AF677285

ADVANCES IN THE BIOSCIENCES

Volume 65

PEPTIDES AND PROTEASES: RECENT ADVANCES

ADVANCES IN THE BIOSCIENCES

Latest volumes in the series:

Volume 48: NEUROMODULATION AND BRAIN FUNCTION
Editors: G. Biggio, P. F. Spano, G. Toffano and G. L. Gessa

Volume 49: PSYCHOLOGICAL ASPECTS OF CANCER
Editors: M. Watson and T. Morris

Volume 50: PATHOPHYSIOLOGICAL ASPECTS OF CANCER EPIDEMIOLOGY
Editors: G. Mathé and P. Reizenstein

Volume 51: FRONTIERS IN HISTAMINE RESEARCH
Editors: C. R. Ganellin and J. C. Schwartz

Volume 52: ORO-FACIAL PAIN AND NEUROMUSCULAR DYSFUNCTION
Editors: I. Klineberg and B. Sessle

Volume 53: THE PINEAL GLAND: ENDOCRINE ASPECTS
Editors: G. M. Brown and S. D. Wainwright

Volume 54: OXYGEN TRANSPORT IN RED BLOOD CELLS
Editor: C. Nicolau

Volume 55: GERM CELL TUMOURS II
Editors: W. G. Jones, A. Milford Ward and C. K. Anderson

Volume 56: RECENT TRENDS IN MEDICAL GENETICS
Editors: K. M. Marimuthu and P. M. Gopinath

Volume 57: ADAPTIVE PROCESSES IN VISUAL AND OCULOMOTOR SYSTEMS
Editors: E. L. Keller and D. S. Zee

Volume 58: BRAIN TUMORS: BIOPATHOLOGY AND THERAPY
Editors: M. A. Gerosa, M. L. Rosenblum and G. Tridente

Volume 59: LEARNING & MEMORY: MECHANISMS OF INFORMATION STORAGE IN THE NERVOUS SYSTEM
Editor: H. Matthies

Volume 60: DISORDERS OF EATING BEHAVIOUR: A PSYCHONEUROENDOCRINE APPROACH
Editors: E. Ferrari and F. Brambilla

Volume 61: DYNAMIC PROPERTIES OF GLIA CELLS II CELLULAR AND MOLECULAR ASPECTS
Editors: T. Grisar, G. Frankck, L. Hertz, W. T. Norton, M. Sensenbrenner and D. M. Woodbury

Volume 62: RESEARCH IN RETINITIS PIGMENTOSA
Editors: E. Zrenner, H. Krastel and H-H. Goebel

Volume 63: RETINITIS PIGMENTOSA: PATIENTS' FIGHT FOR SIGHT
Editors: R. Gizycki and F. Brunsmann

Volume 64: ADVANCES IN AGE PIGMENTS RESEARCH
Editors: E. A. Totaro, P. Glees and F. A. Pisanti

PEPTIDES AND PROTEASES: RECENT ADVANCES

Selected Papers Presented at the 2nd International Meeting on the Molecular and Cellular Regulation of Enzyme Activity in Halle, German Democratic Republic, on August 17–23, 1986

Editors:

R. L. SCHOWEN

University of Kansas, USA

and

A. BARTH

Martin Luther University, GDR

PERGAMON PRESS OXFORD · NEW YORK

BEIJING · FRANKFURT · SÃO PAULO · SYDNEY · TOKYO · TORONTO

U.K.	Pergamon Press, Headington Hill Hall, Oxford OX3 0BW, England
U.S.A.	Pergamon Press, Maxwell House, Fairview Park, Elmsford, New York 10523, U.S.A.
PEOPLE'S REPUBLIC OF CHINA	Pergamon Press, Room 4037, Qianmen Hotel, Beijing, People's Republic of China
FEDERAL REPUBLIC OF GERMANY	Pergamon Press, Hammerweg 6, D-6242 Kronberg, Federal Republic of Germany
BRAZIL	Pergamon Editora, Rua Eça de Queiros, 346, CEP 04011, Paraiso, São Paulo, Brazil
AUSTRALIA	Pergamon Press Australia, P.O. Box 544, Potts Point, N.S.W. 2011, Australia
JAPAN	Pergamon Press, 8th Floor, Matsuoka Central Building, 1-7-1 Nishishinjuku, Shinjuku-ku, Tokyo 160, Japan
CANADA	Pergamon Press Canada, Suite No. 271, 253 College Street, Toronto, Ontario, Canada M5T 1R5

Copyright © 1987 Pergamon Journals Ltd.

All Rights Reserved. No part of this publication may be reproduced, stored in a retrieval system or transmitted in any form or by any means: electronic, electrostatic, magnetic tape, mechanical, photocopying, recording or otherwise, without permission in writing from the publishers.

First edition 1987

ISSN 0065–3446

ISBN 0–08–035726–1

In order to make this volume available as economically and as rapidly as possible the authors' typescripts have been reproduced in their original forms. This method unfortunately has its typographical limitations but it is hoped that they in no way distract the reader.

Printed in Great Britain by A. Wheaton & Co. Ltd., Exeter

Contents

Introduction

No field in science is currently more active than the exploding effort to characterize and understand bioactive peptides and the enzymes which process them. The present volume contains reports on many aspects of this research, organized into sections on peptides and proteases in immune and inflammatory processes, the processing and activity of bioactive peptides, peptide transport, peptide synthesis using proteases, the inhibition of proteases, the mechanisms of action of proteases and related enzymes and the catalytic power of proteases.

The papers were selected from those presented by the 243 scientists who attended the 2nd International Meeting on the Molecular and Cellular Regulation of Enzyme Activity at Halle in the German Democratic Republic (August 17-23, 1986), coming from Bulgaria, Canada, Czechoslovakia, Denmark, the Federal Republic of Germany, Finland, France, the German Democratic Republic, Hungary, Italy, Japan, the Netherlands, Poland, Sweden, Switzerland, the United Kingdom, the USA and the USSR. The meeting was jointed sponsored by the Center for Drug Research, Section of Biosciences, of the Martin Luther University in Halle and the Center for Biomedical Research of the University of Kansas (USA), and enjoyed the support of the European Federation of Medicinal Chemistry, the Society of Experimental Medicine of the GDR, the Biochemical Society of the GDR, Agrochemisches Kombinat Piesteritz (GDR), Bayer AG Leverkusen-Bayerwerk (FRG), Chemiekombinat Bitterfeld (GDR), E.I. Dupont de Nemours (USA), Kombinat GERMED (GDR), Merck Institute of Therapeutic Research-Merck, Sharp & Dohme (USA), SERVA Feinbiochemica-Heidelberg (FRG), Syntex Research-Division of Syntex Inc. (USA-Canada), and The Upjohn Company (USA).

© 1987 Pergamon Journals Ltd.
Printed in Great Britain

PEPTIDES AND PROTEASES IN IMMUNE AND INFLAMMATORY PROCESSES

Dipeptidyl Peptidase IV in Human T Lymphocytes: an Approach to the Function of this Peptidase in the Immune System

S. Ansorge and E. Schön

Medical Academy Magdeburg, Department of Internal Medicine, Division of Experimental Immunology, DDR-3090 Magdeburg, Leipziger Str. 44, GDR

ABSTRACT

The paper is aimed towards the role of the membrane enzyme dipeptidyl peptidase IV (DP IV; E.C. 3.4.14.5) in the immune system. Using different methodological approaches we found the following functions to be inhibited in presence of specific inhibitors or antibodies against DP IV:

(1) DNA synthesis of T lymphocytes induced by mitogens or alloantigens;
(2) immunoglobulin secretion of B lymphocytes induced by PWM;
(3) cell cycle progression of mitogen stimulated lymphocytes between G1 and S.

Models of molecular mechanisms of DP IV action are discussed with respect to a suggested key role of this enzyme in fundamental immunological processes.

KEYWORDS

Dipeptidyl peptidase IV; lymphocyte proliferation; cell cycle progression; lymphokine action.

INTRODUCTION

The dipeptidyl peptidase IV (DP IV, E.C. 3.4.14.5) known since about two decades (Hopsu-Havu and Sarimo, 1967) is an enzyme widely distributed in mammalian cells (cf. Walter and others, 1980) yet its physiological role is largely unknown. The activity of this membrane-bound (Kenny and others, 1976) serine peptidase is characterized by its ability to cleave dipeptides from the free amino termini of polypeptides, and of dipeptide derivatives substituted at their carboxyl terminus with amines. The DP IV shows a high preference for substrates with L-proline and L-alanine in the second (P_1) position.

For immunologists the DP IV has become of particular interest since it has been shown that this enzyme is localized exclusively in the T lymphocyte population (Lojda, 1981; Feller and others, 1982), Scholz and others, 1985, and Schön and others, 1984 a and b).

© 1987 Pergamon Journals Ltd.
Printed in Great Britain

The conclusion that T lymphocytes exhibiting DP IV activity belong to the helper cell compartment has been supported by studies of Scholz and co-workers (1985) (cf. also Flad and others, this issue) who showed that the interleukin-2 (IL-2, T cell growth factor) production by T cells is correlated with the expression of DP IV.

An obvious question concerns whether the expression of DP IV on the surface of T cells is an accidental phenomenon, or whether this peptidase is functionally involved directly in the molecular events of regulation of T cell proliferation and differentiation. One methodical approach to answer these questions is to make use of specific inhibitors of DP IV and antibodies against its structure. In the present communication we summarize recent results of our laboratory showing that the membrane DP IV of T lymphocytes might play a key role in different functions of lymphocytes.

METHODICAL STRATEGY

Three different DP IV-specific inhibitors have been examined: (1) N-Ala-Pro-O(nitrobenzoyl-)hydroxylamine, an irreversible suicide (enzyme activated) inhibitor of dipeptidyl peptidases (Fischer and others, 1983), (2) ε-(4'nitro)benzoxycarbonyl-Lys-Pro, a competitive DP IV inhibitor (Barth and others, 1983), and (3) a polyclonal rabbit anti-porcine kidney DP IV immunoglobulin (Schön and co-workers, unpublished).
As test systems we used the mitogen or alloantigen driven DNA synthesis and proliferation of human blood mononuclear cells during 3 days (Schön and others, 1985), the pokeweed mitogen induced immunoglobulin synthesis after 7 days of incubation (Schön and others, 1985, Ansorge and others, 1986).

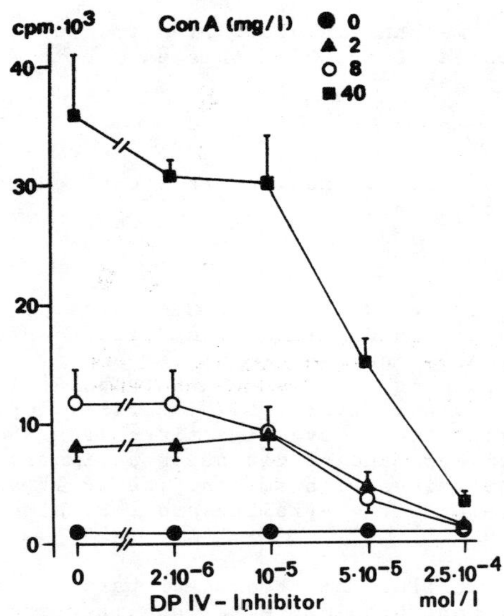

Fig. 1. Inhibition of ConA-induced DNA synthesis of T lymphocytes by the suicide DP IV inhibitor

ROLE OF DP IV IN T AND B LYMPHOCYTE FUNCTION

Using DP IV inhibitors, mitogen-induced DNA synthesis was suppressed in a concentration-dependent manner by both types of inhibitors (Schön and others, 1985). With the suicide inhibitor, 50 % suppression of mitogen-induced proliferation was observed at $5 \cdot 10^{-5}$ (ConA, cf. Fig. 1), 10^{-4} (PWM) and $1.3 \cdot 10^{-4}$ mol/l (PHA). In the case of the competitive inhibitor approx. $2.5 \cdot 10^{-4}$ mol/l (ConA) was necessary to reach half maximal inhibition. The effect on T cell proliferation was not restricted to mitogenic induction but holds also true under allogeneic stimulation (Schön and others, 1985).

It should be pointed out that both inhibitors did not show unspecific or specific cytotoxic effects on lymphocytes in vitro as shown by the unchanged viability and recovery of cultured cells, particularly of the $CD4^+$ T helper lymphocytes as DP IV inhibitor targets. Significant inhibition of mitogen-induced DNA synthesis could also be observed in presence of monospecific polyclonal rabbit anti-DP IV Ig (Fig. 2). This clearly shows that inhibition of DNA synthesis cannot be interpreted as unspecific chemical-induced cytotoxic effects but has to be considered as a specific mechanism acting on membrane DP IV of T lymphocytes.

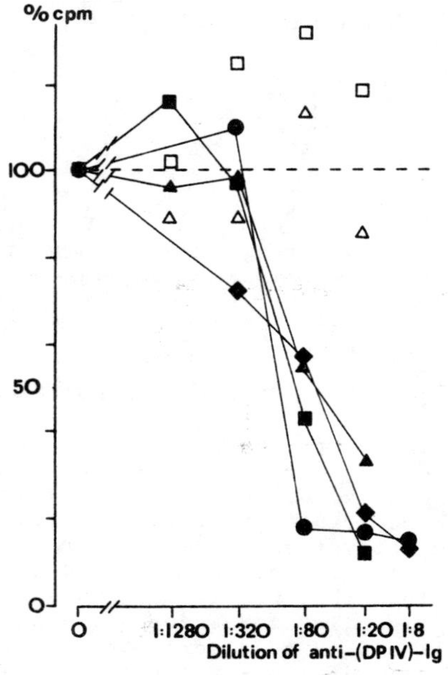

Fig. 2. Inhibition of ConA induced proliferation by a polyclonal rabbit anti-porcine kidney DP IV immunoglobulin. The symbols are representative for separate experiments. The values are expressed as % ^3H-thymidine incorporation related to control cultures without antibody. Full symbols: anti-DP IV. Open symbols: normal rabbit Ig.

As a consequence of the impaired T helper/inducer function, after PWM induction the development of lymphocytes containing cytoplasmic immunoglobulins and/or cytoplasmic thiol-proteindisulfide oxidoreductase (TPO, a typical enzyme marker of Ig producing cells (Ansorge and others, 1986) as well as the secretion of IgG and IgM into the medium was reduced significantly in presence of DP IV inhibitors (Schön and others, 1985) (Fig. 3). These results do not imply a direct action of inhibitors on B cells but have to be considered as an indirect inhibition of B lymphocytes through blocked T helper cells participating normally in B cell growth via helper factors (cf. Fig. 5).

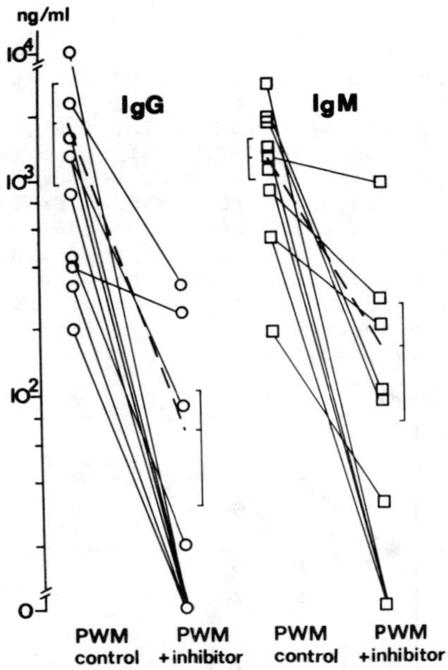

Fig. 3. DP IV inhibitor mediated suppression of PWM stimulated IgG and IgM secretion

DP IV IN CELL CYCLE PROGRESSION

If DP IV is involved in the process of proliferation, then inhibition of that peptidase should result in arresting T cells in a defined cell cycle phase and/or modifying the expression of cycle phased characteristic membrane structures. Investigating the expression of different activation markers after mitogenic stimulation in presence of the suicide inhibitor, we found that the only activation marker under study that was impaired was the HLA class II (DR) antigen. This antigen is expressed on the surface of T lymphocytes after activation and is considered to be a typical marker for the G1T phase, i.e. the postmitotic G1 (Holter and others, 1984). In contrast to this antigen other activation markers as the 4F2 antigen (Kehrl and others, 1984), the interleukin-2 receptor (Tac; Smith, 1980), the formation of blasts (Gibbs and others, 1979) and the induction of cells able to form thermostable SE rosettes (Bryan and others, 1984), all of them characteristics of the early G1a phase, were not influenced by the DP IV inhibitors (Fig. 4, for details cf. communication by Schön and others, this symposium).

Considering these findings, together with the suppression of DNA synthesis by the DP IV inhibitors, the action of DP IV in cell cycle of T lymphocytes appears to be localized between G1a and S phase.

From this it seems reasonable to suggest that a main function of this peptidase lies in overriding the cell cycle restriction point G1. Besides of this putative active regulatory role of DP IV, it is noteworthy that the peptidase activity itself seems to be regulated by external signals. As found recently, PHA stimulated T cells showed an induction of DP IV in the G1 phase up to a 4 fold increase of specific activity (expressed per cell) after a 3 day incubation (Schön and others, this symposium). Moreover, it is highly probable that different molecular forms of DP IV might be involved in these regulatory processes (Kasahara and others, 1984; Smith, this symposium).

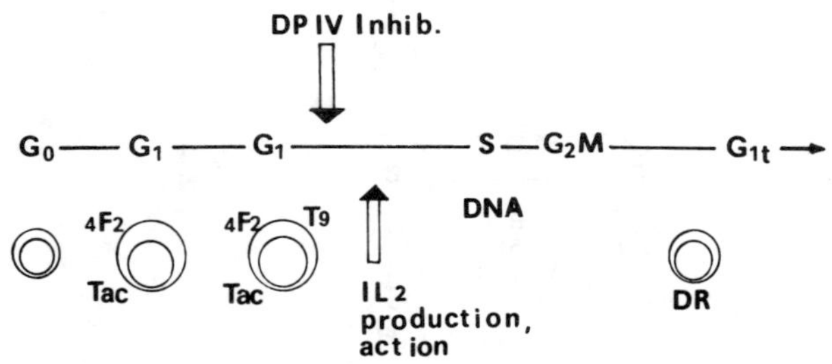

Fig. 4. Schematic representation of the localization of the DP IV action in cell cycle of T cells

MODELS OF MOLECULAR MECHANISM OF DP IV ACTION ON LYMPHOCYTES

Taken together, the data on T and B cell suppression by inhibition of the DP IV activity suggest a very important role for this peptidase in fundamental immunological processes. The detailed molecular mechanisms of the DP IV action, however, are not elucidated as yet.

One of the most probable explanation could be its participation in the signalling process of lymphocytes (Fig. 5) via cytokines (e.g. IL-1, IL-2, interferon-gamma, B cell growth and differentiation factors), namely their posttranslational proteolytic processing and/or inactivation by removing and handling of N-terminal dipeptides. Three different non-exclusive models are possible:
(1) Processing of lymphokine precursors (growth factors and others) on lymphokine producer T cells (T_{act} for IL-2 and T_{res} for IFNγ in Fig. 5), (2) processing and activating of cytokines on cytokine responder T cells (T_{act} for IL-1, T_{res} for IL-2 in Fig. 5), and (3) inactivating of T cell growth inhibiting factors.

Of particular interest concerning model 1 is the study of Flad's group (Scholz and others, 1985) who demonstrated that the bulk of IL-2 production in vitro seems to be restricted to DP IV-positive T lymphocytes. This conclusion has been supported very recently by inhibitor studies in our laboratory. After stimulation of lymphocytes with mitogens and phorbol ester (PHA) in presence of DP IV inhibitors a reduced production of T cell derived lymphokines has been measured (for details cf. communication of Schön and others, this symposium). From these data it seems reasonable to suggest that DP IV is in fact involved in the synthesis, processing and/or release of lymphokines by activated T cells.

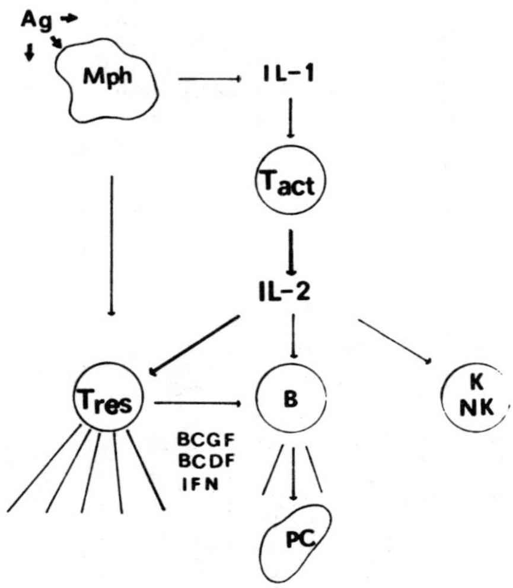

Fig. 5. Schematic representation of communication of immunregulatory cells via cytokines (with reference to Robb 1984)

Conclusive experimental evidences concerning point 2 and 3 have been missing as yet. However, first experimental approaches studying the effect of IL-2 containing cell culture supernatants on DP IV inhibitor-suppressed T proliferation provided first clues that DP IV might also be involved in the processing of active cytokines/lymphokines by T cells (point 2). The growth stimulating activity of these supernatants were suppressed significantly by the competitive DP IV inhibitor (Fig. 6).

Moreover, structural analyses of the N-terminal sequences of cytokines revealed, surprisingly, that most of these factors are characterized by DP IV susceptible sequences (Table 1). Remarkably, also a number of growth factors and other biologically active peptides as some oncogene products (e.g. fes, fms, mos, h-ras-2), not made by and act upon the immune system, share this common property. From these structural considerations it seems highly tempting to speculate that the DP IV may play a key role in modulation of lymphokine action in the immune system by X-Pro-(or X-Ala-) directed limited proteolysis. Moreover, the DP IV catalyzed processing of biologically active peptides, including the formation of special dipeptides, possibly can be envisioned to be of general significance for cell membrane mediated communication regulating normal cell growth and differentiation.

CONCLUSIONS

The data discussed in this paper provide some clues that the DP IV, a characteristic of T helper/inducer lymphocytes, plays an important role in regulation of the immune response. This conclusion has been drawn mainly from studies with different types of DP IV inhibitors

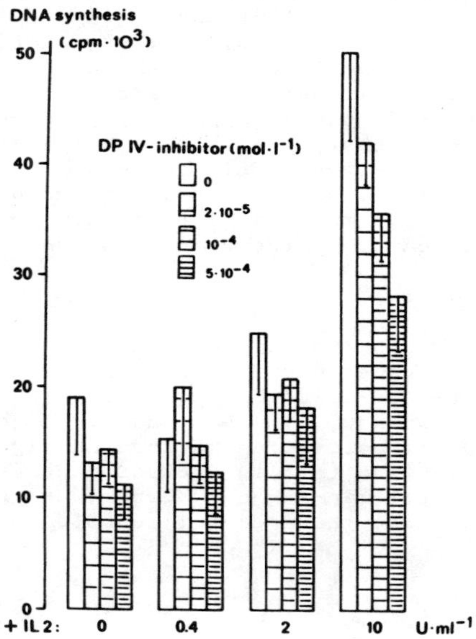

Fig. 6. Inhibition of IL-2 action on PHA prestimulated mononuclear cells (24 h) by the competitive DP IV inhibitor

TABLE 1 Dipeptidyl peptidase IV susceptible structures in the N-terminal sequence of lymphokines, and other biologically active factors regulating cell growth and differentiation (incomplete list)

FACTOR	N-TERMIN. AMINO ACID SEQUENCE	
IL-1 (B)	Ala-Pro	153
IL-2	Ala-Pro	133
IFN-γ	. . . . . Asp-Pro .	146
IFN-a	. . Leu-Pro . . .	155
LYMPHOTOX.	Leu-Pro	171
EGF	. . Tyr-Pro . . .	
STH	Phe-Pro	191
IL-GF I	Gly-Pro	105
IL-GF II	. . Arg-Pro . . .	
SUBST. P	Arg-Pro-Lys-Pro . . .	11
PROLACTIN	Leu-Pro	199
CSF-2	Ala-Pro	144
c-fms	Glu-Pro	950

as well as an anti-DP IV immunoglobulin. Inhibition of DP IV is accompanied by suppression of essential functions of T cells, e.g. proliferation and production and secretion of lymphokines, and B cell activity. Most likely, in cell cycle the peptidase seems to be necessary for T cells in overriding the restriction point at G1.

These data, together with other features of the DP IV, as its localization on the outside of the plasma membrane, its substrate specificity, and its dynamic state of expression (induction after immunostimulation) support the notion that the DP IV plays a key role in the modulation of lymphokine action by X-Pro-(or X-Ala-) directed proteolysis. The molecular mechanisms, however, have not been defined as yet. Available data and structural considerations suggest that the DP IV might act in both, the events of lymphokine production/secretion, or/and in the processing of active cytokines. Moreover, the high frequency of DP IV-susceptible sequences in the N-termini of different growth factors, oncogen products and other biologically active peptides leads to the speculation that this peptidase is of more general significance to the regulation of cellular growth.

ACKNOWLEDGEMENTS

The authors wish to thank the colleagues of different laboratories for their help and suggestions in this work: Drs. A. Barth, H. U. Demuth, and K. Neubert (Halle) for providing the synthetic DP IV inhibitors, Drs. R. Grunow, S. Jahn, and H.-D. Volk (Berlin) for studying expression of activation markers and Ig secretion, Drs. F. Noll, and Evelyn Eichmann (Berlin-Buch) for determining gamma-interferon and Drs. J. Kopp, Ida-Johanna Körner, and W. Malz (Berlin-Buch) for IL-2 assays. We also thank Bärbel Schotte for secretarial expertise.

REFERENCES

Ansorge, S., H.-W. Mansfeld, Ch. Held, L. Broodtaerts, and B. Van Camp (1986). Acta histochem., 78, 65-71.

Barth, A., and G. Kempter (1983). Wiss. Z. Univ. Halle XXXII, 3-50.

Bryan, C. F., S. H. Leech, R. Ducos, C. O. Edwards, J. P. Kushner, M. H. Skolnik, B. Bozelka, J. S. Linn, and R. Gaumer (1984). J. Clin. Immunol., 4, 134-142.

Feller, A. C., C. J. Heijnen, R. E. Ballieux, and M. R. Parwaresch (1982). British J. Haematol., 51, 227-234.

Fischer, G., H.-U. Demuth, and A. Barth (1983). Pharmazie 38, 249-260.

Gibbs, J. H., R. A. Brown, A. J. Robertson, R. C. Potts, and J. S. Beck (1979). J. Immunol. Meth., 25, 147-158.

Holter, W., O. Majdic, K. Liszka, H. Stockinger, O. Köller, and W. Knapp (1984). Immunobiol., 168, 3.

Hopsu-Havu, V. K., and S. R. Sarimo (1967). Hoppe-Seyler's Z. Physiol. Chemie 348, 1540-1550.

Kasahara, Y., G. Leroux-Roels, R. Nakamura, and F. Chirasi (1984). Clin. Chim. Acta 139, 295-302.

Kehrl, J. H., A. Muraguchi, and A. S. Fauci (1984). J. Immunol., 132, 2857-2861.

Kenny, A. J., A. G. Booth, S. G. George, J. Ingram, D. Kershaw, E. J. Wood, and A. R. Young (1976). Biochem. J., 157, 169-182.

Lojda, Z. (1981). J. Histochem. Cytochem., 29, 481-493.

Robb, R. J. (1984). Immunology Today 5, 203-209.

Schön, E., H.-U. Demuth, A. Barth, and S. Ansorge (1984a). Biochem. J., 223, 255-258.

Schön, E., A. Ittenson, K. Klemm, and S. Ansorge (1984b). Acta histochem., 75, 175-182.

Schön, E., H.-W. Mansfeld, H.-U. Demuth, A. Barth, and S. Ansorge (1985). Biomed. Biochim. Acta 44, K9-K15.

Scholz, W., R. Mentlein, E. Heymann, A. C. Feller, A. J. Ulmer, and H.-D. Flad (1985). Cell. Immunol., 93, 199-211.

Walter, R., W. H. Simmons, and T. Yoshimoto (1980). Mol. Cell. Biochem., 30, 111-127.

The Role of Proteolytic Enzymes in the Complement System

M. G. Colomb, G. J. Arlaud, C. Aude and A. Journet

Unité INSERM 238, Département de Recherche Fondamentale, LBIO/ICH, Centre d'Etudes Nucléaires 85 X, 38041 Grenoble Cédex, France

ABSTRACT

The complement system includes five serine proteases responsible for its activation via either the classical or the alternative pathway. With the exception of factor $\overline{D}$, these proteases are part of multiproteic complexes : they all show a trypsin-like specificity, restricted to arginyl bonds and accept only one or two protein substrates. The activity of these proteases is under strict control and leads to the release of polypeptide fragments from their substrates. Small monovalent peptides are mainly involved in anaphylaxis and chemotactism ; large bifunctional peptides are involved in more complex interactions through non specific acceptor and specific receptor sites, playing a significant role at the cellular level, in the early phase of the immune response.

KEYWORDS

Complement system ; limited proteolysis ; serine proteases ; inflammation ; immune response.

INTRODUCTION

The complement system is currently described as a set of proteins in peripheral blood, which exist normally as a resting system (Cooper, 1985 ; Müller-Eberhard, 1986 ; Reid and Porter, 1981). Several stimuli are able to activate complement which, through proteolytic cleavages of several of its proteins, contributes to the humoral immunological defence of the organism.

As shown in Fig. 1, activation occurs via two different pathways, the classical and the alternative pathways, where sequential activation steps lead to a downstream formation of multimolecular proteases, the C3 and C5 convertases.

Biological expression of complement resides in the formation of a cytolytic complex through a series of non-enzymatic reactions following the assembly of the convertases (Müller-Eberhard, 1986). The most significant activities of complement, which will be considered here, are due to the formation of peptides which exert various activities through specific acceptors or receptors on soluble proteins or membranes. The best illustration is the inflammatory process in which peripheral blood complement proteins play a major role. Complement plays also a significant role at the cellular level : cells of the immune system, like macrophages, synthesize complement proteins which are likely to be involved in

© 1987 Pergamon Journals Ltd.
Printed in Great Britain

local interactions : the presentation of antigens by the macrophages or the recognition of antigens by T lymphocytes are believed now to be modulated by non-specific factors such as peptides of the complement system.

STRUCTURE AND ACTIVATION OF THE PROTEOLYTIC ENZYMES OF THE COMPLEMENT SYSTEM

Five proteolytic enzymes, all of the serine proteinase type, participate in the activation of the complement system. Three (C1r, C1s and C2a) are involved in the classical pathway, the other two (factor $\bar{D}$ and Bb) taking part in the alternative pathway (Fig. 1).

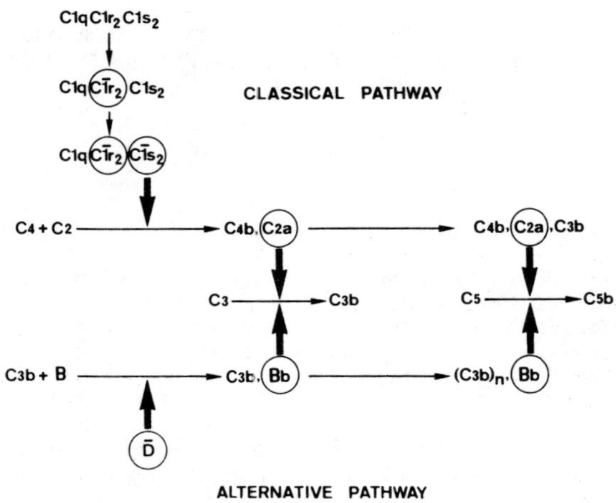

Fig. 1. The two pathways of proteolytic activation of the complement system.
Circles indicate the components that bear a serine proteinase active site. Intermolecular proteolytic cleavages are represented by large arrows.

With the exception of factor $\bar{D}$, which functions as an isolated proteinase, these enzymes all belong to complex proteases : C1, and the C3 and C5 convertases. The structure and activation pattern of each individual serine protease involved in the activation of the complement system is schematized in Fig. 2.

C1r and C1s

Human C1r is a non-covalently linked dimer of apparent Mr 180,000 - 190,000. Each monomer of proenzyme C1r is a single-chain glycoprotein (Mr 92,000), that is split on activation into two disulphide-linked chains, through cleavage of a single Arg-Ile bond (Arlaud and Gagnon, 1985a). Complete sequences of the C-terminal B chain (Arlaud and Gagnon, 1983) and of the N-terminal A chain (Arlaud and others, 1987 ; Gagnon and Arlaud, 1985) have been determined, and confirmed by cDNA sequence analysis (Journet and Tosi, 1986). These are 242 and 446 residues long, respectively, and each contain two N-linked carbohydrates.
Human C1s is isolated as a monomer and dimerizes only in the presence of Ca^{2+} ions. At the monomeric level, its overall structure and activation scheme are

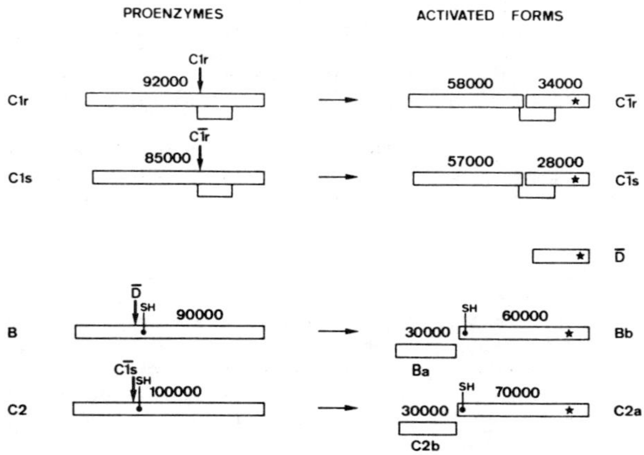

Fig. 2. Structure and activation of the serine proteinases involved in the activation of the complement system.
All structural features are drawn to scale, including the single interchain disulphide bridge in C1r and C1s.
* indicates the serine residues of the active sites and SH represents the free sulphydryl group of C2 and Factor B.
The location of the active site serine residue in C2 is hypothetical.

comparable to those of C1r : the proenzyme form is also a single chain glycoprotein of Mr approx. 85,000 and activation, which is mediated by activated C$\bar{1}$r, occurs through the cleavage of a single Arg-Ile bond (Spycher and others, 1986) and produces two disulphide-bonded chains. Complete sequence of the C-terminal B chain (Carter and others, 1984) and partial sequence of the N-terminal A chain (Spycher and others, 1986) have been obtained by protein sequencing, and the entire sequence of C1s has been recently deduced from nucleotide sequencing (Tosi and others, 1986). Contrary to C$\bar{1}$r; the B chain of C$\bar{1}$s is devoid of carbohydrate, whereas the A chain contains two N-linked carbohydrates. The A chain of both proteases contains a sequence segment homologous to the "epidermal growth factor-like" domains described in other proteins (see Doolittle and others, 1984). In this segment, a residue encoded in both C1r and C1s as asparagine (Journet and Tosi, 1986 ; Tosi and others, 1986) undergoes post-translational modifications likely resulting, at least in C1r, into β-hydroxyasparagine (Arlaud and Gagnon, 1985b ; Arlaud and others, unpublished data). For C$\bar{1}$r and C$\bar{1}$s, the serine proteinase active site is located in the B chains which, except the lack of the "histidine-loop" disulphide bridge (Arlaud and Gagnon, 1981), show extensive homology with other mammalian serine proteinases, indicating that, in both cases , active site formation involves a mechanism similar to that described for other serine proteinases, e.g. chymotrypsin.
Recent studies (Villiers and others, 1985) have established the domain structure and associated functions of C1r and C1s, showing that each monomeric protein consists of two separate domains connected by a short polypeptide strand : (i) an interaction domain (α) corresponding to the N-terminal half of the A chain and associated to specific interaction functions, i.e. calcium binding and C1r-C1s calcium-dependent interactions. (ii) a catalytic domain (γ-B) comprising the C-terminal region of the A chain (γ) disulphide-linked to the B chain and carrying the serine proteinase active site.

A model of the calcium-dependent C1s-C1r-C1r-C1s catalytic subunit of C1 has been proposed, in which the (C1r)2 dimer forms a core, and the C1s molecules are located at each extremity of the complex (Villiers and others, 1985). Upon interaction with C1q, this linear C1s-C1r-C1r-C1s complex is thought to adopt a "8"-shaped conformation in which each C1s catalytic domain is brought into contact with the central C1r catalytic domains (Colomb and others, 1984a, 1984b). Further characterization of C1r catalytic domains at the molecular level led to the conclusion that they interact with each other in a "head to tail" configuration, in which the γ region of one domain faces the B chain of the other domain (Arlaud and others, 1986). This concept is compatible with C1r autolytic activation through an intramolecular cross-mechanism (Arlaud and others, 1985) and with subsequent activation of C1s by activated C1r, thus providing a coherent scheme of C1 activation.
The current concept of C1 function is that its natural tendency is to self-activate, this tendency being submitted in vivo to a permanent negative control by C$\bar{1}$-inhibitor (Ziccardi, 1982). This control could be exerted at the level of the interface between the catalytic domains of C1r and the corresponding domains of C1s, C$\bar{1}$-inhibitor being able to prevent the contact between these domains, thereby preventing C1r activation. This control would be suppressed by activators, acting as positive effectors. Recently, another mechanism of control of C1 activation has been proposed, whereby nascent C3b and C4b inhibit C1 turnover by immune complexes (Ziccardi, 1986).

Factor $\bar{D}$

Human factor $\bar{D}$ is a serine proteinase of Mr 24,000 which, at least in blood, is never found in a proenzyme state, and is presently considered to be synthesized as an active enzyme (Barnum and Volanakis, 1985). The amino acid sequence of Factor $\bar{D}$ has been nearly completed (Johnson and others, 1984 ; Niemann and others, 1984), showing that it is a single-chain protease of 218-222 amino acids which contains no carbohydrate.

C2 et Factor B

C2 and Factor B are unusual serine proteases, with a catalytic chain of molecular weight twice that of typical serine proteinases, and probably with a different activation mechanism (Gagnon, 1984). In its zymogen form, Factor B is a single-chain glycoprotein of Mr about 90,000. Activation of Factor B requires prior formation of a Mg^{2+}-dependent complex C3b,B and is mediated by Factor $\bar{D}$, which splits a single Arg-Lys bond. This cleavage yields two non covalently-linked fragments, Bb (Mr 60,000) which contains the serine active site, and Ba (Mr 30,000), which dissociates from the $\overline{C3b,Bb}$ complex. The complete amino acid sequence of human Factor B has been determined (Gagnon, 1984 ; Mole and others, 1984). The zymogen consists of 739 amino acids and has four N-linked carbohydrates, equally distributed in the Ba et Bb moieties. Although important sequence homology exists between the serine proteinase segment of Bb and that of other serine proteinases, the former shows major differences, such as the lack of the characteristic N-terminal sequence found in these proteinases, including C$\bar{1}$r, C$\bar{1}$s and Factor $\bar{D}$. This suggests that, although the active site of Factor B is similar to that of typical serine proteinases, its activation mechanism is different (Christie and others, 1980). Ba (234 amino acids) contains three regions of internal homology (Morley and Campbell, 1984), which are also related to similar repetitive sequences found in C4b-binding protein (Chung and others, 1985), complement protein H (Kristensen and others, 1986) and β2 glycoprotein I (Lozier and others, 1984). Bb likely consists of two separate globular domains, one of them carrying the serine proteinase active site (Lambris and Müller-Eberhard, 1984 ; Smith and others, 1982).
The overall structure of human C2 is similar to that of Factor B, and its activation follows a similar pattern. Unactivated C2 is a single-chain

glycoprotein of Mr approx. 100,000, and its cleavage by C$\overline{1}$s yields two non-covalently linked polypeptides of Mr 30,000 and 70,000, C2a and C2b, the former carrying the catalytic site. The primary structure of human C2 has been recently derived from cDNA sequence analysis (Bentley and Campbell, 1985), indicating that C2b contains a triple repetitive structure homologous to that found in Ba, and confirming that C2 and Factor B share the same unusual activation mechanism. Although C2 and B both contain a single free -SH group, located close to the N-terminus of fragments Bb and C2a , these groups are not homologous (Christie and Gagnon, 1982 ; Parkes and others, 1983), and the free -SH group of C2 plays an important role in the stability of the classical pathway C3 convertase, $\overline{\text{C4b,C2a}}$, whereas it does not appear to be essential to the activity of Factor B (Kerr, 1980 ; Parkes and others, 1983).

Specificity of the Proteolytic Enzymes

One of the most remarkable properties of the serine proteinases involved in the activation of the complement system is a highly restricted specificity which undoubtedly takes an important part in the control of their activity.
C1r has a double proteolytic role which is confined in the C1 complex. As detailed above, each monomer of the proenzyme is able to activate its neighbour by cleavage of a single Arg-Ile bond (Arlaud and Gagnon, 1985a). Once activated, C$\overline{1}$r retains the same specificity, as it also cleaves an Arg-Ile bond, although in a different substrate, C1s (Spycher and others, 1986). Activated C$\overline{1}$s also has a double role but, contrary to C$\overline{1}$r, it is a true enzyme and its role is extrinsic, as it specifically cleaves the two substrates of C$\overline{1}$, C4 and C2. Both cleavages likely occur at a single site, which is an arginyl bond in C4 (Moon and others, 1981) and probably an Arg-Lys bond in C2 (Kerr and Gagnon, 1982). Factor $\overline{\text{D}}$ can be considered as the most specific of the five proteinases, as it cleaves a single substrate (Factor B), only when it is bound to C3b, and at a single site that is the Arg-Lys bond at position 234 of Factor B (Gagnon, 1984). Bb and C2a, share the same substrates, C3 and C5, in which they cleave the same bonds, and their proteolytic activity is expressed only within Mg^{2+}-dependent complex enzymes, the convertases. The C3 convertases $\overline{\text{C3b,Bb}}$ and $\overline{\text{C4b,C2a}}$, both cleave C3 at a single Arg-Ser bond (De Bruijn and Fey, 1985 ; Tack and others, 1979). The binding of one or more C3b molecules to both $\overline{\text{C4b,C2a}}$ and $\overline{\text{C3b,Bb}}$ changes their substrate specificity from C3 to C5 convertases. The mechanism of this change is not clear (see review by Reid and Porter, 1981), but the active site specificity itself is not greatly modified, as both C5 convertases again cleave an arginyl bond in C5 (Fernandez and Hugli, 1978). Thus, the five serine proteinases involved in the activation of the complement system, although of different types, all are trypsin-like enzymes with a specificity restricted to arginyl bonds.

BIOLOGICAL ACTIVITIES ENSUING ACTIVATION OF THE PROTEOLYTIC ENZYMES OF THE COMPLEMENT SYSTEM

Activities Due to Small Diffusible Peptides

The C3 convertases hydrolyse C3 to generate a Mr 9,000 C3a peptide and a Mr 186,000 C3b molecule which will be considered below. Similarly, the C5 convertases cleave C5 to release a Mr 11,000 glycopeptide, C5a and a Mr 186,000 C5b molecule which participates in the formation of a membrane attack complex (Reid and Porter, 1981). In serum the two small peptides are rapidly transformed in desArg derivatives due to the action of carboxypeptidase-N which removes the C-terminal arginine residues of C3a and C5a. This transformation leads to total inactivation of C3a, but only to partial inactivation of C5a (Hugli,1984).
Other small peptides : C4a, a Mr 8,000 peptide similar to C3a, produced by cleavage of C4 by C$\overline{1}$, and C2k (C2 kinin) yielded by stepwise cleavage of C2 by C$\overline{1}$ and by plasmin, will not be considered here.
C3a and C5a bind to specific receptors on mast cells, basophils, eosinophils,

neutrophils and monocytes (Fearon and Wong, 1983). C3a biological activity, like smooth muscle contraction, is largely mediated through histamine release by these target cells. A major contribution of this peptide to humoral defences is also mediated through an increase of vascular permeability. Recent observations also tend to support a suppressive role of C3a on certain lymphocyte T functions, although no specific receptor has yet been demonstrated on these cells. C5a has very similar activities as C3a. An additional activity of chemotaxis towards neutrophils is mediated by C5a, and its C5a desArg form. Chemotaxis is largely involved in the recruitment of cells for local defence reactions.

Activities Due to Large Bifunctional Molecules

The major proteolytic fragments C4b and C3b, produced by the proteolysis of C4 by $C\overline{1}$ and of C3 by the C3 convertases, have a transient capacity to bind to amino or hydroxyl groups on soluble or membrane acceptors. A small number of C4b or C3b molecules participate in the formation of the convertases, while the vast majority bind covalently to the acceptor (which is, most of the time, the activator of complement) and become available for a second binding to specific receptors through non covalent interactions.

C3b receptors (CR1) are the most investigated among complement receptors (for reviews see Fearon, 1983 ; Fearon and Wong, 1983 ; Ross and Medof, 1985) although significant biological activities are also due to the interaction of other fragments of C3 with their specific receptors. Stepwise proteolysis of bound C3b by factor I, a serine protease, in the presence of cofactors (factor H, CR1) leads to C3bi (Mr 183,000) and C3d,g (Mr 40,000) which exhibit specific non-covalent binding to CR3 and CR2 receptors, respectively. Further proteolysis of C3d,g by plasma or cellular proteases leads to fragment C3d (Mr 35000) able to bind to CR2.

TABLE 1 Cellular Receptors for Bivalent Complement Fragments.

Receptor	Ligand	Cellular location	Biological activity (due to ligand binding)
CR1	C3b C4b	erythrocytes	Factor I cofactor, clearance of immune complexes
		polymorphonuclear cells monocytes monocytes	Opsonisation
		B lymphocytes T lymphocytes K Cells	Enhancement of antibody-dependent cytotoxicity.
CR2	C3d,g	B lymphocytes	Regulation of lymphocyte activation, Epstein-Barr receptor.
		K cells	Enhancement of antibody-dependent cytotoxicity
		lymphoblastoid cells	Epstein-Barr receptor.
CR3	C3bi	same location as CR1 except erythrocytes	Opsonisation, induction of respiratory burst, secretion reactions, enhancement of antibody-dependent cytotoxicity

C3b-receptor (Table 1) fulfills two unique functions : 1. Regulation of complement activation, through the inhibition of the convertases (Fries and others, 1985) ; 2. processing of immune complexes, mainly through the binding of C3b-coated immune complexes to erythrocyte CR1 (Fearon and Wong,1983). Another role, which is not specific to CR1, resides in the enhancement of antibody-dependent cellular activities : human polymorphonuclear cells and monocytes phagocytose erythrocytes coated with IgG antibodies more avidly when the targets also carry C3b. Target cell-bound C3b has also been found to enhance antibody-dependent cellular toxicity by effector lymphocytes carrying CR1 (Fearon and Wong, 1983).

CR2, the receptor for C3d, is predominantly expressed on B lymphocytes and lymphoblastoid cells and has been clearly identified with the Epstein-Barr virus receptor (Mold and others, 1985 ; Nemerow and others, 1985). CR2 has been found to regulate both the effector cell function and proliferative responses of lymphocytes. Recently, an insulin-like growth factor activity has been described for cross-linked C3d, which controls the cell cycle of activated B cells in mouse (Melchers and others, 1985).

CR3 activity, due to the specific binding of C3bi, appears very similar to CR1 activity. However, since C3bi is a more stable fragment than C3b, CR3 may be the effective receptor which regulates in vivo responses and also appears to be a major opsonin (Rothlein and Springer,1985).

The expression of C3 receptors on cell membranes is modulated by physiological factors such as γ-interferon, which has been recently shown to depress ligand binding by CR1 and CR3 on cultured human monocytes. Data are available which show that binding and signaling activities of C3 receptors are regulated separately (Wright and others, 1986). An assessment of the activities of these receptors has been proposed, based on a specific phosphorylation observed for CR1 and CR2 but, unexpectedly, not for CR3 (Changelian and Fearon, 1986).

CONCLUSION

Significant biological activities are due to peptide fragments produced by complement proteases. Although highly specific, these proteases are mimicked by extrinsic proteases e.g. of the contact phase of coagulation, from bacteria or from lysosomes, which are also able to produce active complement fragments or even to trigger the proteolytic cascade of complement. All these interactions clearly show that the complement system cannot be isolated from its environment. Activity of the complement system is controlled by mechanisms not described here, which contribute efficiently to the homeostasis of its potential of proteolytic defence.

REFERENCES

Arlaud, G.J. and Gagnon, J. (1981) Biosci. Rep., 1, 779-784.
Arlaud,G.J. and Gagnon,J. (1983) Biochemistry, 22, 1758-1764.
Arlaud,G.J. and Gagnon,J. (1985a) FEBS Lett., 180, 234-238.
Arlaud,G.J. and Gagnon,J. (1985b) Complement, 2, 4.
Arlaud,G.J., Colomb, M.G. and Villiers,C.L. (1985) Biosci. Rep., 5, 831-837.
Arlaud,G.J., Gagnon,J., Villiers,C.L. and Colomb,M.G. (1986) Biochemistry , 25, 5177-5182.
Arlaud, G.J., Willis, A.C. and Gagnon, J. (1987) Biochem. J. (in the press).
Barnum,S.R. and Volanakis,J.E. (1985) Eur.J.Immunol., 15, 1148-1151.
Bentley,D.R. and Campbell, R.D. (1985) Complement, 2, 9.
Carter,P.E., Dunbar,B. and Fothergill,J.E. (1984) Phil. Trans. R. Soc. Lond. B 306, 219-227.
Changelian,P.S. and Fearon,D.T. (1986) J.Exp.Med, 163, 101-115.
Christie,D.L., Gagnon,J. and Porter,R.R. (1980) Proc. Natl. Acad. Sci. USA, 77, 4923-4927.
Christie,D.L. and Gagnon,J. (1982) Biochem.J., 201, 555-567.
Chung,L.P., Gagnon,J. and Reid,K.B.M. (1985) Mol. Immunol., 22, 427-435.

Colomb,M.G., Arlaud,G.J. and Villiers,C.L. (1984a) Phil. Trans. R. Soc. Lond. B, 306, 283-292.
Colomb,M.G., Arlaud,G.J. and Villiers,C.L. (1984b) Complement, 1, 69-80.
Cooper,N.R. (1985) Adv. Immunol., 37 151-216.
De Bruijn,H.L. and Fey,G.H. (1985) Proc. Natl. Acad. Sci. USA, 82, 708-712.
Doolittle,R.F., Feug,D.F. and Johnson,M.S. (1984) Nature, 307, 558-560.
Fearon,D.T. (1983) Springer Semin. Immunopathol., 6 , 159-172.
Fearon,D.T. and Wong W.W. (1983) Ann. Rev. Immunol., 1, 243-271.
Fernandez,H.N. and Hugli,T.E. (1978) J.Biol.Chem., 253, 6955-6969.
Fries,L.F., Prince,G.M., Gaither,T.A. and Franck M.M.(1985) J.Immunol., 135, 2673-2679.
Gagnon,J. (1984) Phil. Trans. R. Soc. Lond. B, 306, 301-309.
Gagnon,J. and Arlaud,G.J. (1985) Biochem.J., 225, 135-142.
Hugli,T.E. (1984) Springer Semin. Immunopathol., 7, 193-219.
Johnson,D.M.A., Gagnon,J. and Reid,K.B.M. (1984) FEBS Lett., 166, 347-351.
Journet, A. and Tosi, M. (1986) Biochem.J. (in the press).
Kerr,M.A. (1980) Biochem.J., 189, 173-181.
Kerr,M.A. and Gagnon,J. (1982) Biochem.J., 205, 59-67.
Kristensen, T., Wetsel, R.A. and Tack, B.F. (1986) J. Immunol., 136, 3407-3411.
Lambris,J.D. and Müller-Eberhard,H.J. (1984) J.Biol.Chem., 259, 12685-12690.
Lozier,J., Takahashi,N. and Putnam,F.W. (1984) Proc.Natl.Acad.Sci.USA, 81, 3640-3644.
Melchers,F., Erdei,A., Schulz,T. and Dierich,M.(1985) Complement, 2, 54.
Mold,C., Cooper,N.R. and Nemerow G.R.(1985) Complement, 2, 55.
Mole,J.E., Anderson,J.K., Davison,E.A. and Woods,D.E. (1984) J.Biol.Chem., 259, 3407-3412.
Moon,K.E., Gorski,J.P. and Hugli,T.E. (1981) J.Biol.Chem., 256, 8685-8692.
Morley,B.J. and Campbell,R.D. (1984) EMBO J., 3, 153-157.
Müller-Eberhard H.J. (1986) Ann. Rev. Immunol., 4 503-528.
Nemerow,G.R., Siaw,M. and Cooper,N.R.(1985) Complement, 2, 57.
Niemann,M.A., Bhown,A.S., Bennett,J.C. and Volanakis,J.E.(1984) Biochemistry 23, 2482-2486.
Parkes,C., Gagnon,J. and Kerr,M. (1983) Biochem.J., 213, 201-209.
Reid,K.B.M. and Porter,R.R. (1981) Ann. Rev. Biochem., 50, 433-464.
Ross, G.D. and Medof, M.E. (1985) Adv. Immunol., 37, 217-267.
Rothlein,R. and Springer,T.A. (1985) J. Immunol., 135 2668-2672.
Smith,C.A., Vogel,C.-W. and Müller-Eberhard,H.J. (1982) J. Biol. Chem., 257, 9879-9882.
Spycher,S.E., Nick,H. and Rickli,E.E. (1986) Eur. J. Biochem., 156, 49-57.
Tack,B.F., Morris,S.C. and Prahl,J.W. (1979) Biochemistry, 18, 1497-1503.
Tosi,M., Journet,A., Lyonnet,D., Colomb,M. and Meo,T. (1986) In Protides of the Biological Fluids, 34 (H.Peeters, ed.) Brussels.
Villiers,C.L., Arlaud,G.J. and Colomb;M.G. (1985) Proc. Natl. Acad. Sci USA 82, 4477-4481.
Wright,S.D., Detmers,P.A., Jong,M.T.C. and Meyer,B.C. (1986) J. Exp.Med., 163, 101-115.
Ziccardi,R.J. (1982) J. Immunol., 128, 2505-2508.
Ziccardi,R.J. (1986) J. Immunol., 136, 3378-3383.

Phenotypic and Functional Characterization of Human T-Lymphocytes Defined by Monoclonal Antibody Against Dipeptidyl Peptidase IV (DPP IV)

H.-D. Flad*, M. Gruber*, W. Scholz*, A. C. Feller[†], A. J. Ulmer* and E. Heymann[‡]

*Forschungsinstitut Borstel, Parkallee 1-40, D-2061 Borstel;
[†]Universität Kiel, Abt. Pathologie, Hospitalstraße 42, D-2300 Kiel;
[‡]Institut für Med. Chemie, Universität Osnabrück, D-4500 Osnabrück, Federal Republic of Germany

ABSTRACT

Human T-lymphocytes bearing DPP IV on their surface were recognized by a monoclonal anti-DPP IV antibody. When isolated by cell sorting and stimulated with PHA or PWM, DPP IV^{+} cells produced IL-2 in substantial amounts, whereas DPP IV^{-} did not. Further characterization of DPP IV bearing T-cells was achieved by double labelling with monoclonal antibodies against DPP IV and Leu3a, Leu2 or Leu1, and cell sorting. After addition of adherent cells we observed: 1) DPP IV^{+}/Leu3a^{+} and DPP IV^{+}/Leu3a^{-} subsets produced IL-2; 2) DPP IV^{-}/Leu3a^{-} were unable to produce substantial amounts of IL-2; 3) DPP IV^{-}/Leu3a^{+} cells produced some IL-2 probably due to contaminant high producers (DPP IV^{+}/Leu3a^{+}). Corresponding data were obtained with double labelled subsets using anti-DPP IV, anti-Leu2 and anti-Leu1. In a pokeweed mitogen driven B-cell differentiation system DPP IV^{-}/Leu3a^{+} cells were more efficient helper cells than DPP IV^{+}/Leu3a^{+}. We conclude that IL-2 production by T-cells is associated with the expression of DPP IV and that the subsets DPP IV^{+}/Leu3a^{+} and DPP IV^{-}/Leu3a^{+} differ with regard to their capacity to produce B-cell growth and differentiation factors other than IL-2.

KEYWORDS

Human lymphocyte subpopulations; dipeptidyl peptidase IV; interleukin 2.

INTRODUCTION

Dipeptidyl peptidase IV (DPP IV, E.C. 3.4.14.5) is an ubiquitous enzyme located in the plasma membrane of a variety of mammalian cells (Püschel, Mentlein and Heymann, 1982; Mentlein and co-workers, 1984; for review see Walter, Simmons and Yoshimoto, 1980), and first described in 1967 (Hopsu-Havu and Sarimo). Using Gly-Pro-4-methoxy-ß-naphthyl-amide as substrate for histochemical demonstration, DPP IV was also detected in smears of isolated lymphocytes (Lojda, 1977), and several groups provided evidence for the T-cell origin of DPP IV positive lymphocytes (Feller and co-workers, 1982; Chilosi and co-workers, 1982; Schön and co-workers, 1984).
Human T-lymphocytes have been divided into two categories based on the expression of the T8 and T4 antigen, i.e. the cytotoxic/suppressor and the helper/

© 1987 Pergamon Journals Ltd.
Printed in Great Britain

inducer subset, respectively (Reinherz and co-workers, 1979, 1980). We, therefore, decided to investigate the association of DPP IV expression of T-lymphocytes with their functional capacity to produce interleukin 2 (IL-2) and to provide help for B-lymphocytes. Two color immunofluorescence combined with cytofluorometry and cell sorting by means of a single laser system (Loken, Parks and Herzenberg, 1977; Oi, Glazer and Stryer, 1982; Lanier and co-workers, 1983) were applied for the analysis of T-lymphocyte subsets.
Here we report on the detection by monoclonal anti-DPP IV antibody of lymphocytes bearing DPP IV on their surface, the relationship between DPP IV and other known differentiation antigens like Leu1, Leu2, and Leu3, and the capacity of the cells to produce IL-2 and to provide helper activity for the differentiation of B-lymphocytes.

METHODS

Antibodies

A monoclonal anti-DPP IV antibody (II-19-4-7) was prepared as described previously (Scholz and co-workers, 1985; Feller and co-workers, 1986). Specificity of anti-DPP IV antibodies was proven by isoelectric focusing of purified DPP IV by immunoblotting with anti-DPP IV monoclonal antibody and by the reaction with the substrate Gly-Pro-4-methoxy-ß-naphthylamide (Lojda, 1977).
The following monoclonal antibodies were used for analyses of surface phenotypes: phycoerythrin conjugated Leu1, Leu2a and Leu3a from Becton Dickinson (Heidelberg, FRG). For detection of surface bound monoclonal antibodies, biotinylated $F(ab')_2$ fragments of goat anti-mouse IgG (Medac, Hamburg, FRG) and FITC conjugated Avidin (Vector Laboratories, Burlingame, USA) were used. For isolation of B-cells the mouse monoclonal antibody B1 (Coulter Electronics, Krefeld, FRG) was used.

Isolation and Subfractionation of Cells

Human peripheral blood mononuclear cells (PBMC) or purified T-lymphocytes were isolated by discontinuous gradient centrifugation on Percoll-Hank's as described previously (Ulmer and co-workers, 1984). In indicated experiments the cells were stimulated with an optimal concentration (8 µg/ml) of PHA (reagent grade, Wellcome, Beckenham, England), pokeweed mitogen (PWM, 3 µg/ml, Biochrom, Berlin, FRG), or 1:10,000 dilution of Staphylococcus aureus Cowan I suspension (10% w/v, Pansorbin™; Calbiochem, Frankfurt, FRG). Non-adherent, non-phagocytic cells were isolated by incubation in AB serum coated plastic flasks and by depletion of carbonyl-iron phagocytosing cells (Scholz and co-workers, 1985). B-cells were isolated by depletion of E-rosette forming cells and subsequent positive selection of B1-binding cells by rosetting with ox erythrocytes coated with goat anti-mouse IgG and IgM (Medac, Hamburg, FRG) (Gruber and co-workers, 1984). Adherent cells were collected by soft treatment with a rubber policeman. The purity of T-cells, B-cells and monocytes was $<$95%.

Immunofluorescence and Flow Cytometry

The techniques employed were described previously (Scholz and co-workers, 1985, 1986). Cells were analyzed and sorted in the cytofluorograf (System 50 H, Ortho Diagnostic Systems Inc., Raritan, N.J., USA).

Determination of IL-2 Activity

T-cells and their subsets were stimulated with PHA, and the IL-2 activity in the culture supernatants was determined by measuring the proliferation of the IL-2 dependent CTL-6 line (Ulmer and co-workers, 1984; Scholz and co-workers, 1985, 1986).

Culture of B-cells and Measurement of Immunoglobulin Production

Purified B-cells and equal numbers of T-cells were cultured in the presence of pokeweed mitogen or Staphylococcus aureus COWAN I (SAC) using an ultra-micro-culture system of 2 µl and 1000 B-cells as described (Gruber, Ulmer and Flad, 1983). The amount of immunoglobulin produced by B-cells was determined using a micro-ELISA system.

RESULTS and DISCUSSION

IL-2 Production of DPP IV$^+$ and DPP IV$^-$ Cells after Mitogenic Stimulation

Labelling of T-lymphocytes with anti-DPP IV revealed 50-60% DPP IV$^+$ cells. To test the capacity of IL-2 production, the resulting populations were sorted and stimulated with three different mitogens. As shown in a typical experiment (Table 1), DPP IV$^+$ cells secreted IL-2 in response to PHA and PWM, although in different amounts. In contrast, supernatants of DPP IV$^-$ cells showed no significant IL-2 activity. The same populations stimulated with the B-cell mitogen SAC suspension did not lead to IL-2 production.

TABLE 1 Production of IL-2 by DPP IV$^-$ and DPP IV$^+$ T-cells

	Exp.	1	2	3	4
T-cell subset	Mitogen	Interleukin 2 (units/ml)			
DPP IV$^-$	0	$<$2.5	$<$2.5	$<$2.5	$<$1.0
	SAC	$<$2.5	$<$2.5	$<$2.5	$<$1.0
	PWM	$<$2.5	$<$2.5	$<$2.5	$<$1.0
	PHA	$<$2.5	$<$2.5	8.1	$<$1.0
DPP IV$^+$	0	$<$2.5	$<$2.5	$<$2.5	$<$1.0
	SAC	$<$2.5	$<$2.5	$<$2.5	$<$1.0
	PWM	5.5	16.9	10.5	4.3
	PHA	3.9	31.0	26.7	12.1

1×10^5 T-cells/200 µl in RPMI 1640 + 2% FCS were stimulated with PHA (8 µl/ml), PWM (3 µg/ml), or Staphylococcus aureus (1:10000) in presence of 5×10^3 ANAE-positive adherent cells (monocytes) for 24 h. Aliquots of 50 µl culture supernatants were added to 2×10^3 CTL-6 cells (50 µl, RPMI 1640/10% FCS) and incubated for further 24 h. Proliferation of CTL-6 cells was determined by a 5 h pulse with ^{3}HdT. IL-2 units/ml were calculated from CPM/culture using a laboratory IL-2 standard.

IL-2 Production of Subsets Defined by the DPP IV and Leu3 Antigens

By double labelling the following four subsets of T-cells were quantified: DPP IV$^+$/Leu3$^+$ 38 $\pm$ 6%; DPP IV$^-$/Leu3$^-$ 32 $\pm$ 7%; DPP IV$^+$/Leu3$^-$ 20 $\pm$ 4%; DPP IV$^-$/Leu3$^+$ 10 $\pm$ 5% (Table 2). The subsets were isolated by sorting with a purity of >98% and tested for their IL-2 producing capacity in response to PHA. The addition of 10% adherent cells ($\geq$ 95% esterase positive monocytes) enhanced the IL-2 producing capacity and revealed for DPP IV$^+$/Leu3$^+$ cells the highest level of IL-2 production. This requirement for monocytes for optimal IL-2 production is well known (Chouaib and Fradelizzi, 1982; Solbach and co-workers, 1982). In contrast, DPP IV$^-$/Leu3$^-$ cells did not show substantial IL-2 production even in presence of adherent cells. With DPP IV$^+$/ Leu3$^-$ cells we found intermediate IL-2 production after addition of adherent cells. Within the small fraction of DPP IV$^-$/Leu3$^+$ cells, IL-2 production was strongly reduced but could be restored by the addition of 10% adherent cells.

TABLE 2 IL-2 Production of Double Fluorescence Labelled Lymphocytes Highly Purified by Cell Sorting

Phenotype[a)]	% of cells[b)]	IL-2 production[c)] Addition of adherent cells 0 %	10 %
DPP IV$^+$/Leu3a$^+$	38 ± 6	8 ± 14 (5)[d)]	159 ± 30 (6)
DPP IV$^+$/Leu3a$^-$	20 ± 4	11 ± 23 (4)	83 ± 40 (6)
DPP IV$^-$/Leu3a$^+$	10 ± 5	ND (2)	85 ± 18 (2)
DPP IV$^-$/Leu3a$^-$	32 ± 7	3 ± 5 (4)	15 ± 15 (6)
unseparated	-	31 ± 17 (4)	112 ± 29 (6)

a) Phagocyte depleted nonadherent lymphocytes were stained with anti-DPP IV and Leu3a, and the indicated subpopulations were isolated by cell sorting.
b) Phagocyte depleted nonadherent PBMC were stained with anti-DPP IV + biotin-GAM/avidin-FITC (green) and Leu3a - PE (red). Only forward light scatter gated lymphocytes were analyzed for green or red fluorescence. Mean ± SD of 9 experiments.
c) IL-2 units/ml compared with a standard preparation containing 100 units.
d) Values in paranthesis indicate the number of experiments (n). In some experiments only part of the possible subsets was examined.
ND = not detectable

Relationship between Expression of DPP IV, Leu2, Leu1, and IL-2 Production

In further experiments we examined the coexpression of DPP IV, Leu2, and Leu1. As shown in Table 3, there were more DPP IV$^+$/Leu2a$^-$ cells than DPP IV$^+$/Leu2a$^+$ cells, and these subsets corresponded to the respective Leu3a-positive or -negative subset. Also, the majority of Leu1$^+$ cells were DPP IV$^+$.
With regard to IL-2 producing capacity, DPP IV$^+$/Leu2a$^-$ cells were the highest IL-2 producers, which confirms the data obtained with the DPP IV$^+$/Leu3a$^+$ subset (Table 2). However, also DPP IV$^+$/Leu2a$^+$ cells were capable of producing IL-2, while DPP IV$^-$ cells of either Leu2a or Leu1 phenotype were producing less IL-2. To exclude that the monoclonal anti-DPP IV antibody itself activated lymphocytes, separated and unseparated cells were cultured with various amounts of the antibody. No effect on proliferation and IL-2 production was noted which suggested that the anti-DPP IV was not activating or mitogenic per se (data not shown).

Helper Activity of DPP IV$^+$/Leu3a$^+$ and DPP IV$^-$/Leu3a$^+$ Subsets in Differentiation of B-cells

To test the helper activity of DPP IV$^+$/Leu3a$^+$ and DPP IV$^-$/Leu3a$^+$ T-cells, different numbers of these cells were added to purified B-cells, and the cultures were stimulated with PWM. Immunoglobulin production was determined after seven days of culture. Surprisingly, DPP IV$^-$/Leu3a$^+$ cells proved to be more efficient

TABLE 3 IL-2 Production of Lymphocytes Highly Purified by Cell Sorting after Double Labelling with anti-DPP IV and Leu2a or Leu1

Phenotype[a)]	% of cells[b)]	IL-2 production[c)] Addition of adherent cells 0 %	10 %
DPP IV$^+$/Leu2a$^+$	14 ± 2	1	142
DPP IV$^+$/Leu2a$^-$	51 ± 6	1	205
DPP IV$^-$/Leu2a$^+$	9 ± 3	NT	NT
DPP IV$^-$/Leu2a$^-$	27 ± 5	1	79
unseparated	-	31	184
DPP IV$^+$/Leu1$^+$	63 ± 7	3	48
DPP IV$^+$/Leu1$^-$	2 ± 1	NT	NT
DPP IV$^-$/Leu1$^+$	20 ± 7	1	6
DPP IV$^-$/Leu1$^-$	15 ± 3	1	4
unseparated	-	4	33

a) Phagocyte depleted nonadherent lymphocytes were stained with anti-DPP IV and Leu2a or Leu1, and the indicated subpopulations were isolated by cell sorting. The purity of the sorted cells was greater than 95% as shown in reanalyses, and the adherent cells consisted of 97% esterase positive monocytes.

b) Phagocyte depleted nonadherent PBMC were stained with anti-DPP IV + biotin-GAM/avidin-FITC (green) and Leu2a-PE or Leu1-PE (red.) Only forward light scatter gated lymphocytes were analyzed for green or red fluorescence. Mean ± SD of 3 experiments.

c) IL-2 units/ml compared with a standard preparation containing 100 units.

NT = not tested

helper cells than DPP IV$^+$/Leu3a$^+$ when the cultures were stimulated with PWM. In contrast, there was no significant difference in helper activity of both subsets when cultures were stimulated with SAC (Fig. 1). Since the DPP IV$^+$/Leu3a$^+$ cells had been shown to have higher IL-2 producing capacity than DPP IV$^-$/Leu3a$^+$ cells, we concluded that these subsets differ with regard to their capacity of producing B-cell growth and differentiation factors other than IL-2. Studies along these lines are presently conducted in our laboratory.

ACKNOWLEDGEMENTS

The technical assistance of Mrs. F. Klützke and Mrs. R. Bergmann is gratefully acknowledged.

This work was supported by grants of German Leprosy Relief Association and Deutsche Forschungsgemeinschaft (Fl 104/4-2).

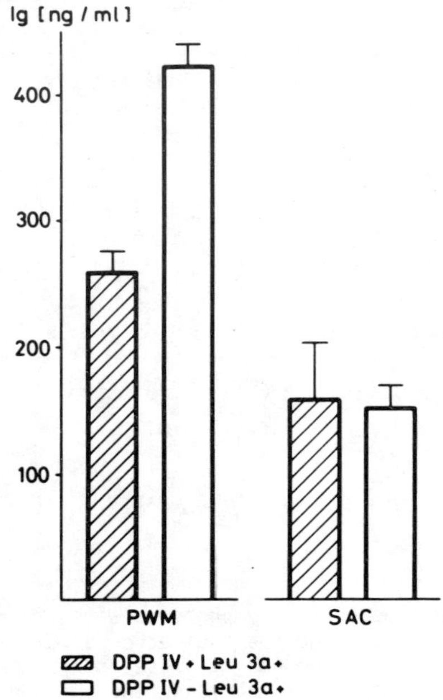

Fig. 1. Immunoglobulin production of purified B-cells cocultured with DPP IV$^+$/Leu3a$^+$ or DPP IV$^-$/Leu3a$^-$ T-cells and stimulated with PWM or SAC. Results are from one representative experiment (mean ± SD of three cultures.

REFERENCES

Chilosi, M., G. Pizzolo, F. Menestrina, A. M. Iannucci, F. Bonetti, and L. Fiore-Donati (1982). Am. J. Clin. Pathol. 77, 714-720

Chouaib, S. and D. Fradelizzi (1982). J. Immunol. 129, 2463-2468

Feller, A. C., H. J. Razun, E. Heymann, H. Haas, W. Scholz, and M. R. Parwaresch (1986). Virch. Arch. Path. Anat. 409, 263-273

Gruber, M., A. J. Ulmer, and H.-D. Flad (1983). Immunobiology 165, 267

Gruber, M., A. J. Ulmer, W. Scholz, E. Brandt, and H.-D. Flad (1984). Scand J. Immunol. 20, 171

Hopsu-Havu, V. K. and S. R. Sarimo (1967). Hoppe-Seyler's Z. Physiol. Chem., 348, 1540-1550

Lanier, L. L., A. M. Le, J. H. Phillips, N. L. Warner, and G. F. Babcock (1983). J. Immunol. 131, 1789-1796

Lojda, Z. (1977). Histochemistry 54, 299-309

Loken, M. R., D. R. Parks, and L. A. Herzenberg (1977). J. Histochem. Cytochem. 25, 899-907

Mentlein, R., R. Heymann, W. Scholz, A.C. Feller, and H.-D. Flad (1984). Cell. Immunol. 89, 11-19

Oi, V. T., A. N. Glazer, and L. Stryer (1982). Cell Biol. 93, 981-986
Püschel, G., R. Mentlein, and E. Heymann (1982). Eur. J. Biochem. 126, 359
Reinherz, E. L., P. C. Kung, G. Goldstein, and S. F. Schlossman (1979). Proc. Nat. Acad. Sci. USA 76, 4061-4065
Reinherz, E. L., P. C. Kung, G. Goldstein, and S. F. Schlossman (1980). J. Immunol. 124, 1301-1307
Schön, E., H. U. Demuth, A. Barth, and S. Ansorge (1984). Biochem. J. 223, 255-258
Scholz, W., R. Mentlein, E. Heymann, A. C. Feller, A. J. Ulmer, and H.-D. Flad (1985). Cell. Immunology 93, 199-211
Scholz, W., A. C. Feller, A. J. Ulmer, and H.-D. Flad (1986). In: E. L. Reinherz, B. F. Haynes, L. M. Nadler, and I. D. Bernstein (Eds.), Leukocyte Typing, Vol. 1, Springer, pp. 489-501
Solbach, W., S. Barth, M. Röllinghoff, and H. Wagner (1982). Clin. Exp. Immunol. 49, 167-175
Ulmer, A. J., W. Scholz, M. Ernst, E. Brandt, and H.-D. Flad (1984). Immunobiology 166, 238-250
Walter, R., W. H. Simmons, and T. Yoshimoto T. (1980). Mol. Cell. Biochem. 30, 111-127

Cathepsin B Activity in Unfixed and Unrecalcified Cryostat Sections of Normal and Arthritic Mouse Knee Joints

C. J. F. van Noorden and I. M. C. Vogels

Laboratory of Histology and Cell Biology, University of Amsterdam, Academic Medical Centre, Meibergdreef 15, 1105 AZ Amsterdam, The Netherlands

ABSTRACT

The possible involvement of cathepsin B in the destruction of articular cartilage in arthritic lesions was investigated by histochemical means. Unfixed and undecalcified cryostat sections of normal and antigen-induced arthritic knee joints were used. Sections with good morphology were obtained by attaching adhesive tape to the cutting surface of the tissue blocks. Cathepsin B activity was demonstrated by a post-azo-coupling method. Final reaction product indicating cathepsin B activity was localized in small granules within the cytoplasm of synoviocytes, fibroblasts, chondrocytes and osteoclasts. Leupeptin and E-64 inhibited the formation of final reaction product completely. Induction of arthritis resulted in a vast increase of cathepsin B activity in synoviocytes, fibroblasts and chondrocytes and therefore, cathepsin B could be involved in the destruction of articular cartilage.

KEYWORDS

Cathepsine B; histochemistry; proteinase; arthritis; articular cartilage; knee joint; synovium; fibroblast; chondrocyte; osteoclast.

Unfixed and undecalcified cryostat sections of mouse knee joints have been used for the histochemical detection of cathepsin B activity in view of its possible role in the genesis of arthritic lesions. In the literature, several attempts have been reported to include enzyme histochemical methods in the study of the development of arthritic lesions but have been hampered by difficulties in obtaining cryostat sections with a good morphology from undecalcified joints (Henderson, 1983; Van Noorden and Vogels, 1986). Fixation and decalcification result in a better preservation of the morphology but also lead to rapid inactivation of many enzymes. In the present study, unfixed and undecalcified cryostat sections with a good morphology were prepared by using adhesive tape on the cutting surface (Ullberg, 1954).

Non-inflamed knee joints and knee joints of mice with antigen-induced arthritis have been used in our study. Arthritis was induced by intra-articular injection of methylated bovine serum albumin (mBSA) into the right knee joint of animals having been immunized previously with mBSA in saline according to Brackertz,

© 1987 Pergamon Journals Ltd.
Printed in Great Britain

Mitchell and Mackay (1977). The left knee joint received saline only and served as control. 5 days after the induction of arthritis, when the right knee joint was severely affected by inflammation (Lens, Van den Berg and Van de Putte, 1984), both knee joints were dissected in toto and skin and superficial layers were removed subsequently. The joints were embedded in an aquous solution of 8% (w/v) gelatin white and frozen in liquid nitrogen (Altman, 1978). Sections (thickness, 10 μm) were cut at low speed with a motor-driven cryostat fitted with a tungsten carbide knife at an obtuse angle (10-20°; Pearse and Gardner, 1972; Johnstone, Bitensky and Chayen, 1972). The sections were attached to transparant tape (Scotch tape 800) to keep the integrity of the tissue (Fig. 1). The tape was fastened with a brush onto the section surface of the block. The microtome knife then cut underneath the tape and sections attached to the tape were prepared. Pieces of tape with adherent sections were fixed on glass slides with ordinary tape.

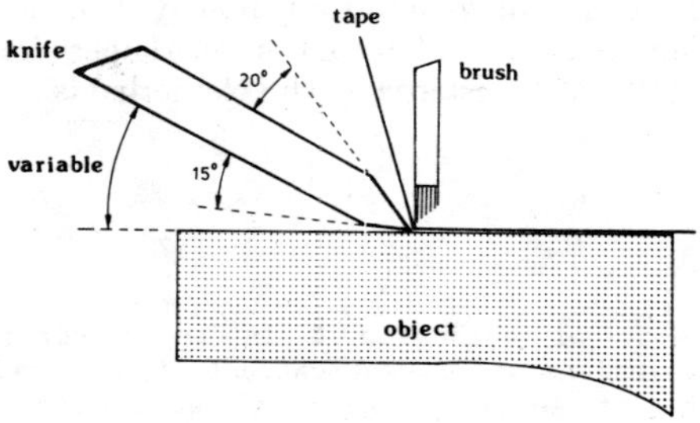

Fig. 1. Schematic representation of sectioning of cryostat sections attached to adhesive tape. Reprinted with permission from Rijntjes and others (1979).

Cathepsin B activity was demonstrated with a post-azo-coupling method which has been developed in our laboratory (Van Noorden and Vogels, 1986). Simultaneous coupling methods are not applicable for the histochemical demonstration of the activity of cysteine proteinases such as cathepsin B. Thiol compounds are essential for these proteinases whereas diazonium salts are destroyed by thiol compounds (Barrett, 1979; Kugler, 1982; Lojda, 1984). Moreover, diazonium salts strongly inhibit the activity of cysteine proteinases (Barrett, 1979; Graf, Baici and Sträuli, 1979; Gossrau, 1981, 1983).

Sections were incubated for 15 min at 37°C in a medium consisting of 100 mM phosphate buffer (pH 6.0), 1 mg/ml N-CBZ-L-alanyl-L-arginyl-L-arginyl-L-arginine-4-methoxy-2-naphthylamide.2HCl, 1.3 mM EDTA, 10 mM dithiothreitol and 2.7 mM L-cysteine. Control reactions were performed in the presence of 5 μM leupeptin or 5 μM E-64 (Barrett and others, 1982). After incubation, the sections were rinsed thoroughly for 5 min at 37°C with 10 mM N-ethylmaleimide in 100 mM phosphate buffer (pH 8.0) in order to remove all SH-groups. Post-coupling was carried out at room temp for 15 min using 0.5 mM Fast Blue BB dissolved in 100 mM phosphate buffer (pH 8.0). At pH 8.0, naphthylamine derivatives are very little water-soluble (Raap, 1983), whereas the capture reaction with Fast Blue BB is much faster at an alkaline pH. After coupling, the sections were rinsed with 0.9% NaCl solution and treated with 100 mM copper sulphate for 15 min at room temp in order to enhance and further insolubilize the final reaction product (McCabe and Chayen, 1965). Afterwards, the sections were rinsed and embedded in glycerol-gelatin jelly.

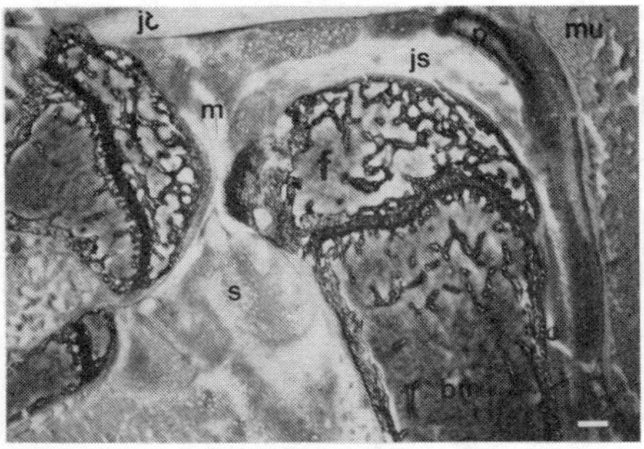

Fig. 2. Photomicrograph of a cryostat section of a mouse knee joint with induced arthritis after staining with Giemsa. js, joint space; m, meniscus; p, patella; s, synovium (inflamed); bm, bone marrow; mu, muscle; jc, joint capsule; f, femur; t, tibia. Bar = 500 μm.

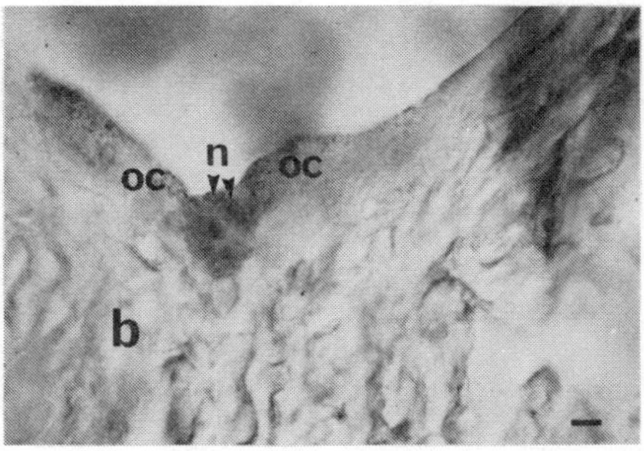

Fig. 3. Osteoclast-like cell in a cryostat section attached to tape of a control mouse knee joint and stained for cathepsin B activity. Final reaction product is precipitated in small lysosome-like granular structures within the cytoplasm. b, bone; oc, osteoclast; n, nuclei. Bar = 10 μm.

Figure 2 shows a low power photomicrograph of an arthritic knee joint after Giemsa staining. The integrity of the sections was well kept and the morphological aspect of the joint tissue was good; dislocation of especially the softer tissues such as synovium or bone marrow did not occur. Serial sections were obtained without any difficulty and nonspecific staining of the adhesive tape was never observed.

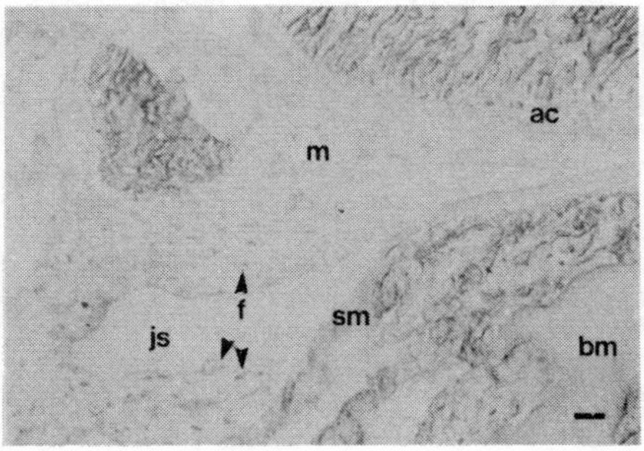

Fig. 4. Cryostat section of a control mouse knee joint after staining for cathepsin B activity. m, meniscus; js, joint space; sm, synovial membrane; ac, articular cartilage; bm, bone marrow, f, fibroblast. Bar = 40 μm.

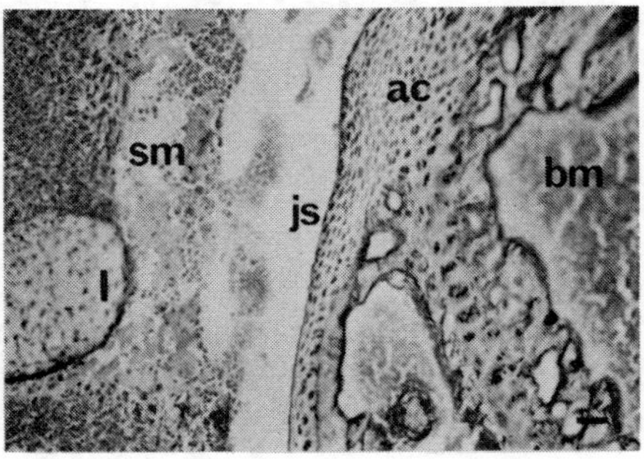

Fig. 5. Cryostat section of a mouse knee joint with induced arthritis after staining for cathepsin B activity. js, joint space; sm, synovial membrane; l, ligament; ac, articular cartilage; bm, bone marrow. Bar = 40 μm.

Final reaction product indicating cathepsin B activity was localized in small granules within the cytoplasm of synoviocytes, fibroblasts, chondrocytes and osteoclasts (Fig. 3). Osteoclasts contained a considerable activity under normal conditions whereas synoviocytes, fibroblasts and chondrocytes showed a moderate activity (Fig. 4). Induction of arthritis lead to a vast increase in cathepsin B activity in synoviocytes, fibroblasts and chondrocytes but the activity in osteoclasts did not seem to be affected very much (Fig. 5). The reaction appeared to be specific since proteinase inhibitors such as leupeptin and E-64 inhibited the reaction completely in all cells of both control and inflamed knee joints (Fig. 6).

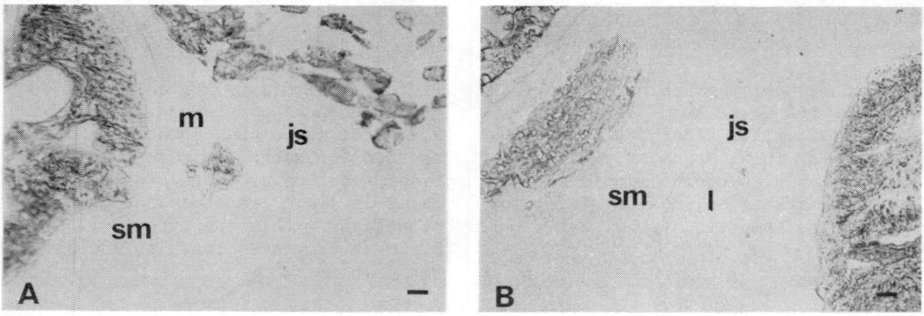

Fig. 6. Cryostat sections of mouse knee joints stained for cathepsin B activity. A, control joint incubated in the presence of leupeptin; B, inflamed joint incubated in the presence of E-64. m, meniscus; js, joint space; sm, synovial membrane; l, ligament. Bar = 40 μm.

It is concluded that tape-attached cryostat sections allow the application of enzyme histochemistry to tissue structures of varying hardness without previous fixation or embedding in order to prevent dislocation of structures or disturbance of histology (Van Noorden and Vogels, 1986). Moreover, localization of cathepsin B activity in these cryostat sections is acceptable at the cellular level. In this connection we like to state that localization of activities of cysteine proteinases is always troublesome and recrystallization of the final reaction product occurs readily (Dolbeare and Vanderlaan, 1979; Graf, Baici and Sträuli, 1981; Kugler, 1982; Gossrau, 1983; Lojda, 1984). The final reaction product is due to cathepsin B activity and may not only be used for a topological demonstration of its activity, but also merits further investigation for quantitative purposes. The high activity of cathepsin B in synoviocytes, fibroblasts and chondrocytes in arthritic lesions strongly suggests that cathepsin B activity could be involved in the destruction of articular cartilage.

ACKNOWLEDGEMENTS

The authors wish to thank mrs T.M.S. Pierik for the careful preparation of the manuscript.

REFERENCES

Altman, F.P. (1978). A gelatin embedding technique as an aid in the preparation of unfixed cryostat sections. Histochem. J., 10, 617-620.

Barrett, A.J. (1979). Cathepsin B and other thiol proteinases. In A.J. Barrett (Ed.) Proteinases in mammalian cells and tissue. North Holland, Amsterdam, p.p. 182-208.

Barrett, A.J., Kembhavi, A.A., Brown, M.A., Kirschke, H., Knight, C.G., Tamai, M., and Hanada, K. (1982). L-trans-Epoxysuccinyl-leucylamido(4-guanidino)-butane (E-64) and its analogues as inhibitors of cysteine proteinases including cathepsins B, H and L. Biochem. J., 201, 189-198.

Brackertz, D., Mitchell, G.F., and Mackay, T.R. (1977). Antigen-induced arthritis in mice. Arthr. Rheum., 20, 841-850.

Dolbeare, F., and Vanderlaan, M. (1979). A fluorescent assay of proteinases in cultured mammalian cells. J. Histochem. Cytochem., 27, 1493-1495.

Gossrau, R. (1981). Investigation of proteinases in the digestive tract using 4-methoxy-2-naphthylamine (MNA) substrates. J. Histochem. Cytochem., 29, 464-480.

Gossrau, R. (1983). Fluorescence histochemical detection of hydrolases in tissue sections and culture cells. Histochemistry, 79, 87-94.

Graf, M., Baici, A., and Sträuli, P. (1981). Histochemical localization of cathepsin B at the invasion front of the rabbit V2 carcinoma. Lab. Invest., 45, 587-596.

Henderson, B. (1983). The application of quantitative cytochemistry to the study of diseases of the connective tissues. Progr. Histochem. Cytochem., 15/1, 1-86.

Johnstone, J., Bitensky, L., and Chayen, J. (1972). Cryostat sections of undemineralized bone. J. Clin. Path., 25, 742.

Kugler, P. (1982). Fluorescent histochemical demonstration of cathepsin B in the rat yolk sac. Histochemistry, 75, 215-218.

Lens, J.W., Van den Berg, W.B., and Van de Putte, L.B.A. (1984). Quantitation of arthritis by ^{99m}Tc-uptake measurements in the mouse knee joint: correlation with histological joint inflammation scores. Agents Actions, 14, 723-728.

Lojda, Z. (1984). Die Histochemie der Proteasen. Acta Histochem., Suppl. Bd. 30, 9-28.

McCabe, M., and Chayen, J. (1965). The demonstration of latent particulate aminopeptidase activity. J. Roy. Micr. Soc., 84, 361-371.

Pearse, A.D., and Gardner, D.L. (1972). Preparation of unfixed undemineralized bone sections: the Bright bone cryostat. J. Clin. Path., 25, 26-29.

Raap, A.K. (1983). Localization mechanisms in enzyme histochemistry. PhD Thesis, University of Leiden, pp. 115-131.

Rijntjes, N.V.M., Van de Putte, L.B.A., Van der Pol, M., and Guelen, P.J.M. (1979). Cryosectioning of undecalcified tissues for immunofluorescence. J. Immunol. Methods, 30, 263-268.

Ullberg, S. (1954). Studies on the distribution and fate of S^{35} labeled benzylpenicellin in the body. Acta Radiol., Suppl. 118.

Van Noorden, C.J.F., and Vogels, I.M.C. (1986) Enzyme histochemical reactions in unfixed and undecalcified cryostat sections of mouse knee joints with special reference to arthretic lesions. Histochemistry, 86, in press.

PROCESSING AND ACTIVITY OF BIOACTIVE PEPTIDES

Human β-Casomorphin-7 Immunoreactive Material in the Plasma of Women During Pregnancy and After Parturition

G. Koch*, K. Wiedemann*, E. Drebes*, W. Zimmermann[+], G. Link[+] and H. Teschemacher*

*Rudolf Buchheim-Institut für Pharmakologie, Frankfurter Straße 107, 6300 Gießen; Zentrum für Frauenheilkunde und Geburtshilfe, Klinikstraße 32, 6300 Gießen, Federal Republic of Germany

ABSTRACT

Human ß-casein contains fragments with opioid activity, termed "human ß-casomorphins". The present study was undertaken to investigate whether ß-casomorphins or fragments thereof may be cleaved from ß-casein in mammary tissue or elsewhere in the organism of women during pregnancy or after delivery and may eventually appear in the women's blood. Thus, a method for the extraction of $ß_H$-casomorphin-7 immunoreactive material from human plasma and a radioimmunoassay for its determination in the plasma extracts were developed. Blood samples were collected from 35 pregnant women and from 138 women after delivery and in addition, samples were taken from four men and sixteen nonpregnant women at different stages of the ovarian cycle as controls.

No $ß_H$-casomorphin-7 immunoreactive material was detectable in the plasma of the four men and the sixteen nonpregnant women. However, $ß_H$-casomorphin-7 immunoreactive material was present in the plasma of 10 out of 35 pregnant women and in the plasma of 43 out of 138 women after parturition.

It is suggested that the presence of $ß_H$-casomorphin-7 immunoreactive material in the plasma of women during pregnancy and after delivery reflects the proliferative and secretory changes which the mammary gland undergoes during that time.

KEYWORDS

ß-casein; ß-casomorphins; milk; opioid peptides; radioimmunoassay; breast-feeding; pregnancy.

INTRODUCTION

Recently, bovine ß-casein has been shown to contain segments with opioid activity, $ß_B$-casein (60-63), (60-64), (60-65) and (60-66); they have been termed "ß-casomorphins" (Brantl and colleagues, 1979; Henschen and colleagues, 1979; Lottspeich and colleagues, 1980).

© 1987 Pergamon Journals Ltd.
Printed in Great Britain

Opioid effects of these peptides have been reported (for review see Brantl and Neubert, 1986). Very recently, fragments of human ß-casein, located in positions corresponding to those bovine ß-casein fragments, have been demonstrated to behave like opioids, as well (Brantl, 1984; Yoshikawa and colleagues, 1984; Koch and colleagues, 1985). They were named human ß-casomorphins (Brantl, 1984).

The present study was undertaken to investigate whether human ß-casomorphins or ß-casomorphin-related peptides may be cleaved from ß-casein and may occur in the blood of women during pregnancy and after delivery.

MATERIALS AND METHODS

Plasma Extraction

Plasma samples were obtained from 35 pregnant women, from 138 women after delivery, from sixteen nonpregnant women at different stages of the ovarian cycle and from four men.

For extraction of plasma a method described by Wiedemann and Teschemacher (1986) was used in a modified form. Octadecasilyl silica cartridges (SEP PAK C 18 cartridges, Waters Ass., Milford, U.S.A.) were prepared by successive application of 5 ml methanol, 5 ml 8 M urea and 10 ml distilled water. Then 5 ml aliquots of acidified plasma were applied to each cartridge followed by 10 ml distilled water and 10 ml 4% acetic acid (Cahill and Akil, 1982). Retained peptides were eluted with 5 ml of a mixture of ethanol and acetic acid (96/4; v/v). The organic solvent of the eluate was evaporated using a SPEED VAC concentrator (Savant, New York, U.S.A.); the remaining aqueous phase was lyophilized and the residue stored at -20°C until further processing. For radioimmunoassay or chromatography, the residue was reconstituted in 500 µl of buffer 3 (see below) and is henceforth referred to as "extract".

The extraction recovery rates, the intra- and inter-assay coefficients of variation for the extraction method and the linearity of recovery were determined as described by Wiedemann and Teschemacher (1986).

Radioimmunoassay

For the radioimmunoassay, antisera against $ß_H$-casomorphin-7 were raised in rabbits according to a method described by Weber and colleagues (1982) with modifications (Wiedemann and Teschemacher, 1986). $ß_H$-casomorphin-7 was radioiodinated (^{125}I; Amersham-Buchler, Braunschweig, F.R.G.) and purified as described (Wiedemann and Teschemacher, 1986) using the chloramine T method (Hunter and Greenwood, 1962). For purification of the labelled peptide a Bio-Gel P2 column was used.

The assay procedure was performed at 0°C. The buffers used were: buffer 1 (0.1 M bis-tris-propane buffer, pH 7.0), buffer 2 (buffer 1 with 0.0005% gamma-globulin; w/v) and buffer 3 (buffer 2 with 0.1% triton X-100; w/v).

For the assay, to 200 µl of buffer 1 were added, in succession, 50 µl buffer 3 ("zero-standard") or 50 µl $ß_H$-casomorphin-7 in buffer 3 ("standard") or 50 µl plasma extract ("sample") and 100 µl antiserum solution (rabbit $ß_H$-casomorphin-7 antiserum diluted 1:5000 with buffer 2). Radioiodinated $ß_H$-casomorphin-7 (50 µl) was added 24 hours later. Again 24 hours later the incubation was terminated by addition of 300 µl of a charcoal suspension (1 g charcoal, 0.5 g bovine serum albumin in 100 ml buffer 1). The samples were incubated for 15 minutes, centrifuged (12 000 g, 5 min) and 550 µl of the supernatant were transferred to another tube. Radioactivity in the supernatant was determined in a gammacounter (Kontron, Zürich, Switzerland).

The intra-assay coefficient of variation for one radioimmunoassay was calculated from the determination of immunoreactivity in 10 aliquots from a sample; the inter-assay coefficient of variation was calculated from the IC_{50} values of the standard curves in ten differnt radioimmunoassays.

The minimum detectable amount of $ß_H$-casomorphin-7 in one assay tube was derived from 10 standard curves for 10% inhibition of tracer binding.

RESULTS

Determination of $ß_H$-Casomorphin-7 Immunoreactive Material in Human Plasma: Characteristics of the Method

The recovery rate for extraction of $ß_H$-casomorphin-7 from plasma was 45%. The correlation coefficient (r) for the linearity of recovery of $ß_H$-casomorphin-7 was 0.98. Both, the intra- and the inter-assay coefficient of variation for the extraction recovery rate were 3.5%. The intra- and inter-assay coefficients of variation for the radioimmunoassay were 3.4% and 10.6%, respectively. The minimum detectable amount of $ß_H$-casomorphin-7 in the radioimmunoassay was found to be 2 ± 0.6 fmol per assay tube.

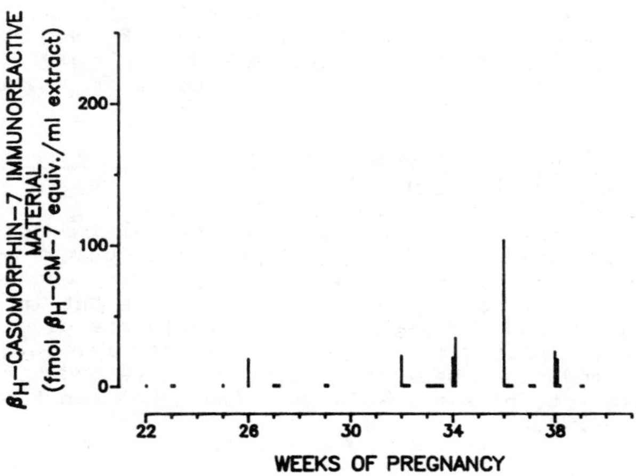

Fig. 1. $ß_H$-casomorphin-7 immunoreactive material in the plasma of pregnant women. Mean values from duplicate determinations in each plasma sample are shown.

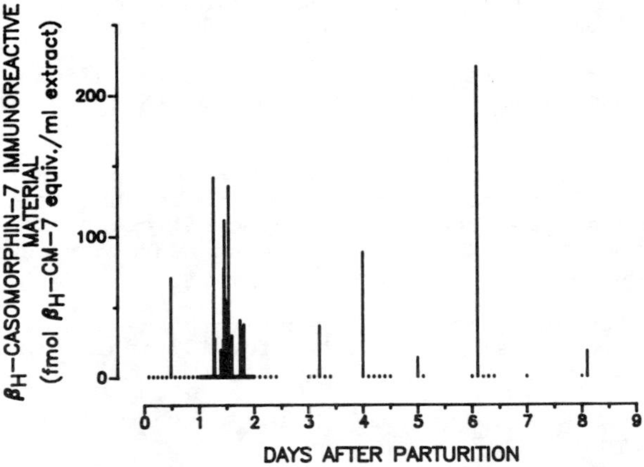

Fig. 2. $ß_H$-casomorphin-7 immunoreactive material in the plasma of women after parturition. Mean values from duplicate determinations in each plasma sample are shown.

The cross reactivities of the $ß_H$-casomorphin-7 antiserum used for the radioimmunoassay with various peptides were calculated as 100 times the ratio of the IC_{50} of unlabelled $ß_H$-casomorphin-7 to the IC_{50} of the tested peptide. The cross reactivities were 0.34% for $ß_H$-casomorphin-8, 0.02% for $ß_H$-casomorphin-5 and -9 and less than 0.01% for $ß_H$-casomorphin-4, $ß_H$-casomorphin-(1-2), -(2-3), -(3-4), -(4-5), -(5-6), -(6-7), -(7-8), -(8-9), -(4-9), $ß_H$-casein, $ß_B$-casomorphin-4, -5, -7, -8, $ß_B$-casomorphin-4 amide, human ß-endorphin, leu-enkephalin, angiotensin and adrenocorticotropic hormone.

Human ß-Casomorphin-7 Immunoreactive Material in the Plasma of Women During Pregnancy and after Parturition

Using the developed methods, no $ß_H$-casomorphin-7 immunoreactive material could be demonstrated in the plasma of four men and sixteen nonpregnant women at various stages of the ovarian cycle. However, $ß_H$-casomorphin-7 immunoreactive material was present in the plasma of 10 out of 35 pregnant women (Fig. 1) and in the plasma of 43 out of 138 women after parturition (Fig. 2). Plasma levels of immunoreactive $ß_H$-casomorphin-7 seem to rise from the second trimester of pregnancy; highest plasma levels were found between the first and sixth day postpartum.

DISCUSSION

In this study $ß_H$-casomorphin-7 immunoreactive material has been demonstrated to occur in the plasma of women during pregnancy and after delivery. The material was absent in the plasma of men and nonpregnant women at different stages of the ovarian cycle.

Plasma levels of immunoreactive $ß_H$-casomorphin-7 seem to rise from the second trimester of pregnancy until parturition. Coincidentally from the second trimester on the breasts are capable of milk secretion provided that the pregnancy is terminated (Weiss, 1977). Interestingly, immunoreactive $ß_H$-casomorphin-7 as a mammary gland product appears in the blood at a time when no milk is secreted. Possibly this rise of immunoreactive $ß_H$-casomorphin-7 in plasma reflects the proliferative changes which the mammary gland undergoes during that time. Between the first and sixth postpartum days an increase above the plasma levels at the end of pregnancy is observed. This might reflect the onset of copious milk secretion after delivery. After the sixth day postpartum plasma levels of immunoreactive $ß_H$-casomorphin-7 fall again.

So far we can only speculate on the factors which cause the appearance of this material in the pregnant and nursing women's blood and on its destination . After delivery the mechanical strain, which is put upon the mother's breast when nursing the newborn, an increased intramammary pressure, caused by the first onset of milk secretion and the disappearance of the desmosomes in the alveolar cells during lactation (Pitelka, 1978) may allow milk proteins to enter the blood stream through the intercellular junctions.

But also hormonally controlled release of milk proteins into the blood is conceivable. Hormonal control mechanisms for the regulation of alpha-lactalbumin levels in plasma, as suggested by Martin and colleagues (1980), might also be considered for an eventual release of immunoreactive $ß_H$-casomorphin-7 since its plasma levels obviously run in parallel to those of alpha-lactalbumin. It is tempting to speculate that $ß_H$-casomorphin-7 or related peptides could, thus, be released to participate in the regulatory control of milk production. Indeed, intraperitoneal administration of $ß_B$-casomorphin-7 led to an increase of prolactin levels in rat plasma (Nedvidkova and colleagues, 1985).

Furthermore ß-casomorphins might also have pathologic significance. ß-Casomorphin-related peptides have been suggested to occur in the cerebrospinal fluid of women with postpartum psychosis (Lindström and colleagues, 1984).

To conclude with, after having demonstrated the presence of $ß_H$-casomorphin-7 immunoreactive material in the plasma of women during pregnancy and after delivery the possible biological significance of this material remains to be elucidated.

ACKNOWLEDGEMENTS

The excellent technical assistance of Ms. E. Drebes is greatly acknowledged and we wish to thank Dr. M.L. Groves for the gift of human ß-casein. This work was supported by Deutsche Forschungsgemeinschaft (Te.73/7-2)

REFERENCES

Brantl, V. (1984). Novel opioid peptides derived from human ß-casein: human ß-casomorphins. Eur. J. Pharmacol., 106, 213-214.
Brantl, V., H. Teschemacher, A. Henschen and F. Lottspeich (1979).

Novel opioid peptides derived from casein (ß-casomorphins). I. Isolation from bovine casein peptone. Hoppe-Seyler's Z. Physiol. Chem., 360, 1211-1216.

Brantl, V. and K. Neubert (1986). Opioid peptides derived from food proteins. Trends Pharmacol. Sci., 7, 6-7.

Cahill, C. and H. Akil (1982). Plasma ß-endorphin-like immunoreactivity, self reported pain perception and anxiety levels in women during pregnancy and labor. Life Sci., 31, 1871-1874.

Henschen, A., F. Lottspeich, V. Brantl and H. Teschemacher (1979). Novel opioid peptides derived from casein (ß-casomorphins). II. Structure of active components from bovine casein peptone. Hoppe-Seyler's Z. Physiol. Chem., 360, 1217-1224.

Hunter, W.M. and F.C. Greenwood (1962). Preparation of iodine-131 labelled human growth hormone of high specific activity, Nature, 194, 495-496.

Koch, G., K. Wiedemann and H. Teschemacher (1985). Opioid activities of human ß-casomorphins. Naunyn-Schmiedeberg's Arch. Pharmacol., 331, 351-354.

Lindström, L.H., F. Nyberg, L. Terenius, K. Bauer, G. Besev, L.M. Gunne, S. Lyrenäs, G. Willdeck-Lund and B. Lindberg (1984). CSF and plasma ß-casomorphin-like opioid peptides in post-partum psychosis. Am. J. Psychiatry, 141, 1059-1066.

Lottspeich, F., A. Henschen, V. Brantl and H. Teschemacher (1980). Novel opioid peptides derived from casein (ß-casomorphins). III. Synthetic peptides corresponding to components from bovine casein peptone. Hoppe-Seyler's Z. Physiol. Chem., 361, 1835-1839.

Martin, R.H., M.R. Glass, C. Chapman, G.D. Wilson and K.L. Woods (1980). Human alpha-lactalbumin and hormonal factors in pregnancy and lactation. Clin. Endocrinol., 13, 223-230.

Nedvidkova, J., E. Kasafirek, A. Dlabac and V. Felt (1985). Effect of ß-casomorphin and its analogues on serum prolactin in the rat, Exp. Clin. Endocrinol., 85, 249-252.

Pitelka, D. (1978). Cell contacts in the mammary gland. In M. Peaker (Ed.), Lactation: A Comprehensive Treatise, Vol. 4, Academic Press, New York. pp. 41-66.

Weber, E., C. Evans, J.-K. Chang and J.D. Barchas (1982). Antibodies specific for alpha-N-acetyl-ß-endorphins: Radioimmunoassays and detection of actylated ß-endorphins in pituitary extracts. J. Neurochem., 38, 436-447.

Weiss, G., E.M.O. Facog, J.A. Hochman, L.T. Goldsmith, I. Rifkin and B.G. Steinetz (1977). Secretion of progesterone and relaxin by the human corpus luteum at midpregnancy and at term. Obstet. Gynecol. 50, 679-681.

Wiedemann, K. and H. Teschemacher (1986). Determination of ß-endorphin and fragments thereof in human plasma using high-performance liquid chromatography and a multiple radioimmunoassay system. Pharm. Res., 3, 142-149.

Yoshikawa, M., T. Yoshimura and H. Chiba (1984). Opioid peptides from human ß-casein. Agric. Biol. Chem., 48, 3185-3187.

Casein-derived Opioid Peptides: Physiological Significance?

H. Teschemacher

Rudolf Buchheim-Institut für Pharmakologie der Justus Liebig-Universität, Frankfurter Str. 107, D-6300 Gießen, Federal Republic of Germany

ABSTRACT

Recently, bovine α-casein as well as bovine or human ß-casein have been demonstrated to contain segments which display opioid activity.

Several studies have been performed to obtain information, whether ß-casein segments with opioid activity, named "ß-casomorphins", might be cleaved from ß-casein under in vivo conditions to elicit opioid or non-opioid effects. Indeed, ß-casomorphins or ß-casomorphin-immunoreactive materials have been found in small intestine contents of mini-pigs after oral administration of a bovine casein preparation or in small intestine contents of adult humans after bovine milk ingestion, respectively. Further, ß-casomorphin immunoreactive materials have been demonstrated in the plasma of new-born calves after their first milk intake and in the plasma of pregnant women and in lactating or non-lactating mothers. Thus, ß-casomorphins obviously are present or are likely to be generated in adult or in new-born mammals after milk intake or even in pregnant or nursing females to participate eventually in the regulation of gastrointestinal or other functions of the respective organism.

In summary, although a physiological role of casein-derived opioid peptides remains to be elucidated, all data raised so far seem to be compatible with a regulatory role of these peptides in the frame of mammalian nutrition or reproduction.

KEYWORDS

α-Casein-derived exorphins; ß-casein; ß-casomorphins; opioid peptides; food peptides; mammalian reproduction.

INTRODUCTION

Milk contains numerous substances with well-known significance in the mammalian organism, such as hormones (Koldovsky, 1980) or immunocompetent compounds (Hanson, 1982); however, it is not clear so far, which physiological role many of these substances play as milk con-

© 1987 Pergamon Journals Ltd.
Printed in Great Britain

stituents. Recently, segments of α- and ß-casein have been demonstrated to display opioid activity and a series of in vitro and in vivo studies have been performed to obtain information, whether these milk constituents might be nutrients with regulatory functions or might even be of significance in the frame of mammalian reproduction as suggested for other opioid peptides (Zagon and colleagues, 1982).

CASEIN-DERIVED OPIOID PEPTIDES: ORIGIN, AMINO ACID SEQUENCES AND EFFECTS

Recently, bovine α-casein (Mercier and colleagues, 1970) has been shown to contain fragments with opioid activity (Loukas and colleagues, 1983; Zioudrou and Klee, 1979; Zioudrou and colleagues, 1979), bovine α-casein (90-96), ARG-TYR-LEU-GLY-TYR-LEU-GLU, as well as fragments thereof, i.e. bovine α-casein (90-95), (91-96) or (91-95). These "α-casein-derived exorphins" have been shown to interact with opioid receptors in rat brain membranes. Further, they can inhibit adenylate cyclase in NG 108-15 cells and they also inhibit the contractions of the electrically stimulated mouse vas deferens; the effects on adenylate cyclase and mouse vas deferens are antagonized by the specific opioid antagonist naloxone. All opioid peptides so far known except two of the α-casein-derived opioids, bovine α-casein (90-95) and (90-96), carry a TYR residue at their N-terminus which interacts with opioid receptors. These two peptides carry an ARG residue in N-terminal position and, amazingly, display even higher opioid activities than their analogues with N-terminal TYR residues (Loukas and colleagues, 1983).

Furthermore, bovine (Ribadeau-Dumas and colleagues, 1972) and human ß-casein (Greenberg and Groves, 1984) have been demonstrated to contain fragments with opioid activity, as well. Opioid peptides derived from bovine (Brantl and colleagues, 1979; Henschen and colleagues, 1979; Lottspeich and colleagues, 1980) or human ß-casein (Brantl, 1984; Yoshikawa and colleagues, 1984; Koch and colleagues, 1985) have been termed "bovine" (Brantl and colleagues, 1979) or "human ß-casomorphins" (Brantl, 1984) respectively; opioid activity has been shown to be displayed by bovine ß-casomorphin-8, corresponding to bovine ß-casein (60-67), TYR-PRO-PHE-PRO-GLY-PRO-ILE-PRO, or by human ß-casomorphin-8, corresponding to human ß-casein (51-58), TYR-PRO-PHE-VAL-GLU-PRO-ILE-PRO, and their respective fragments. It should be mentioned that the bovine ß-casomorphin amino acid sequence is also contained in the amino acid sequence of buffalo (Petrilli and colleagues, 1983) or ovine ß-casein (Richardson and Mercier, 1979).

Characteristic opioid activities such as interaction with opioid receptors in rat brain membranes or effects on the guinea pig ileum "strip" have been demonstrated for bovine (Brantl and colleagues, 1981) as well as for human ß-casomorphins (Yoshikawa, 1984; Brantl 1984; Koch and colleagues, 1985). Further opioid effects of bovine ß-casomorphins on the central nervous system, such as analgetic (Brantl and colleagues, 1981; Grecksch and colleagues, 1981; Chang and colleagues, 1982), electrophysiological (Reymann and colleagues, 1985) or behavioural effects (Panksepp and colleagues, 1984) have been reported as well as opioid effects on the endocrine system, e.g. on the release of insulin (Schusdziarra and colleagues, 1983a), somatostatin (Schusdziarra and colleagues, 1983b), pancreatic polypeptide (Schusdziarra and colleagues, 1983c), thyrotropin (Mitsuma and colleagues, 1984) or prolactin (Nedvidkova and colleagues, 1985). Opioid effects in non-mammals have also been shown (Karabensch and colleagues, 1985).

Non-opioid interactions of bovine ß-casomorphins with the cardiovascular (Hoffmann and colleagues, 1986), or the neuroendocrine system (Zadina and Kastin, 1985) or of a human ß-casomorphin fragment with the immune system (Parker and colleagues, 1984) have also been demonstrated. Some of these as well as further opioid effects have been reviewed (Brantl and Teschemacher, 1983; Brantl and Neubert, 1986).

POSSIBLE PHYSIOLOGICAL SIGNIFICANCE OF CASEIN-DERIVED OPIOID PEPTIDES

Milk is a characteristic of the mammalian species and milk constituents may be assumed to play an important role as "food hormones" (Morley, 1982); these food hormones might participate in the regulation of gastrointestinal functions or, moreover, might participate in the regulation of reproductive processes in the mammalian species. Thus, various efforts have been undertaken to obtain information about such a physiological significance of milk constituents with opioid activity, i.e. casein-derived opioid peptides. A prerequisite for such a significance would be their cleavage from α- or ß-casein, respectively, their presence for a certain time and, finally, their degradation in the respective mammalian organism, i.e. the adult or the newborn mammal after milk intake or even the pregnant or nursing female. Thus, in vitro studies have been performed which may demonstrate that there is in fact a chance for casein fragments with opioid activity to be present in the respective organism for a limited time. Conclusive evidence for the presence of casein-derived opioid peptides under in vivo conditions can only result from in vivo studies, however.

Evidence for Release and Degradation of Casein-derived Opioid Peptides Under In Vivo Conditions: In Vitro Studies

α-Casein-derived exorphins. In pepsin hydrolysates of bovine α-casein, fragments with opioid activity, α-casein (90-95) and (90-96), have been found (Zioudrou and colleagues, 1979; Loukas and colleagues, 1983). These peptides proved resistant to trypsin or proteolytic enzymes of rat brain membranes, but are degraded by chymotrypsin, subtilisin and pronase. Thus, there is a chance that α-casein-derived opioid peptides are liberated after milk intake in the stomach of mammals to elicit opioid effects there; effects in more caudally located sections of the gastrointestinal tract are less likely in view of the peptides' sensitivity to chymotrypsin etc.

ß-Casomorphins. Upon incubation of buffalo ß-casein with proteolytic enzymes adapting gastrointestinal digestion, "ß-casomorphin-8" was released at the "brush border stage" (Petrilli and colleagues, 1984). From enzymatic bovine casein digests ß-casomorphins or derivatives thereof, respectively, have been isolated (Brantl and colleagues, 1979; Chang and colleagues, 1985) and in a trypsin digest of delipidated human casein a human ß-casomorphin fragment was found (Parker and colleagues, 1984). Finally, upon digestion of bovine milk designed as an imitation of the natural gastrointestinal digestion process, ß-casomorphin immunoreactive materials were demonstrated at the "pancreatic" and the "intestinal mucosa digestion stage", whereas no such material was found at the "stomach digestion stage" (Svedberg and colleagues, 1985).

Bovine ß-casomorphins seem to be resistant to pronase, thermolysin and carboxypeptidases A or B (Henschen and colleagues, 1979). However, they have been shown to be degraded by carboxypeptidase Y (Henschen and colleagues, 1979) and, in particular, by a group of enzymes specifically cleaving amino acid sequences after proline residues (which are contained in the ß-casomorphin sequence in several positions), i.e. dipeptidylpeptidase IV, proline-specific endopeptidase or postproline cleaving enzyme (Hartrodt and colleagues, 1982a; Hartrodt and colleagues, 1982b); one or several of these enzymes might be present in mammalian plasma, since bovine ß-casomorphins have been shown to be degraded in bovine plasma with a half life of about 10 min (Kreil and colleagues, 1983).

Thus, although the reports on the release of ß-casomorphins look somewhat anecdotal, a release of ß-casomorphins or precursors thereof in the small intestine of adult or new-born mammals after milk intake seems to occur. ß-Casomorphins might elicit any effects before being degraded; degradation, in particular, can be expected in presence of peptidases specifically cleaving proline-containing amino acid sequences.

Evidence for Release and Degradation of ß-Casomorphins Under in Vivo Conditions: In Vivo Studies

Whereas in vitro studies may indicate the possibility that casein-derived opioid peptides occur under in vivo conditions, conclusive evidence for release and degradation in adult or new-born mammals after milk intake or in pregnant or nursing females can be expected from in vivo studies, only. Thus, several studies have been performed to elucidate, whether ß-casomorphins occur under in vivo conditions - a prerequisite for any physiological significance.

In fact, after bovine milk ingestion in adult humans, ß-casomorphin immunoreactive materials have been demonstrated in small intestine contents. Apparently these materials consisted of ß-casomorphins as well as, to their major part, of precursors thereof (Svedberg and colleagues, 1985). However, no such materials were found in the plasma of adult humans after ingestion of bovine milk or cheese (Teschemacher and colleagues, 1986). The results of Svedberg and colleagues (1985) found support by data of Meisel (1986), who demonstrated the presence of bovine ß-casomorphin-11 in the duodenum of minipigs after administration of a bovine casein preparation; probably, except this "ß-casomorphin precursor", also ß-casomorphins were contained in the duodenal contents.

Thus, although no ß-casomorphins seem to reach the circulation after milk intake in adult mammals, they are apparently released in the small intestine from ß-casein and may elicit any opioid or non-opioid effects in the gastrointestinal wall before being degraded. Influencing, e.g. gastrointestinal motility (Kromer and colleagues, 1982), they might participate in the regulation of gastrointestinal functions as "food hormones" (Morley, 1982).

Further, in the plasma of new-born calves, a ß-casomorphin-7 immunoreactive material was found after the calves' very first milk intake after birth; the material had not been present prior to the milk intake (Umbach and colleagues, 1985). However, as shown by chromatography, the material proved not identical with ß-casomorphin-7, but rather represents a precursor thereof. In addition, in contrast to ß-casomorphin-7, the material proved very stable against enzymatic attack in the new-born calves' plasma (Umbach and colleagues, 1985). Thus, in new-born mammals after milk intake, from the gastrointestinal tract a ß-casomorphin precursor is apparently absorbed, from which ß-casomorphins may be cleaved at any site in the new-born's organism to elicit opioid or non-opioid effects. Preferred candidates would be central effects such as induction of behavioural phenomena in new-borns (Panksepp and colleagues, 1984). In principle, a permeation through the blood brain barrier appears to be possible (Ermisch and colleagues, 1983).

In order to obtain information, whether ß-casomorphins might already occur in the organism of the mother, plasma samples from pregnant women and from lactating or non-lactating mothers were analysed for the presence of ß-casomorphin-7 and -8 immunoreactive materials. In fact, such materials were found in most of these women, whereas no such material was found in men and nonpregnant women included in the study as controls (Koch and colleagues, 1986). Interestingly, a material eventually related to ß-casomorphins has been found in the plasma of some women suffering from postpartum psychosis (Lindström and colleagues, 1984). Thus, ß-casomorphins or, at least, their precursors, from which they might be cleaved proved to be present already in the organism of pregnant and nursing women. ß-Casomorphins might, thus, participate in the endocrinic regulation of pregnancy, e.g. by modulation of prolactin release (Nedvidkova, 1985); prolactin or gonadotropin release, at least in part, seem to be controlled by endogenous opioids (Yen and colleagues, 1985; Leadem and Kalra, 1985) as well as the milk-ejection reflex (Wright, 1985).

Finally, after intracerebroventricular injection in rats, ß-casomorphin-5 has been shown to be degraded rapidly (Stark and colleagues, 1986). Thus, ß-casomorphins can be degraded rapidly under in vivo conditions, which is a prerequisite for the flexibility of a signal transfer system using transmitters or hormones as messengers.

Conclusion

In summary, although a physiological significance of casein-derived opioid peptides remains to be elucidated, all data so far raised are compatible with a nutritive function or even a role in the frame of mammalian reproduction.

REFERENCES

Brantl, V. (1984). Novel opioid peptides derived from human ß-casein. Eur. J. Pharmacol. 106, 213-214.

Brantl, V., H. Teschemacher, A. Henschen and F. Lottspeich (1979). Novel opioid peptides derived from casein (ß-casomorphins): I. Isolation from bovine casein peptone. Hoppe-Seyler's Z. Physiol. Chem. 360, 1211-1216.

Brantl, V., H. Teschemacher, J. Bläsig, A. Henschen and F. Lottspeich (1981). Opioid activities of ß-casomorphins. Life Sci. 28, 1903-1909.

Brantl, V. and H. Teschemacher (1983). Opioids in milk. Trends Pharmacol. Sci. 4, 193.

Brantl, V. and K. Neubert (1986). Opioid peptides derived from food proteins. Trends Pharmacol. Sci. 7, 6-7.

Chang, K.-J., P. Cuatrecasas, E. T. Wei and J.-K. Chang (1982). Analgesic activity of intracerebroventricular administration of morphiceptin and ß-casomorphins: Correlation with the morphine (μ) receptor binding affinity. Life Sci. 30, 1547-1551.

Chang, K.-J., Y. F. Su, D. A. Brent and J.-K. Chang (1985). Isolation of a specific μ-opiate receptor peptide, morphiceptin, from an enzymatic digest of milk proteins. J. Biol. Chem. 260, 9706-9712.

Ermisch, A., H.-J. Rühle, K. Neubert, B. Hartrodt and R. Landgraf (1983). On the blood-brain barrier to peptides: ^{3}H-ß-casomorphin-5 uptake by eighteen brain regions in vivo. J. Neurochem. 41, 1229-1233.

Grecksch, G., C. Schweigert and H. Matthies (1981). Evidence for analgesic activity of ß-casomorphin in rats. Neurosci. Lett. 27, 325-328.

Greenberg, R., M. L. Groves and H. J. Dower (1984). Human ß-casein. Amino acid sequence and identification of phosphorylation sites. J. Biol. Chem. 259, 5132-5138.

Hanson, L. A. (1982). The mammary gland as an immunological organ. Immunol. Today 3, 168-172.

Hartrodt, B., K. Neubert, G. Fischer, U. Demuth, T. Yoshimoto and A. Barth (1982a). Degradation of ß-casomorphin-5 by prolin-specific endopeptidase (PSE) and post-proline cleaving enzyme (PPCE). Pharmazie 37, 72-73.

Hartrodt, B., K. Neubert, G. Fischer, H. Schulz and A. Barth (1982b). Synthese und enzymatischer Abbau von ß-Casomorphin-5. Pharmazie 37, 165-169.

Henschen, A., F. Lottspeich, V. Brantl and H. Teschemacher (1979). Novel opioid peptides derived from casein (ß-casomorphins). II. Structure of active components from bovine casein peptone. Hoppe-Seyler's Z. Physiol. Chem. 360, 1217-1224.

Hoffmann, S., E. Günther, C. Liebmann, P. Mentz, K. Neubert and A. Barth (1986). Casomorphin - Strukturaspekte. Z. Chem. 26, 254-255.

Karabensch, K.-H., D. Otto, K. Neubert, I. Born, B. Hartrodt, H.-D. Jakubke, P. Kuhl, S. Hoffmann, P. Walther, H. Matthies, W. Gründler und A. Barth (1985). Zur Zeitabhängigkeit der Wirkung von ß-Casomorphin-5 und zwei Analoga auf das Nervensystem von Insekten. Biol. Zentralbl. 104, 539-550.

Koch, G., K. Wiedemann and H. Teschemacher (1985). Opioid activities of human ß-casomorphins. Naunyn-Schmiedeberg's Arch. Pharmacol. 331, 351-354.

Koch, G., K. Wiedemann and W. Zimmermann (1986). Human ß-casomorphin -8 immunoreactive materials in the plasma of nursing mothers. Naunyn-Schmiedeberg's Arch. Pharmacol. 332, R 85.

Koldovsky, O. (1980). Hormones in milk. Life Sci. 26, 1833-1836.

Kreil, G., M. Umbach, V. Brantl and H. Teschemacher (1983). Studies on the enzymatic degradation of ß-casomorphins. Life Sci. 33, Suppl. I: 137-140.

Kromer, W., W. Pretzlaff and R. Woinoff (1981). Regional Distribution of an opioid mechanism in the guinea-pig isolated intestine. J. Pharm. Pharmacol. 33, 98-101.

Leadam, C. A. and S. P. Kalra (1985). Effects on endogenous opioid peptides and opiates on luteinizing hormone and prolactin secretion in ovariectomized rats. Neuroendocrinology 41, 342-352.

Lindström, L. H., F. Nyberg, L. Terenius, K. Bauer, G. Besev, L. M. Gunne, S. Lyrenäs, G. Willdeck-Lund and B. Lindberg (1984). CSF and plasma ß-casomorphin-like opioid peptides in post-partum psychosis. Am. J. Psychiatry 141, 1059-1066.

Lottspeich, F., A. Henschen, V. Brantl and H. Teschemacher (1980). Novel opioid peptides derived from casein (ß-casomorphins). III. Synthetic peptides corresponding to components from bovine casein peptone. Hoppe-Seyler's Z. Physiol. Chem. 361, 1835-1839.

Loukas, S., D. Varoucha, C. Zioudrou, R. A. Streaty and W. A. Klee (1983). Opioid activities and structures of α-casein-derived exorphins. Biochemistry 22, 4567-4573.

Meisel, H. (1986). Chemical characterization and opioid activity of an exorphin isolated from in vivo digests of casein. FEBS Lett.

196, 223-227.
Mercier, J. C., F. Grosclaude and B. Ribadeau-Dumas (1972). Primary structure of bovine caseins. A Review. Milchwissenschaft 27, 402-408.
Mitsuma, T., T. Nogimori and M. Chaya (1984). ß-Casomorphin inhibits thyrotropin secretion in rats. Exp. Clin. Endocrinol. 84, 324-330.
Morley, J. E. (1982). Food peptides: A new class of hormones? J. Am. Med. Assoc. 247, 2379-2380.
Nedvidkova, J., E. Kasafirek, A. Dlabac and V. Felt (1985). Effect of beta-casomorphin and its analogue on serum prolactin in the rat. Exp. Clin. Endocrinol. 85, 249-252.
Panksepp, J., L. Normansell, S. Siviy, J. Rossi III and A. J. Zolovick (1984). Casomorphins reduce separation distress in chicks. Peptides 5, 829-831.
Parker, F., D. Migliore-Samour, F. Floch, A. Zerial, G. H. Werner, J. Jollès, M. Casaretto, H. Zahn and P. Jollès (1984). Immunostimulating hexapeptide from human casein: amino acid sequence, synthesis and biological properties. Eur. J. Biochem. 145, 677-682.
Petrilli, P., F. Addeo and L. Chianese (1983). Primary structure of water buffalo casein: Tryptic and CNBr peptides. Ital. J. Biochem. 32, 336-344.
Petrilli, P., D. Picone, C. Caporale, F. Addeo, S. Auricchi and G. Marino (1984). Does casomorphin have a functional role? FEBS Lett. 169, 53-56.
Reymann, K. G., A. N. Chepkova, K. Schulzeck and T. Ott (1985). Effects of ß-casomorphin on dentate hippocampal field potentials in freely moving rats. Biomed. Biochim. Acta 44, 749-754.
Ribadeau-Dumas, B., G. Brignon, F. Grosclaude and J.-C. Mercier (1972). Structure primaire de la caséine ß bovine: Séquence complète. Eur. J. Biochem. 25, 505-514.
Richardson, B. C. and J. C. Mercier (1979). The primary structure of the ovine ß-caseins. Eur. J. Biochem. 99, 285-297.
Schusdziarra, V., A. Holland, R. Schick, A. de la Fuente, M. Klier, V. Maier, V. Brantl and E. F. Pfeiffer (1983a). Modulation of postprandial insulin release by ingested opiate-like substances in dogs. Diabetologia 24, 113-116.
Schusdziarra, V., R. Schick, A. de la Fuente, A. Holland, V. Brantl and E. F. Pfeiffer (1983b). Effect of ß-casomorphins on somatostatin release in dogs. Endocrinology 112, 1948-1951.
Schusdziarra, V., R. Schick, A. Holland, A. de la Fuente, J. Specht, V. Maier, V. Brantl and E. F. Pfeiffer (1983c). Effect of opiate-active substances on pancreatic polypeptide levels in dogs. Peptides 4, 205-210.
Stark, H., B. Lössner and H. Matthies (1986). Degradation of ß-casomorphin in the rat brain in vivo. Biomed. Biochim. Acta 45, 557-563.
Svedberg, J., J. de Haas, G. Leimenstoll, F. Paul and H. Teschemacher (1985). Demonstration of ß-casomorphin immunoreactive materials in in vitro digests of bovine milk and in small intestine contents after bovine milk ingestion in adult humans. Peptides 6, 825-830.
Teschemacher, H., M. Umbach, U. Hamel, K. Praetorius, G. Ähnert-Hilger, V. Brantl, F. Lottspeich and A. Henschen (1986). No evidence for the presence of ß-casomorphins in human plasma after ingestion of cows' milk or milk products. J. Dairy Res. 53, 135-138.
Umbach, M., H. Teschemacher, K. Praetorius, R. Hirschhäuser and H. Bostedt (1985). Demonstration of a ß-casomorphin immunoreactive material in the plasma of newborn calves after milk intake. Reg. Peptides 12, 223-230.
Wright, D. M. (1985). Evidence for a spinal site at which opioids may act to inhibit the milk-ejection reflex. J. Endocrinol. 106, 401-407.

Yen, S. S. C., M. E. Quigley, R. L. Reid, J. F. Ropert and N. S. Cetel (1985). Neuroendocrinology of opioid peptides and their role in the control of gonadotropin and prolactin secretion. Am. J. Obstet. Gynecol. 152, 485-493.
Yoshikawa, M., T. Yoshimura and H. Chiba (1984). Opioid peptides from human ß-casein. Agric. Biol. Chem. 48, 3185-3187.
Zadina, J. E. and A. J. Kastin (1985). Interactions between the anti-opiate Tyr-MIF-1 and the Mu opiate morphiceptin at their respective binding sites in brain. Peptides 6, 965-970.
Zagon, I. S., P. J. McLaughlin, D. J. Weaver and E. Zagon (1982). Opiates, endorphins and the developing organism: A comprehensive bibliography. Neurosci. Biobehav. Rev. 6, 439-479.
Zioudrou, C. and W. A. Klee (1979). Possible roles of peptides derived from food proteins in brain functions. In R. J. Wurtman and J. J. Wurtman (Eds.), Nutrition and the Brain. Raven Press, New York, pp. 125-158.
Zioudrou, C., R. A. Streaty and W. A. Klee (1979). Opioid peptides derived from food proteins. J. Biol. Chem. 254, 2446-2449.

Studies of Dipeptidyl Peptidases I–IV of the Human Placenta

K. Lalu, S. Lampelo and T. Vanha-Perttula

Department of Anatomy, University of Kuopio,
P.O. Box 6, 70211 Kuopio, Finland

ABSTRACT

Homogenates of human term placentae were utilized to demonstrate the presence of four dipeptidyl peptidases (DPP I-IV) after sequential fractionation with gel filtration and anion exchange chromatography. DPP I was assayed with SerTyrNA as substrate and showed the characteristics of a thiol-dependent serine enzyme with optimum at pH 4.5 and a molecular weight of 210 kd. DPP II analysed with LysAlaNA as substrate had an optimum at pH 5.5 and a molecular weight of 130 kd with no dependence on thiol groups or divalent ions. DPP III was studied with ArgArgNA as substrate. It was optimally hydrolysed at pH 8.8 and had a molecular weight of 84 kd. The enzyme was highly suppressed by chelating agents but could be reactivated by Co^{2+} and some other divalent ions. DPP IV with GlyProNA as substrate displayed an optimum at pH 8.2. The fractionations displayed multiple forms of the enzyme possibly indicating the presence of proto- and polymeric DPP IV activities. The protomer had a molecular weight of 260 kd and showed no dependence on thiol groups but was sensitive to Zn^{2+}.

KEYWORDS

Dipeptidyl peptidases; fractionation; biochemical characteristics

INTRODUCTION

Our previous studies have revealed that the human term placenta contains at least five aminopeptidases with distinct enzymic properties and subcellular distribution (Lampelo and Vanha-Perttula, 1979, 1980; Lampelo, 1982; Lampelo, Lalu and Vanha-Perttula, 1982, 1983a,b). Substantial evidence has been obtained that only one of them, cystine aminopeptidase or oxytocinase, is released into the maternal circulation (Lalu, Lampelo and Vanha-Perttula, 1983), although the pregnant serum contains two other well-defined aminopeptidases (Lalu, Lampelo and Vanha-Perttula, 1985). Four dipeptidyl peptidases (DPP) have been identified in a variety of mammalian tissues (for ref. see McDonald and Schwabe, 1977), but only one of them, DPP IV, has been isolated and characterized biochemically from the human placenta (Püschel, Mentlein and Heymann, 1982).

In this report we present the separation and characterization of four DPP activities from the term human placenta.

Advances in the Biosciences Vol. 65
© 1987 Pergamon Journals Ltd.
Printed in Great Britain

MATERIALS AND METHODS

ArgArgNA, GlyProNA, LysAlaNA and SerTyrNA as well as antipain and leupeptin were obtained from Bachem Feinchemikalien (Bubendorf, Switzerland). Amastatin, bestatin, leupeptin, pepstatin and sulfhydryl modifiers were purchased from Sigma Chemical Co. (St Louis, MO, USA). Sephacryl S-300, Mono Q column and protein standards were obtained from Pharmacia Fine Chemicals (Uppsala, Sweden). All other chemicals used were of analytical grade quality from Merck AG (Darmstadt, W. Germany).

Normal full term human placentae were washed in the pre-cooled (4 °C) TBS (20 mM Tris-HCl, pH 7.0, containing 0.15 M NaCl) to remove blood clots. A 20-g sample of tissue was homogenized with Bamix Blendor for 5 min in 40 ml of TBS containing 0.5 % (v/v) Triton X-100. The mixture was stirred for 1 h at 0 °C, rehomogenated in Potter - Elvehjem homogenizer and centrifuged at 105 000 xg for 1 h. The supernatant fluid was applied to a gel filtration (GF) column (2.6 x 95 cm, V_0=151 ml, V_t=504 ml) of Sephacryl S-300, which had been equilibrated with TBS. The same buffer was used in the elution and fractions of 4.0 ml/10 min were obtained at 4 °C with a fraction collector. The active fractions were pooled, concentrated and dialyzed in a Diaflo concentrator. The samples were then separately fractionated in Mono Q anion exchange chromatography (AEC), which was eluted with individually programmed NaCl-gradient created with an HPLC controller and pumps. Fractions of 1 ml/min were collected and the active fractions were pooled and used in the biochemical characterization of the enzymes.

The hydrolysis of β-naphthylamide derivatives was assayed in the standard reaction mixture consisting of 0.15 ml of 0.2 M buffer [Na-acetate, pH 4.0, containing 3 mM dithiothreitol (DTT) for DPP I with SerTyrNA; Na-acetate, pH 5.5, for DPP II with LysAlaNA; Tris-HCl, pH 8.0, for DPP III with ArgArgNA and for DPP IV with GlyProNA], 0.05 ml of enzyme preparation and 0.1 ml of substrate (final conc. 0.4 mM). The reaction was stopped by adding 0.5 ml of p-dimethylaminobenzaldehyde solution (equal volumes of p-dimethylaminobenzaldehyde, 10 g/l, in methanol and 1 M sodium acetate buffer, pH 1.4, mixed immediately before the use). The absorbances were read at 450 nm in a spectrophotometer.

The effect of pH on the hydrolysis was studied in a series of 0.2 M Na-acetate and Tris-HCl buffers in the ranges from pH 3 to 7 and from pH 7 to 10, respectively.

Various modifiers were dissolved in distilled water. In the standard reaction mixture, the buffer was partially replaced with the modifier so that the volume of the buffer was at least 0.1 ml. The enzyme was preincubated with the modifier at 37°C for 30 min if not otherwise indicated. After preincubation the substrate was added and the reaction mixture incubated at 37°C. Means and standard errors (± s.e.m.) were calculated and statistical analysis was performed by Student's t-test.

The approximate molecular weights of the enzyme samples were determined by gel filtration in a column of Sephadex G-200 (1.5 x 144.4 cm). The column was calibrated with Blue Dextran 2000 (V_0), catalase (232 kd), aldolase (158 kd), bovine serum albumin (67 kd), ovalbumin (43 kd), chymotrypsinogen (25 kd) and ribonuclease A (13.7 kd).

RESULTS

The distribution of DPP I-IV activities with the four model substrates after GF on Sephacryl S-300 Superfine is shown in Fig. 1. DPP I with SerTyrNA as substrate at pH 4.0 with dithiothreitol showed one major activity peak at

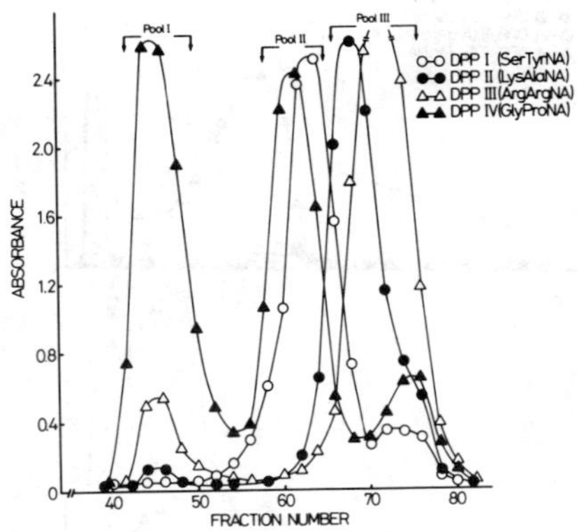

Fig. 1. Elution of dipeptidyl peptidases (DPP) I-IV with the model substrates as indicated after gel filtration of human placental homogenates on Sephacryl S-300 Superfine column at 4°C. The activities are given as absorbances at 450 nm after an incubation of 10-60 min at 37°C.

fractions 58-68. Similarly only one activity area at fractions 65-75 was obtained for DPP II with LysAlaNA as substrate at pH 5.5. The major DPP III activity with ArgArgNA as substrate at pH 8.0 eluted in fractions 68-78, while a minor peak was found in fractions 42-48. With GlyProNA as substrate at pH 8.0 two major activity areas at fractions 41-51 and 57-65, respectively, were found and a minor hydrolysis peak at fractions 72-78.

The fractions 42-50 after Sephacryl GF were pooled (Pool I) and eluted in AEC with a linear (0-0.3 M) NaCl-gradient (Fig. 2, upper panel). One major activity peak was obtained with GlyProNA for DPP IV at 0.25 M NaCl, while only a small peak was obtained with the other substrates using 2 M NaCl after the gradient. The pooled peak was used in further analyses of DPP IV activity.

The Sephacryl fractions 58-66 resulted in a minor activity peak for DPP II at 0.1 M NaCl and an enrichment of DPP I at 0.18 M NaCl in AEC (Fig. 2, middle panel). Slightly later and superimposed with this activity, a hydrolytic activity for DPP IV was also found. The fractions containing the high DPP I activity were pooled for further characterization of this enzyme.

The pooled Sephacryl fractions 68-78 showed a symmetrical sharp activity peak for DPP II at 0.1 M NaCl and a separate activity area for DPP III at 0.24 M NaCl in AEC with only minor contributions by DPP I and IV in the fractions (Fig. 2. lower panel). The former peak was used for the detailed analysis of DPP II and the latter for the characterization of DPP III activities.

The pooled enzyme preparations after AEC gave a single optimum at pH 4.5, 5.5, 8.8 and 8.2 for DPP I-IV, respectively.

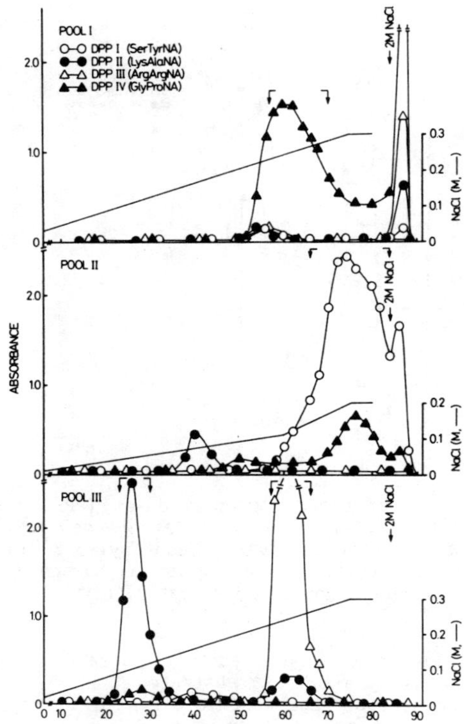

Fig.2. Elution of enzyme pools I-III after gel filtration (See Fig. 1) from Mono Q anion exchange chromatographic column with NaCl gradients as indicated. Each DPP was measured separately from the fractions. For the studies on enzyme characteristics pooled preparations were collected as indicated by the arrows for DPP IV (upper panel), DPP I (middle panel) and DPP II and III (lower panel).

Table 1 summarizes the effects of thiol modifiers on the pooled enzyme activities. It is obvious that DPP II and IV are not influenced by the reducing agents (cysteine, dithiothreitol) or thiol reagents (iodoacetamide, N-ethylmaleimide, p-chloromercuribenzoate). DPP I and III are highly sensitive to p-CMB, while clear differences between the enzymes were found in the response to the other modifiers. In particular, the reducing agents were mainly activating for DPP I, while a marked suppression was obtained with DPP III. Also the responses of these two enzymes to IAA and NEM were significantly different.

In Table 2 the effects of various proteinase inhibitors are shown. It is seen that DPP I is suppressed by the higher concentrations of antipain and leupeptin. None of the modifiers has any obvious effect on DPP II. On the other hand, DPP III seems to be highly sensitive to o-phenanthroline. In contrast, none of these modifiers were able to influence the hydrolysis rate of DPP IV.

TABLE 1. Effect of Thiol Modifiers on the Activities of Separated DPP I - DPP IV

Thiol modifier	Fin. conc. (mM)	DPP I	DPP II	DPP III	DPP IV
Cysteine	1	113 ± 5.8	106 ± 7.9	67 ± 16.0	123 ± 7.2
DTT	1	168 ± 1.7	105 ± 1.9	12 ± 1.7	118 ± 4.1
IAA	1	39 ± 0.7	109 ± 2.5	114 ± 1.3	116 ± 5.0
NEM	1	81 ± 4.6	110 ± 1.2	28 ± 4.2	95 ± 8.8
p-CMB	0.05	8 ± 1.7	115 ± 1.0	1 ± 0.7	105 ± 9.6

The enzymes were preincubated at 37°C for 30 min with the modifiers in the appropriate reaction mixtures for the enzymes before the addition of the substrates. The values in percentages are mean± s.e.m. of 3-5 duplicate experiments with the control values of 100 % obtained without added modifiers.

TABLE 2. Effect of Proteinase Inhibitors and o-Phenanthroline on the Activities of the Separated DPP I - DPP IV

Inhibitor	Fin. conc. mM or μM*	DPP I	DPP II	DPP III	DPP IV
Amastatin	0.5*	99 ± 0.9	101 ± 1.2	107 ± 3.9	100 ± 0.3
	1*	96 ± 2.0	99 ± 1.2	105 ± 3.5	99 ± 0.7
Antipain	0.01	81 ± 1.9	105 ± 3.2	101 ± 14.0	116 ± 11.9
	0.05	55 ± 2.1	111 ± 3.8	130 ± 10.0	113 ± 6.4
	0.1	53 ± 5.8	118 ± 2.0	141 ± 14.7	113 ± 4.7
Benzamidine	5	113 ± 4.3	102 ± 4.5	57 ± 12.8	117 ± 8.1
Bestatin	0.5*	119 ± 3.6	115 ± 1.0	72 ± 10.4	123 ± 3.8
Leupeptin	0.01	50 ± 7.5	105 ± 4.6	109 ± 6.4	108 ± 3.2
	0.05	16 ± 5.1	110 ± 3.5	125 ± 9.3	108 ± 5.4
	0.1	16 ± 4.3	115 ± 2.3	128 ± 7.6	114 ± 5.4
Pepstatin	0.01	108 ± 6.1	102 ± 2.1	96 ± 0.6	101 ± 1.0
	0.05	79 ± 5.5	100 ± 3.2	91 ± 2.1	106 ± 0.9
	0.1	63 ± 1.5	94 ± 1.3	108 ± 6.2	104 ± 1.3
Puromycin	1.0	117 ± 1.0	88 ± 2.8	77 ± 13.1	124 ± 7.9
o-Phenan-	1	97 ± 2.7	126 ± 5.5	24 ± 6.0	136 ± 8.9
throline	2	97 ± 4.5	139 ± 9.3	22 ± 6.7	140 ± 9.9

The enzymes were preincubated at 37°C for 30 min with the modifiers in the appropriate reaction mixtures for the enzymes before the addition of the substrates. The values in percentages are mean± s.e.m. of 3-5 duplicate experiments with the control values of 100 % obtained without added modifiers.

The effect of some divalent ions on the enzyme activities are given in Table 3. DPP I and II were not markedly influenced by these ions. On the other hand, both DPP III and IV were markedly suppressed by Ni^{2+} and Zn^{2+}. With DPP IV a somewhat lower decline of activity was found also with Co^{2+} and Mn^{2+}. In contrast, DPP III displayed a marked activation by Co^{2+} and a minor activation also by Ca^{2+} and Mn^{2+}. A further study on the effects of divalent ions was carried out after an initial treatment of the enzyme samples with 1 mM EDTA (Fig. 3). These analyses clearly showed that DPP I, II and IV are not influenced by EDTA and even in its presence DPP IV is sensitive to Ni^{2+} and Zn^{2+}. DPP III, on the other hand, is highly sensitive to EDTA but can be reactivated by Co^{2+} as well as by certain concentrations of Ca^{2+}, Mn^{2+} and Zn^{2+}.

TABLE 3. Effect of Divalent Metal Ions on the Separated Activities of DPP I - DPP IV

Divalent metal ion	Fin. conc. (mM)	DPP I	DPP II	DPP III	DPP IV
Ca^{2+}	1	97 ± 4.1	102 ± 4.1	118 ± 6.6	93 ± 1.9
	2	99 ± 2.9	100 ± 4.1	117 ± 7.1	88 ± 1.9
Co^{2+}	1	90 ± 1.8	102 ± 4.5	432 ± 80.9	53 ± 8.7
	2	93 ± 3.9	100 ± 5.0	353 ± 67.3	30 ± 4.1
Mg^{2+}	1	98 ± 3.4	103 ± 1.7	97 ± 2.6	93 ± 1.9
	2	90 ± 5.6	102 ± 1.4	100 ± 4.8	92 ± 2.1
Mn^{2+}	1	80 ± 6.9	101 ± 0.8	144 ± 20.4	62 ± 5.7
	2	82 ± 6.7	99 ± 0.7	133 ± 21.0	28 ± 14.6
Ni^{2+}	1	77 ± 3.2	94 ± 2.0	9 ± 1.0	18 ± 8.3
	2	73 ± 2.9	87 ± 1.9	7 ± 1.5	7 ± 6.5
Zn^{2+}	1	66 ± 5.2	101 ± 6.3	1 ± 0.7	3 ± 2.0
	2	60 ± 5.2	94 ± 5.3	0 ± 0.3	2 ± 1.7

The enzymes were preincubated at 37°C for 15 min before the addition of the substrates. Experimental details are described in Materials and Methods. The values in percentages are mean ± s.e.m. of 3-5 duplicate experiments with the control value of 100 % obtained without added metal ions.

The apparent molecular weights for DPP I-IV by gel filtration on Sephadex G-200 resulted in the tentative values of 210 kd, 130 kd, 84 kd and 260 kd, respectively. A second peak for DPP III and IV was found at the void volume of this fractionation and may represent aggregated or particulate forms of these enzymes.

DISCUSSION

In the present study we have been able to demonstrate that the human placenta contains enzyme activities, which hydrolyse the model substrates developed to measure dipeptidyl peptidases I-IV. Two sequential fractionations, gel filtration and anion exchange chromatography, resulted in a separation good enough for their identification and characterization. Previously these enzymes have been found in many mammalian tissues but only DPP IV has been purified and characterized from the human placenta (Püschel, Mentlein and Heymann, 1982). The cellular localization of the enzymes in the placenta has also been obscure, except for a recent immunohistochemical demonstration of DPP IV in myofibroblasts and stromal cells of the villi (Feller and co-workers, 1986).

DPP I has been characterized as a thiol-dependent enzyme of lysosomal origin (for ref. see McDonald and Schwabe, 1977). The enzyme has been regarded identical with cathepsin C and is able to degrade also many biologically active compounds (McDonald and co-workers, 1969; Callahan, McDonald and Ellis, 1972). The placental enzyme with optimum at acid pH was closely similar to the activities previously reported in various other tissues (McDonald and Schwabe, 1977).

DPP II has been regarded as a lysosomal activity with acid pH-optimum (McDonald and Schwabe, 1977; 1980; Sannes, 1983). It has also been demonstrated that DPP II is a serine enzyme (McDonald and Schwabe, 1980; Fukasawa and co-workers, 1983). Our observations with the placental activity did not support the latter concept, since the inhibitors of serine enzymes (antipain, leupeptin) were not suppressive. The other biochemical properties, however, corresponded closely to those reported for DPP II in other tissues (for ref. see McDonald and Schwabe, 1977).

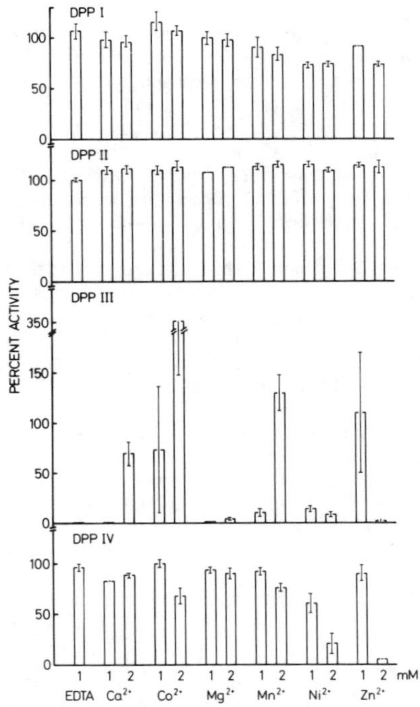

Fig. 3. The effect of EDTA (1 mM) alone was studied by preincubating the enzymes with EDTA at 37°C for 1 h before adding a substrate. The effect of metal ions on the EDTA-pretreated enzymes was studied by adding them into the assay medium for 15 min at 37°C before the substrate. Final concentrations of the effectors are shown in the Figure. The results are given as the percentage (mean ± s.e.m.) from the control (100 %) of five duplicate analyses.

DPP III has been regarded as a cytosolic serine enzyme with some thiol activation (Lee and Snyder, 1982). It has typically an alkaline pH-optimum and displays the ability to split some biologically active compounds (McDonald and Schwabe, 1977; Lee and Snyder, 1982). The placental enzyme was suppressed by some thiol reagents (p-CMB, NEM) but not all (IAA). A similar response to the thiol reagents has been found with the placental aminopeptidase II_1 (Lampelo, Lalu and Vanha-Perttula, 1983b), but the basic mechanism of such a reaction remains obscure. The most characteristic property of DPP III was the high sensitivity to the chelating agents. After EDTA the enzyme could be reactivated by Co^{2+} and to some extent also by Ca^{2+}, Mn^{2+} and Zn^{2+}. These characteristics are typical of a metalloenzyme. Our observations with the inhibitors (antipain, leupeptin) of the serine enzymes do not support the role of serine in the active center. It is possible that the placental DPP III is distinct from the enzyme(s) characterized in other organs and tissues with ArgArgNA as substrate. However, previous studies have given conflicting results regarding the response of DPP III to chelating agents and serine enzyme inhibitors (Ellis and Nuenke, 1967; Hopsu-Havu, Jansén and Järvinen, 1970; Lee and Snyder, 1982; Hazato and co-workers, 1984). Therefore, further studies with a purified enzyme preparation are needed to disclose the detailed characteristics of this placental enzyme, since some of the suppressive effects, e.g., with bestatin and puromycin may be encountered by the contribution of aminopeptidases sequentially splitting the substrate (Lampelo, Lalu and Vanha-Perttula, 1983a,b).

DPP IV has been found as a microsomal enzyme with distinct post-proline cleavage site. The enzyme has previously been purified and characterized from the human placenta (Püschel, Mentlein and Heymann, 1982). This study has disclosed its heterogeneity, substrate specificity and modifier characteristics. Our observations are consistent with these results including the sensitivity to Zn^{2+}. The elution of a substantial amount of the enzyme at the void volume in gel filtration indicates that the detergent treatment of the homogenate was not able to solubilize completely DPP IV from the membranes, in which it is known to be attached (for ref. see McDonald and Schwabe, 1977).

Dipeptidyl peptidases are enzymes, which have strict requirements for their substrates and are able to split only dipeptide residues from peptides and proteins with unprotected NH_2-terminus. It is obvious that these enzymes belong to a cellular machinery, which is responsible for processing of macromolecules in different cellular compartments. They are also able to process a variety of biologically active peptides (e.g. ACTH, glucagon, secretin, insulin, angiotensin II by DPP I: McDonald and co-workers, 1969, McDonald and Schwabe, 1977; ACTH, glucagon, angiotensin II, β-endorphin, β-lipotropin and enkephalin by DPP III: Lee and Snyder, 1982, Hazato and co-workers, 1984; substance P, casomorphin, kentsin and renin inhibitor by DPP IV: Püschel, Mentlein and Heymann, 1982). The human placenta is a site for very active metabolic functions, synthetic as well as degradative. We can only assume that the four dipeptidyl peptidases are actively engaged in these processes.

REFERENCES

Callahan, P.X., J.K. McDonald, and S. Ellis (1972). Fed. Proc. 31, 1105-1113.

Ellis, S. and J.M. Nuenke (1967). J. Biol. Chem. 242, 4623-4629.

Feller, A.C., H.-J. Radzun, E. Heymann, H. Haas, W. Scholz, and R.M. Parwaresch (1986). Virch. Arch., A409, 263-273.

Fukasawa, K., K.M. Fukasawa, B.Y. Hiraoka, and M. Harada (1983). Biochim. Biophys. Acta 745, 5-11.

Hazato, T., M. Shimamura, A. Achimura, and T. Katayama (1984). J. Biochem. 95, 1265-1271.

Hopsu-Havu, V.K., C.T. Jansén, and M. Järvinen (1970). Arch. klin. Derm. 236, 267-281.

Lalu, K., S. Lampelo, and T. Vanha-Perttula (1983). Progr. Clin. Enzymol. 2, 299-309.

Lalu, K., S. Lampelo, and T. Vanha-Perttula (1985). Int. J. Biochem. 17, 1227-1235.

Lampelo, S. (1982). Med. Biol. 60, 278-281.

Lampelo, S., K. Lalu, and T. Vanha-Perttula (1982). Placenta 3, 379-394.

Lampelo, S., K. Lalu, and T. Vanha-Perttula (1983a). Int. J. Biochem. 15, 709-714.

Lampelo, S., K. Lalu, and T. Vanha-Perttula (1983b). Placenta 4, 499-514.

Lampelo, S., and T. Vanha-Perttula (1979). J. Reprod. Fertil. 56, 285-296.

Lampelo, S., and T. Vanha-Perttula (1980). J. Reprod. Fertil. 58, 225-235.

Lee, C.-M., and S.H. Snyder (1982). J. Biol. Chem. 257, 12043-12050.

McDonald, J.K., and C. Schwabe (1977). Intracellular exopeptidases. In A.J. Barrett (Ed.), Proteinases in Mammalian Cells and Tissues, North Holland, Amsterdam, pp. 311-391.

McDonald, J.K., and C. Schwabe (1980). Biochim. Biophys. Acta 616, 68-81.

McDonald, J.K., B.B. Zeitman, T.J. Reilly, and S. Ellis (1969). J. Biol. Chem. 244, 2693-2709.

Püschel, G., R. Mentlein, and E. Heymann (1982). Eur. J. Biochem. 126, 359-365.

Sannes, P.L. (1983). J. Histochem. Cytochem. 31, 684-690.

Studies on the Secretion of Salivary Kallikren[1]

J. R. Garrett and R. E. Smith

Department of Oral Pathology, The Rayne Institute,
King's College School of Medicine and Dentistry,
London SE5 9NU, UK
and Lawrence Livermore National Laboratory, Livermore
California, USA

ABSTRACT

The substrate D-Val-Leu-Arg-4-methoxy-2-naphthylamide has been found very useful for the histochemical demonstration of active glandular kallikrein in the parenchyma of submandibular glands of the cat. Its 7-amino-4-trifluoromethyl coumarin analogue affords an extremely sensitive method for corresponding biochemical assay of saliva and glandular homogenates. Both inhibitor studies and the use of zymographic membrane overlays, impregnated with the substrate and applied after isoelectric focussing, have demonstrated that the substrate is monospecific for glandular kallikrein in cat submandibular saliva. Stimulation experiments have confirmed that sympathetic nerves provide the main drive for the secretion of kallikrein but we have found that this is normally a mixed alpha- plus beta-adrenergic response and not just an alpha-activity. The speed with which kallikrein reached remarkably high concentrations in individual drops of saliva suggest that similar rapid mobilization may occur in life and that a special separate population of sympathetic fibres may exist for this purpose. Denervation experiments indicated that parasympathetic rather than sympathetic impulses normally influence resynthesis. Analyses of blood samples have shown that glandular kallikrein does not normally pass in significant amounts to the glandular interstitium. The enzyme has been found extremely hardy and this may have an importance for its exocrine activities, the functions of which are still not known.

KEYWORDS

Glandular kallikrein; salivary glands; submandibular glands; saliva; secretion; naphthylamide substrates; trifluoromethyl coumarin substrates; zymographic membrane overlays.

INTRODUCTION

Although Werle and Von Roden (1936) discovered glandular kallikrein in

[1] A more complete account of the work is given by Garrett and others (1986).

© 1987 Pergamon Journals Ltd.
Printed in Great Britain

salivary glands 50 years ago, its function in the glands remains obscure and many problems still exist about its formation and secretion. The conventional role of glandular kallikrein is to produce lysyl bradykinin from low molecular weight kininogen by limited proteolysis, i.e. it is a kininogenase. The enzyme is now classed as an endopeptidase of the serine protease type; it has a special affinity for cleaving arginyl-amide bonds and, as with other enzymes of this group, it displays strong esteratic activity.

A false boost to the presence of salivary kallikrein was given in the 50's when Hilton & Lewis (1957) claimed that nerve-induced vasodilatation in salivary glands was due entirely to a local release of kallikrein causing the formation of kinin in the glandular interstices. This view is no longer held, but many people still hang on to the belief that kallikrein in the glands exists there to fulfill a role on the blood vascular side. Most people also consider that glandular kallikrein has the exclusive function of acting on kininogen to form kinin. Actually, we have no real idea about the true function of kallikrein in salivary glands, or why it usually occurs in far greater amounts in submandibular glands than in parotid glands.

An interest in glandular kallikrein emerged during some experimental studies on the effects of stimulating the autonomic nerves to submandibular glands in cats (Garrett & Kidd, 1975). At that time the localization of kallikrein in the gland was unknown and much debated. Our work showed that stimulation of the sympathetic nerve caused the secretion of secretory granules from the striated ducts but parasympathetic nerve stimulation did not. Based on this finding and other correlations with known physiological changes in submandibular kallikrein, we predicted that glandular kallikrein occurs in the secretory granules of the striated ducts (Garrett & Kidd, 1975). Within a short time this view was supported immunocytochemically by Hojima and others (1977). However, we have preferred to pursue our investigations of this enigmatic enzyme by harnassing its catalytic activity. Our first attempts at developing a specialised esteratic procedure for its detection histochemically were not very successful. The ideal we were hoping to achieve was a

SINGLE SPECIFIC SUBSTRATE for detecting GLANDULAR KALLIKREIN both

a) HISTOCHEMICALLY in GLANDULAR SECTIONS, and

b) BIOCHEMICALLY in SALIVA and GLANDULAR HOMOGENATES

If this aim could be realised it would overcome a common problem experienced in the correlation of histochemical and biochemical results on the same tissues, since the methods used for either procedure are usually very different.

In 1979 Lexow and colleagues successfully used N-acetyl-L-Methionine α-Naphthyl ester for the histochemical detection of esteroproteases in submandibular glands of mice. Professor von Deimling kindly sent us some of this substrate and we obtained encouraging results on sections of cat submandibular glands (Al-Gailani and others, 1980a). There was strong staining in striated ducts that was not inhibited by E-600. Subsequently we found that sympathetic stimulation caused extensive depletion of the reaction in the striated ducts (Al-Gailani and others, 1980b). However, we were in no doubt from the outset that there were problems of specificity with this substrate. A chance encounter then occurred in 1980 between the present authors, at the Meeting of the Histochemical Society in New Orleans. As a result it was agreed to test some 4-methoxy-2-napthylamide (MNA) compounds of oligopeptides for the histochemical detection of the protease activity of glandular kallikrein.

CHOICE OF A SUITABLE HISTOCHEMICAL SUBSTRATE

Based on the known cleavage of kininogen by kallikrein, the first substrate we tested was Z-Ser-Pro-Phe-Arg-MNA but the results on cat submandibular glands were not very promising. So we went back to the drawing board and tested a number of MNA substrates biochemically with purified glandular kallikrein using fluorimetric assessment of the released MNA leaving group. The most selective reaction, by far, occurred with D-Val-Leu-Arg-MNA (Garrett and co-workers, 1985). Surprisingly the use of the N-terminal blocking procedure Z-Val-, rather than D-Val-, caused a much slower cleavage of the substrate by glandular kallikrein. This probably relates to the configuration of the active sight of the enzyme but the details are not known.

Encouraged by the biochemical results we therefore tested the substrate D-Val-Leu-Arg-MNA histochemically on aldehyde-fixed preparations of submandibular glands, using Fast Blue B as the coupling agent, and obtained some very satisfactory results on cat glands and in a number of other species (Garrett and co-workers, 1982), including man (Garrett and co-workers, 1984). The reaction product was localised precisely in a periluminal rim in striated ducts. This activity in human glands was inhibited by low concentrations of Aprotinin, but the periluminal activity in striated ducts of cat submandibular glands was not inhibited by high concentrations of Aprotinin. These differences are in agreement with known biochemical results on the inhibition of kallikrein (see Vogel & Werle, 1970). It does suggest, however, that a good degree of specificity for kallikrein was being afforded by the substrate with sections of human glands.

EFFECTS OF TISSUE FIXATION

The kallikrein in salivary glands is very soluble in aqueous solutions and unfixed sections gave unsatisfactory histochemical results by standard procedures; the activity could readily be monitored in the supernatant fluid. Therefore, for conventional histochemistry, fixation is required. Fixatives containing both glutaraldehyde and formaldehyde gave stronger and better localized reactions in cryostat sections than when formaldehyde alone was used. The enzyme showed remarkable resistance to aldehyde fixation and crude EM grade of glutaraldehyde was as useful as the more expensive distilled glutaraldehyde. The best results were undoubtedly achieved by perfusion fixation but satisfactory results were obtained after immersion fixation, so long as enough time had been allowed for immobilization of the enzyme. With small blocks 3-4 mm thick, fixation for at least 4 hours was required and the preferred time was 6 hours. Some inhibition occurred thereafter but reasonable results were still obtainable with fixation for 24 hours in 4% formaldehyde - 5% glutaraldehyde. Providing sufficient time for adequate fixation had been allowed, the enzyme would stand up moderately well to paraffin embedding and interpretable results were achieved, but the appearances were never so good as in cryostat sections.

CHOICE OF THE BIOCHEMICAL SUBSTRATE

Catalytic cleavage of the same MNA leaving group from oligopeptide substrates can be assessed fluorimetrically in solutions, as was done for the selection of the most suitable histochemical substrate (see previously). Nevertheless, we prefer to use the fluorimetric leaving group 7-amino-4-trifluoromethyl-coumarin (AFC) for biochemical assessment because it provides a more sensitive reaction and there is far better colour discrimination (Smith and co-workers, 1980). The substrates emit a blue light in the bound form but the AFC leaving group, when released by enzymatic cleavage, gives a green fluorescence. Purified kallikrein was tested with a variety of AFC substrates. The dynamics of the reactions were found to be closely similar to those with MNA substrates. Table 1 - (previously unpublished) gives the results, comparing glandular kallikrein with plasma kallikrein.

TABLE I

Comparative Rates of Hydrolysis of Different Peptide Substrates Caused by Purified Glandular Kallikrein (G.K) and Plasma Kallikrein (P.K).

	SUBSTRATE	G.K	P.K
1)	Arg-AFC	0	1.4
2)	Z-Ala-Arg-Arg-AFC	0	4.4
3)	Boc-Phe-Arg-AFC	0	4.7
4)	D-Val-Lys-Lys-Arg-AFC	0	5.7
5)	Boc-D-Ser-Pro-Phe-Arg-AFC	0	16.5
6)	D-Val-Lys(Z)-Lys(Z)-Arg-AFC	0	29.6
7)	Boc-D-Phe-L-Phe-Arg-AFC	0	36.4
8)	BZ-Val-Lys-Lys-Arg-AFC	0.2	7.6
9)	Acetyl-Phe-Arg-AFC	0.5	43.2
10)	D-Ser-Pro-Phe-Arg-AFC	1.3	36.0
11)	D-Phe-L-Phe-Arg-AFC	1.5	100.0
12)	Leu-Arg-AFC	2.4	0.9
13)	D-Ala-Leu-Lys-AFC	9.0	0.3
14)	D-Val-Leu-Lys-AFC	13.6	1.3
15)	D-Val-Leu-Arg-AFC	100.0	30.8

It will be noted that our preferred peptide group D-Val-Leu-Arg also gave a moderate reaction with plasma kallikrein. However, this reaction could be totally inhibited by low concentrations of soya-bean trypsin inhibitor (SBTI) without affecting the reaction with glandular kallikrein. The substrate was also cleaved by trypsin, albeit more slowly than glandular kallikrein, (Garrett and co-workers, 1985) but this reaction was also totally inhibited by low concentrations of SBTI. Therefore, addition of SBTI to the substrate mixture imparts an improved specificity to it for detecting glandular kallikrein. We have tested samples of sympathetic saliva from cat submandibular glands and found that inclusion of 200μg/ml SBTI caused no inhibitory effects on their reactions with D-Val-Leu-Arg-AFC. This suggests that there are no other trypsin like enzymes in this particular saliva, that are capable of hydrolysing the substrate. Thus, the reaction of cat submandibular saliva with D-Val-Leu-Arg-AFC per se is likely to be specific for glandular kallikrein. Nevertheless, as a precaution, all routine biochemical assessments have been carried out in the presence of 100μg/ml of SBTI.

Assessment of glandular kallikrein in saliva indicated that it is a very hardy enzyme. It was not affected by repeated freezing and thawing. Furthermore, it showed no change when a sample was left on the shelf for a month at room temperature.

USE OF ZYMOGRAPHIC OVERLAY MEMBRANES FOR ASSESSING SPECIFICITY

A useful zymographic method for identifying protease-containing bands after isoelectric focussing has been developed (Smith, 1984). A removable overlay membrane, previously impregnated with the appropriate AFC substrate is applied for a short while to the surface of the electrophoretic gel after the electrophoretic procedure. On removal and drying of the overlay membrane, the sites of reaction can be identified as fluorescent bands under blue light. Membranes, impregnated with D-Val-Leu-Arg-AFC, have been used to test reactions after isoelectric focussing of samples of saliva and to compare them with those from purified glandular kallikrein from cat submandibular glands (a gift from Prof. M. Schachter). The results showed 4-6 reactive bands in the pI range 4.2-5.0 in both types of specimen with no fewer bands in the purified glandular kallikrein (Garrett and co-workers, 1985). This confirms that in the cat saliva D-Val-Leu-Arg-AFC is mono-specific for glandular kallikrein.

Furthermore, the bands in sympathetic and parasympathetic saliva were identical but parasympathetic saliva had to be less diluted to enable detection. In addition, homogenates or unfixed cryostat sections (eluted onto the gel with a triton mixture) gave similar banding. Cryostat sections from sympathetically stimulated glands showed gross reduction in the fluorescent reaction compared with that seen in sections from control glands. Unfixed cryostat sections, mounted and dried on filter papers, retained electrophoretically detectable activity indefinitely.

In cat submandibular saliva, the fortunate absence of other proteases, capable of cleaving the substrate, imbues a monospecificity for D-Val-Leu-Arg substrates that permits them to be used satisfactorily for identifying glandular kallikrein activity per se. This is not the case in rat submandibular glands which contain many enzymes capable of hydrolysing the substrate (Arnold and co-workers, 1983, Garrett co-workers, 1985). Other glands will have to be tested accordingly for the degrees of specificity that D-Val-Leu-Arg substrates provide, as the need arises.

NERVE-STIMULATION STUDIES ON CAT SUBMANDIBULAR GLANDS

This work was undertaken in vivo to assess the influence of the secretory nerves on the secretion of glandular kallikrein into saliva.

A) Parasympathetic Nerve Stimulation

This produced a copious flow of saliva that would continue for very long times. The saliva, however, only contained low concentrations of glandular kallikrein, usually in the range of 0.2-2µg equivalents/ml with low frequency stimulation (1-8Hz). At higher frequency stimulation small increases in the concentrations of kallikrein occurred but they never reached high levels. The maximum that was obtained was 36 µg equivalents/ml. Nevertheless, a secretion of low concentrations of glandular kallikrein continued indefinitely in parasympathetic saliva. However, no histochemical changes were detected in the glands despite prolonged stimulation.

B) Sympathetic Nerve Stimulation

This caused an erratic flow of saliva that could be improved by applying the stimulation intermittently. Despite the poor flow of saliva there was always a considerable secretion of glandular kallikrein into the saliva, reaching levels up to 3000 µg equivalents/ml in early samples. Correspondingly there was an extensive depletion of kallikrein activity in homogenates, falling by as much as 90%. Sections of the glands showed a similar depletion histochemically. The high concentrations of kallikrein in the saliva permitted assessment of secretion in individual drops. Activity rapidly reached a high peak, between the 4th and 9th drop, then gradually declined. The highest concentration encountered was 5.7 mg equivalents/ml in the peak drop from one animal. Despite prolonged stimulation on occasions, the concentration in the saliva was still found to be greater than in parasympathetic saliva.

C) Mixed Parasympathetic and Sympathetic Nerve Stimulation

Was undertaken for short periods at low frequency, in order to obtain a steadier output of kallikrein. This was done with the objective of allowing adrenergic blocking drugs to be assessed more purposefully than is possible when sympathetic stimulation is applied in isolation, with the rapidly changing concentrations of kallikrein that occur. Furthermore, conditions of mixed stimulation are more likely to reflect naturally occurring secretory events in the glands (see Emmelin, 1986). The same protocol was used on both glands from each animal. The responses in the first gland were tested without blocking drugs, and compared with the subsequent effects of blocking drugs on the contralateral gland. Addition of sympathetic stimulation at 2-5 Hz on parasympathetic stimulation caused a 10- to 45- fold increase in kallikrein secretion that was moderately well maintained. Alpha-adrenergic blockade totally inhibited the sympathetic secretion of kallikrein under these conditions. However, contrary to general belief, beta-adrenergic blockade per se caused more than 50% reduction in the secretion of kallikrein during mixed nerve stimulation.

DENERVATION EXPERIMENTS

A) Chronic Sympathectomy

Although the sympathetic nerves provide the main drive for the secretion of salivary kallikrein an absence of such impulses appeared not to affect the glandular content of kallikrein.

B) Chronic Parasympathectomy

Loss of parasympathetic impulses, on the other hand, soon caused an extensive depletion (65-80%) of the kallikrein content in the glands. However, sympathetic impulses were still capable of causing most of the remaining kallikrein to be secreted.

TESTING THE VENOUS EFFLUENT FROM THE GLANDS FOR GLANDULAR KALLIKREIN.

The blood supply to the submandibular gland was isolated and after, heparinizing the animal, the blood flow from the glandular vein was monitored by a drop-counter before returning the blood to the animal. Samples of blood between stimulation periods and during parasympathetic stimulation or sympathetic stimulation were removed for assessing kallikrein-like activity with D-Val-Leu-Arg-AFC, in the presence of SBTI (Edwards and co-workers, 1986). All blood samples showed a low amidase reaction (mean value 17 ng equivalents/ml). No changes were detected during any stimulation protocol. Yet, sympathetic stimulation contained up to 1000 µg equivalents/ml and the glandular content of kallikrein was reduced by 60-80%. Furthermore, on testing neat plasma with very diluted sympathetic saliva, no changes were detected in the inhibitory capacity of the plasma during the course of an experiment, thereby indicating that its inhibitory capacity was not being saturated. Thus, it seems unlikely that there is a significant movement of glandular kallikrein to the blood during its secretion into the saliva.

DISCUSSION

Use of MNA or AFC compounds of D-Val-Leu-Arg has proved very helpful for studying the histochemical and biochemical events of the secretion of glandular kallikrein from submandibular glands of cats.

The dramatically rapid secretion of kallikrein into saliva on sympathetic stimulation is thought to reflect a potential to mobilize it similarly in life. The purpose of such a rapid mobilization is not known and can only be speculated about. If we ever gain a glimpse of the natural influences inducing this to occur reflexly, more insight about possible functions of this enigmatic enzyme in saliva should accrue. It seems a distinct possibility that a separate population of sympathetic nerves, under separate central control, pass specially to the striated ducts for the express purpose of enabling sudden large increases in the secretion of glandular kallikrein into the saliva. It is naive to assume that its sole function in the saliva is to cause a release of kinin from kininogen, but other possible activities for it are not yet known. The present work has found it unlikely that salivary kallikrein has a purposeful regular role on the vascular side of the gland. This is not-to-deny that some kallikrein may pass there if secretion occurs against an obstruction. Under such conditions the kallikrein may well make an important contibution to the ensuing vasodilatation, which could be propitious for the removal of foreign material entering the interstices of the gland in this manner.

Our investigations have shown that, contrary to popular belief, beta-adrenergic activation of the cells contributes towards the sympathetic secretion of kallikrein. However, for the beta-influences to be effective the cells require alpha-activation at the same time, as is likely to happen in life.

Denervation studies indicate that parasympathetic impulses are essential for the normal resynthesis of kallikrein in the glands, though these impulses cause only a small secretion of the enzyme. Work by Emmelin & Henriksson (1953), using muscarinic blocking agents, suggest that the acetycholine from the parasympathetic impulses plays an important part in this resynthetic drive.

The enzyme has been found to be very hardy. It stands up to prolonged fixation, remains unchanged in saliva for long periods, withstands drying and remoistening. This hardiness may play a part in its natural function and a role in the licking of fur may be important.

CONCLUSIONS

SALIVARY KALLIKREIN is present for

1) EXOCRINE PURPOSES.
2) ITS FUNCTION IS OBSCURE; it is naive to assume that it acts only on kininogen.
3) SPECIAL RAPID MOBILIZATION INTO SALIVA IS LIKELY AT TIMES, under separate sympathetic control.
4) IT IS AN EXTREMELY HARDY ENZYME and this may afford an important aspect of its function.

ACKNOWLEDGMENTS

This work would not have been possible without the generous help of the Wellcome Trust & N.A.T.O. (Grant number 333/82).
A travel grant from the British Council is gratefully acknowledged.
We wish to thank Dr. A.V. Edwards, The Physiological Laboratory, Cambridge, for providing the facilities for the blood flow experiments.

REFERENCES

Al-Gailani, M., J. R. Garrett, A. Kidd, K. Kyriacou, and P. Leite (1980a). Localisation of esteroproteases in "resting" salivary glands from different species and the effects of the organophosphorous inhibitor E600.

Histochemistry, 66, 59-74.

Al-Gailani, M., J. R. Garrett, and A. Kidd (1980b). Histochemical assessment of factors affecting ductal esteroproteases in submandibular glands of cats. J. Physiol., 307, 40-41P.

Arnold, W.H., T. B. Ørstavik, and M. Holck (1983). A 4-methoxy-2-naphylamide substrate for the histochemical localization of esteroproteases in the submandibular gland of the rat. Histochem. J., 15, 139-146.

Edwards, A. V., J. R. Garrett, J. Liao, and R. E. Smith (1986). Testing for possible movement of kallikrein to the blood from submandibular glands on nerve stimulation in cats. J. Physiol., 371, 97P.

Emmelin, N. (1986). Nerve interactions in salivary glands. J. Dent. Res., (in press).

Emmelin, N., and K. G. Henriksson (1953). Depressor activity of saliva after section of the chorda tympani. Acta Physiol. Scand., 30, (Suppl. 111), 75-82.

Garrett, J. R., and A. Kidd (1975). Effects of nerve stimulation and denervation on secretory material in submandibular striated duct cells of cats, and the possible role of these cells in the secretion of salivary kallikrein. Cell Tiss. Res., 161, 71-84.

Garrett, J. R., A. Kidd, K. Kyriacou, and R. E. Smith (1984). New observations on the localization of kallikrein-like activity in human salivary parenchymal and mast cells by enzyme histochemistry. Histochem. J., 16, 789-791.

Garrett, J. R., A. Kidd, K. Kyriacou, and R. E. Smith (1985). Use of different derivatives of D-Val-Leu-Arg for studying kallikrein activities in cat submandibular glands and saliva. Histochem. J., 17, 805-818.

Garrett, J. R., A. Kidd, K. Kyriacou, and R. E. Smith (1986). Salivary kallikrein and its secretion: structural and functional investigations on submandibular glands in cats. In G. L. Haberland and J. W. Rohen (Eds.), Kininogenases - Kallikrein, Vol 7. Schattauer Verlag, Stuttgart. pp. 33-58.

Garrett, J. R., R. E. Smith, A. Kidd, K. Kyriacou, and R. J. Grabske (1982). Kallikrein-like activity in salivary glands using a new tripeptide substrate, including preliminary secretory studies and observations on mast cells. Histochem. J., 14, 967-979.

Hilton, S. M., and G. P. Lewis (1957). Functional vasodilatation in the submandibular salivary glands. Brit. med. Bull., 13, 189-196.

Hojima, Y., B. Maranda, C. Moriwaki, and M. Schachter (1977). Direct evidence for the location of kallikrein in the striated ducts of the cat's submandibular glands by the use of specific antibody. J. Physiol., 268, 793-801.

Lexow, U., C. Grossarth, and O. Von Deimling (1979). Histochemical demonstration of mouse submandibular esterproteases with a new chromogenic substrate. Histochemistry, 60, 327-334.

Smith, R. E. (1984). Identification of protease isoenzymes after analytical isoelectric focusing using fluorogenic substrates impregnated into cellulose membranes. J. Histochem. Cytochem., 32, 1265-1274.

Smith, R. E., E. R. Bissell, A. R. Mitchell, and K. W. Pearson (1980). Direct photometric or fluorometric assay of proteinases using substrates containing 7-amino-4-trifluoromethyl-coumarin. Thrombosis Res., 17, 393-402.

Vogel, R. and E. Werle (1970). Kallikrein inhibitors. In E. G. Erdos and A. F. Wilde (Eds.), Handbook of Experimental Pharmacology, Vol. XXV. Springer, Heidelberg. pp. 213-249.

Werle, E. and P. Von Roden (1936). Uber das Vorkommen von Kallikrein in den Speicheldrusen und im Mundspeichel. Biochem. Z., 286, 213-219.

Peptidases in Human Cerebrospinal Fluid Converting Dynorphin and Substance P to Bioactive Fragments

Fred Nyberg

Department of Phamacology, University of Uppsala,
P.O. Box 591, S-751 Uppsala, Sweden

ABSTRACT

During the last decade great interest has been focussed on the mechanism behind the termination of neuropeptide signalling. In contrast to the re-uptake systems described for most other neurotransmitters, such as noradrenaline and amino acids, the peptides are likely to be removed by enzyme-induced catabolism. Indeed, several enzymes present in nervous tissue have been reported to be involved in the degradation or conversion of various neuropeptides. This presentation describes the kinetic and biochemical characteristics of enzymes purified from human cerebrospinal fluid (CSF), with high capacity of converting dynorphin A and substance P. It also discusses the biological activity of the generated products.

KEYWORDS

Cerebrospinal fluid; dynorphin; substance P, convertases; peptidases; processing, bioactive fragments.

INTRODUCTION

The heptadecapeptide dynorphin (dyn A; Tyr-Gly-Gly-Phe-Leu-Arg-Arg-Ile-Arg-Pro-Lys-Leu-Lys-Trp-Asp-Asn-Gln) and the undecapeptide substance P (SP; Arg-Pro-Lys-Pro-Gln-Gln-Phe-Phe-Gly-Leu-Met-NH_2) are present in many areas of the central nervous system (CNS). Areas rich in dynorphins are the hypothalamus-pituitary system, the hippocampus and the substantia nigra (Vincent and co-workers, 1982a; Cone and co-workers,1983; Zamir, Palkowitz and Brownstein, 1984a-b). High levels of substance P are found in brain areas associated with primary afferents transmitting nociceptive stimuli (Hökfelt and co-workers, 1977; Pernow, 1983) and in extra-pyramidal brain areas, such as striatum and substantia nigra (Brownstein and co-workers, 1977; Hong and co-workers, 1977), constituting a striato-nigral pathway. Evidence for a striato-nigral pathway containing dynorphin, parallell to that of subtance P, has also been presented (Vincent co-workers, 1982b; Zamir and co-workers, 1984c; McLean and co-workers, 1985). According to their biological activities, dynorphin and substance P belong to two different neuropeptide families: the opioid peptides (the endorphins) and the tachykinins, respectively. The opioid peptides are derived from three genetically distinct precursors with different CNS distribution, proopiomelanocortin, proenkephalin A and proenkephalin B. Each precursor is polyvalent, i.e. capable releasing several biologically active peptides. The enkephalin core

© 1987 Pergamon Journals Ltd.
Printed in Great Britain

Tyr-Gly-Gly-Phe-X (X=Leu or Met) is the minimum sequence necessary for opioid activity. Substance P is released from two different precursors, each with a unique biosynthetic origin, alpha- and beta-preprotachykinin. Peptides belonging to the tachykinin family have similar biological activity and share a common C-terminal sequence, Phe-X-Gly-Leu-Met-NH_2 (X=Phe or Val), in which the C-terminus is amidated. The N-terminal sequence is unique for each peptide and accounts for its specific action. In view of their proposed role in neurotransmission and neuromodulation, dynorphin and substance P have been thoroughly studied with regard to the mechanism of their inactivation. Several enzymes capable of degrading dynorphin (Mizuno and Matsuo, 1984; Devi and Goldstein, 1984) and substance P (Benuck and Marks, 1975; Mc Geer and Singh, 1979; Kato and co-workers, 1980; Lee and co-workers, 1981; Matsas and co-workers, 1983) are known. Recent work performed in this laboratory has demonstrated the presence of dynorphin A and substance P converting activities in human CSF (Nyberg co-workers, 1984; Nyberg, Nordström and Terenius, 1985). This article summarizes our studies of these enzymes.

EXPERIMENTAL

Peptides and chemicals were from sources previously noted (Nyberg and co-workers, 1984; Nyberg, Nordström and Terenius, 1985; Nyberg, LeGrevés and Terenius, 1985; Nyberg, Nylander and Terenius, 1986). Enzyme assays were based on product quantification by radioimmunoassay (RIA) and performed as in (Nyberg and co-workers, 1984, Nyberg, Nordström and Terenius, 1985). Conversion of dynorphin A was guided by a RIA specific for its N-terminal hexapeptide fragment (Leu)enkephalin-Arg^6 (Christenson-Nylander and co-workers, 1985). Substance P conversion was monitored by RIA:s specific for the fragments SP(1-7) (Rimon and co-workers, 1984) and for the C-terminal part of the molecule (Rimon and co-workers, 1984). The latter assay was used for detection of the fragment SP (5-11), which was separated from the parent peptide by ion exchange chromatography on SP-Sephadex (Bergström and co-workers, 1983). In the purification procedure large volumes of pooled human CSF were fractionated on a tandem column system with DEAE-Sepharose CL-6B and phenyl- Sepharose CL-4B (Nyberg, Nordström and Terenius, 1985). After sample application the columns were dismantled and enzyme fractions adsorbed to the respective gel were separately eluted as described (Nyberg and co-workers, 1984; Nyberg, Nordström and Terenius, 1985) and further purified by gel filtration on Sephadex G-100 and Sephacryl S-200. The purified enzymes were characterized by studies of various biochemical and kinetic parameters. High performance liquid chromatography (HPLC) was used for both gel permeation and reversed phase separation as described previously (Nyberg, Nordström and Terenius, 1985). Electrophoreses were run at acidic pH using the agarose suspension procedure (Nyberg and Terenius, 1983).

RESULTS

By the procedures outlined here we were able to partially purify a dynorphin converting endopeptidase (DCE) which cleaves the heptadecapeptide at the Arg^6-Arg^7 bond. Two different activities capable of hydrolyzing substance P were also recovered. One is an endopeptidase, which cleaves substance P at the Phe^7-Phe^8 and Phe^8-Gly^9 bonds; the other acts on substance P like a postproline-dipeptidyl-aminopeptidase (PPDAP). The purification scheme including all these enzymes is summarized in Fig. 1.

The DCE activity is recovered in the fraction eluted from the phenyl-Sepharose column. Upon gel filtration it shows molecular heterogeneity. A major peak, eluting as a protein of molecular weight around 40 kdaltons was collected for kinetic studies. It showed pH and temperature optima in the range of pH 6-8 and 35-45°C,

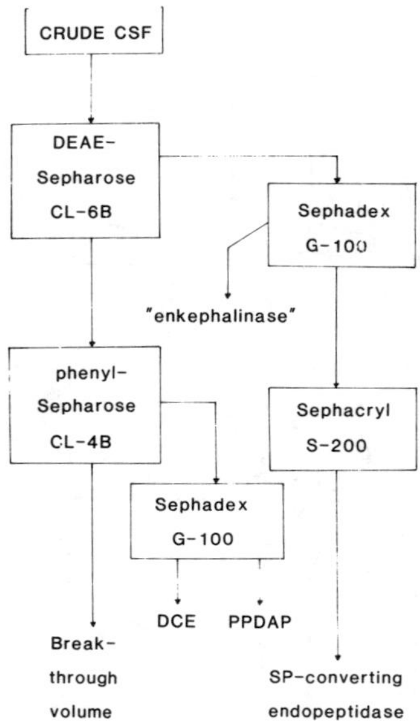

Fig. 1. Schematic representation of the procedures used for the purification of dynorphin and substance P converting enzymes from human CSF.

respectively, and exhibited high specificity for releasing the N-terminal fragment (Leu)enkephalin-Arg^6, not only from dynorphin A but also from other proenkephalin B derived peptides, e.g. alpha-neoendorphin, dynorphin B (Fig. 2) and dynorphin A (1-8). The enkephalyl hexapeptide appeared as a major product in all experiments and the formation of this fragment was confirmed by amino acid analysis. Studies also indicated that dynorphin A and B were converted at a higher rate than alpha-neoendorphin and dynorphin A(1-8). A K_m-value of 0.45 mM was estimated for the conversion of dynorphin B. The DCE activity was not affected by EDTA or iodoacetate but strongly inhibited by phenylmethylsulphonyl fluoride (PMSF) in analogy with enzymes classified as serine proteases.

The substance P converting endopeptidase was initially adsorbed to the DEAE-Sepharose column (Fig. 1). Its further purification on Sephadex G-100 and Sephacryl S-200 was monitored by RIA quantitation of the SP(1-7) fragment. In the Sephadex G-100 fractionation the SP(1-7) generating activity was resolved in at least two separate fractions (Fig. 3); one eluting ahead of albumin (> 68 kdaltons) and the other at a position corresponding to 40 kdaltons or thereabouts. It can further be seen in Fig. 3 that the activity of the large fraction was almost abolished in the presence of phosphoramidon, a potent inhibitor of enkephalinase (EC 3.4.24.11). The 40 kdalton entity was further fractionated on Sephacryl S-200 and analyzed by HPLC gel permeation on a TSK-G3000-SW column. By the latter technique an approximate molecular weight of 43 kdaltons was estimated. The purified endopeptidase showed maximum activity at neutral pH in a temperature interval of 35-45 ^oC. The enzyme releases the SP (1-7) and (1-8) fragments in approximately equimolar proportions (Fig. 4) and is inhibited by EDTA and bacitracin but it is not affected by capto-

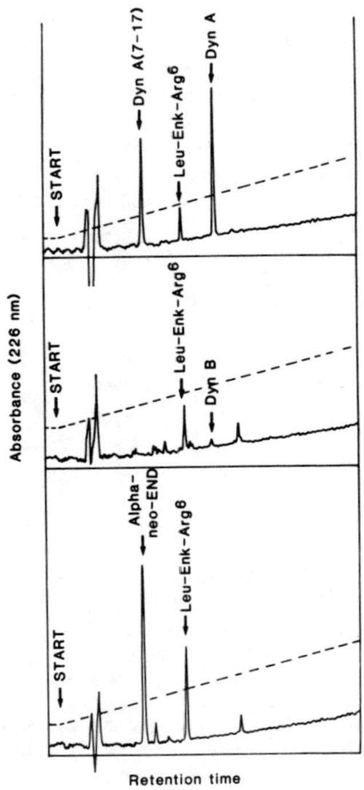

Fig. 2. RP-HPLC separation of incubation mixtures of dynorphin converting enzyme and dyn A (upper panel), dynB (in the middle) and alphaneoendorphin (lower panel). Peptides and products were resolved by a 40 min. linear gradient of acetonitrile (14-45%) containing 0.04% trifluoroacetic acid. The position of the released N-terminal fragment leu-enkephalin-Arg^6 is indicated by an arrow.

pril (an inhibitor of angiotensin converting enzyme, EC 3.4.15.1) phosphoramidon or iodoacetate. Removal of N-terminal amino acids from substance P decreased the rate of conversion (Table 1) and the rate of degradation is also reduced as a result of modifications in the C-terminal region. No or negligible conversion was obtained in experiments with SP(1-8), (MePhe^8, Sarkosin^9)substance P or with other tachykinins or phenylalanine-containing peptides.

The PPDAP-like activity attacking substance P was found in the fraction eluted from the phenyl-Sepharose column (Fig. 1). Following gel filtration on Sephadex G-100 the enzyme activity was detected in fractions eluting near the void volume. Substance P was converted into its C-terminal SP(3-11) and (5-11) fragments. The presence of these fragments was confirmed by electrophoresis. The enzyme was not sensitive to bacitracin but inhibited by proline-containing peptides.

DISCUSSION

The procedure for CSF fractionation described here (see Fig. 1) allows the recovery of several neuropeptide degrading enzymes. The endopeptidase capable of releasing (Leu)enkephalin-Arg^6 from proenkephalin B derived peptides acts with specificity towards consecutive pairs of basic amino acid residues. It exhibited higher rate for converting dynorphin A than e.g. its (1-8) fragment suggesting that the sequences beyond the cleavage site are of importance. Since the conversion of precursors of many neuropeptides occurs at a sequence of two basic amino acids the DCE activity in CSF may represent an enzyme of general processing character. In fact, it was

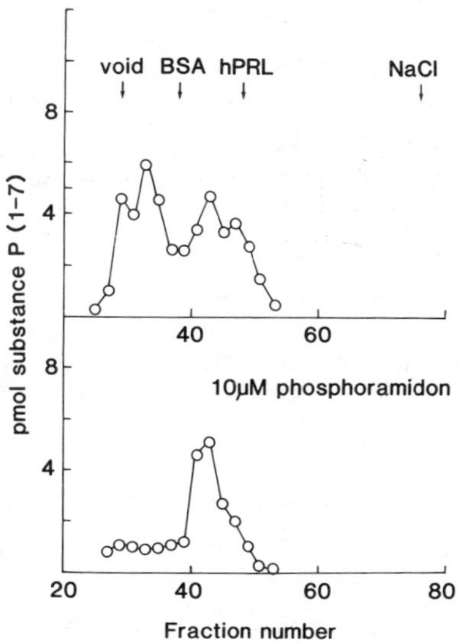

Fig. 3. Gel filtration on Sephadex G-100 of substance P (1-7), generating activity collected from the DEAE-Sepharose column. Enzyme activity was measured in the absence (upper panel) and presence (lower panel) and presence (lower panel) of phosphoramidon, using a RIA procedure for product quantitation. The elution positions of protein standards bovine serum albumin (BSA, MW = 68,000) and human prolactin (hPRL, MW = 22,000) are indicated by arrows.

found that methorphamide, an opioid peptide derived from proenkephalin A is also cleaved by DCE. However, in runs with other opioid or non-opioid peptides, containing pairs of consecutive basic amino acid residues (such as beta-endorphin or ACTH), no or negligible conversion was observed. The formation of (Leu)-enkephalin-Arg^6 from dynorphin A by DCE esults in a change of receptor affinity; from kappa receptor for the parent peptide to delta receptor selectivity for the generated product. The fragment dynorphin A (7-17) has no opioid activity but may still possess bioactivity since the related (6-17) fragment produced rotational behaviour in rats when injected unilaterally into the substantia nigra (Herrera-Marschitz and co-workers, 1984). The effect was not reversible by naloxone.

TABLE 1 Conversion of SP and related peptides by SP(1-7) and SP(1-8) generating enzyme purified from human cerebrospinal fluid

Peptide	Structure	HPLC-detected Conversion[1]
SP	Arg-Pro-Lys-Pro-Gln-Gln-Phe-Phe-Gly-Leu-Met-NH_2	90
SP (3-11)	" " " " "	35
SP (5-11)	" " " "	10
SP (1-8)	" " " " -Phe	<5
SP free acid	" " " " " -Met-COOH	70
SP (Tyr^8)	" " " " -Tyr " -Met-NH_2	50
SP ($MePhe^8$-$Sarkosin^9$)	" " " -MePhe-Sar "	<5
Neurokinin alpha	His-Lys-Thr-Asp-Ser-Phe-Val-Gly-Leu-Met-NH_2	<5
Neurokinin beta	Asp-Met-His-Asp-Phe-Phe-Val-Gly-Leu-Met-NH_2	<5
Leu-enkephalin	Tyr-Gly-Gly-Phe-Leu	<5
Leu-enkephalin-Arg^6	" " " -Arg	<5
Beta-casomorphine	Tyr-Pro-Phe-Pro-Gly-Pro-Ile	<5

[1] Incubations were performed in 50 µl of 20 mM Tris-HCl buffer (pH 7.8) at 37 °C for 90 min. with 5 µg enzyme in each vial. HPLC separated products and unaltered substrates were quantified by calibration runs with peptide standards at various concentration.

The substance P converting endopeptidase exhibits high specificity for the undecapeptide. It is evident that even minor structural modifications of the peptide result in reduced rate of degradation. The released N-terminal fragments SP(1-7) and (1-8) have no tachykinin activity but are still bioactive. For instance, both fragments produce antinociception in mice following central or peripheral administration (Stewart and co-workers, 1982). The C-terminal heptapeptide fragment SP(5-11), which is released by the PPDAP-like activity, also retains neurotropic activity. The fragment acts as a potent substance P agonist in many bioassays (e.g. Nieber and Oehme, 1982) and was shown to compete with the iodinated undecapeptide for binding to brain synaptosomes (Viger and co-workers, 1983).

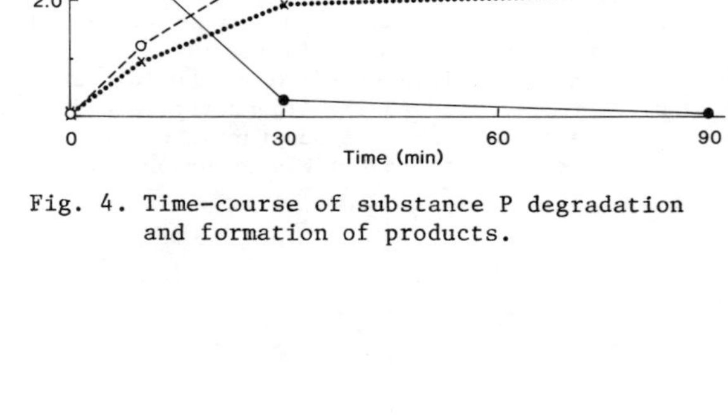

Fig. 4. Time-course of substance P degradation and formation of products.

The presence in CSF of the dynorphin and substance P converting activities has several implications. Undoubtedly, the peptidases are bound to influence the peptide levels in the fluid. In fact, recent studies performed in this laboratory indicated that both dynorphin A(1-6) and SP(5-11) exhibit higher CSF levels than the respective parent compounds (Nyberg, Nylander and Terenius, 1986; Rimon and co-workers, 1984). Whether these enzymes are present due to passive leakage from the synaptic area into the extracellular fluid and then the CSF or actually are secreted into the fluid is unknown. It is clear, however, that the CSF itself, must be considered a compartment in dynamic activity, particularly since the studied peptidases not only seem to have a function in terminating the dynorphin or substance P action but also to release a second signal masked within their structure.

ACKNOWLEDGEMENT

This study was supported by The Swedish Medical Council, Tore Nilsons Fond för Medicinsk Forskning and Åke Wibergs Stiftelse.

REFERENCES

Benuck, M. and Marks, N. (1975) Biochem. Biophys. Res. Commun. 65, 153-160.

Bergström, L., Christensson, I., Folkesson, R., Stenström, B. and Terenius L. (1983) Life Sci. 33, 1613-1619.

Brownstein, M.J., Mroz, E.A., Tappaz M.L. and Leeman, S.E. (1977) Brain Res. 135, 315-323.

Christensson-Nylander, I., Nyberg, F., Ragnarsson, U. and Terenius, L. (1985) Reg. Peptides 11, 65-76.

Cone, R.I., Weber.,E., Barchas J.D. and Goldstein, A. (1983) J. Neurosci. 3, 2146-2152.

Devi, L. and Goldstein. A (1984) Proc. Natl. Acad. Sci. USA 81, 1892-1896.

Herrera-Marschitz, Hökfelt, T., Ungerstedt, U., Terenius, L. and Goldstein, M. (1984) Eur. J. Pharmacol. 102, 213-227.

Hong, J.S., Yang, H.Y., Racagni, G. and Costa, E.(1977) Brain Res. 122, 541-544.

Hökfelt, T., Ljungdahl, G., Terenius, L., Elde, R., and Nilsson, G. (1977) Proc. Natl. Acad. Sci. USA 74, 3081-3185.

Kato, T., Kojima K., Nagatsu, T. and Sakakibara, S. (1980) J. Neurochem 35, 527-535.

Lee, C.M., Sandberg, B.E.B., Hanly, M.R. and Iversen L.L. (1981) Eur. J. Biochem 114, 315-327.

Matsas, R., Fulcher, I.S., Kenny, A.J. and Turner, A.J. (1983) Proc. Natl. Acad. Sci. USA 80, 3111-3115.

Mc Geer, E.G. and Singh E.A. (1979) Neurosci. Lett. 14, 105-108.

Mc Lean; S., Bannon M.J., Zamir, N. and Pert, C.B. (1985) Brain Res. 361, 185-192.

Mizuno, K. and Matsuo, H. (1984) Nature 309, 558-560.

Nieber, K. and Oehme, P. (1982) Pharmazie 37, 368-369.

Nyberg, F. and Terenius, L. (1983) Anal. Biochem. 134, 245-253.

Nyberg, F., Le Greves, P., Sundqvist C. and Terenius, L. (1984) Biochem. Biophys. Res. Commun. 125, 244-250.

Nyberg, F., Nordström, K. and Terenius, L. (1985a) Biochem Biophys Res. Commun 131, 1069-1074.

Nyberg, F., Le Greves, P. and Terenius, L.(1985b) Proc. Natl. Acad. Sci. USA 82, 3921-3924.

Nyberg, F., Nylander, I. and Terenius L. (1986) Brain Res. 371, 278-286.

Pernow, B. (1983) Pharmacol. Reviews 35, 86-141.

Rimon, R., Le Grevés, P., Nyberg, F., Heikkilä, L., Salmela, L. and Terenius, L. (1984) Biol. Psychiatr. 19:509-516.

Stewart, J.M., Hall, M., Harkins, J., Frederickson, R.C.A., Terenius, L., Hökfelt, T. and Krivoy, W.A. (1982) Peptides 3, 851-857.
Vincent, S.R., Hökfelt, T., Christensson, I. and Terenius, L. (1982a) Neurosci. Lett. 85, 185-190.
Vincent, S.R., Hökfelt, T., Christensson, I. and Terenius, L. (1982b) Eur. J. Pharmacol. 85, 251-252.
Viger, K., Beaujouan, J.C., Torrens, Y. and Glowinski, J. (1983) J. Neurochem. 40: 1030-1039.
Zamir, N., Palkovitz, M. and Brownstein, M.J. (1984a) Brain Res. 307, 61-68.
Zamir, N., Palkovitz, M. and Brownstein, M.J. (1984b) J. Neurosci. 4, 1240-1247.
Zamir, N., Palkovitz, M., Weber, E., Mezey, E. and Brownstein, M.J. (1984c) Nature 307, 643-645

Bradykinin Competitive Antagonists: Design and Activities

John M. Stewart and Raymond J. Vavrek

Department of Biochemistry, University of Colorado School of Medicine, Denver, CO 80262, USA

ABSTRACT

Competitive antagonists of the biological actions of the peptide hormone bradykinin have been synthesized and assayed for biological activities on smooth muscles and on blood pressure. The antagonism is specific for bradykinin and related kinins and is reversible. Evidence with the antagonists suggests that several subclasses of "classical" bradykinin receptors may exist.

KEYWORDS

Analogs; antagonists; bradykinin; peptide synthesis; solid phase synthesis.

INTRODUCTION

Understanding the complex biological actions of bradykinin (BK) (Arg-Pro-Pro-Gly-Phe-Ser-Pro-Phe-Arg) requires specific, competitive, and potent sequence-related antagonists, but no suitable antagonists of BK activity have been described previously. The frustrating search for an antagonist of BK, which spanned 25 years from the isolation, structure elucidation and synthesis of the correct BK sequence, entered a new phase with our description of 7-D-Phe-BK, the first true BK antagonist. It antagonized BK action in the *in vitro* guinea pig ileum (GPI) assay, one of the classic assay systems for BK. Antagonist potency was increased and the spectrum of BK antagonist action was broadened to include the *in vitro* rat uterus (RUT) and the *in vivo* rat blood pressure (RBP) by incorporating additionally beta-2-thienylalanine (Thi) at positions 5 & 8 to replace the Phe residues, and by extending the N-terminus with basic amino acid residues (DArg-, Lys-Lys-) (Stewart and Vavrek, 1986; Vavrek and Stewart, 1985). Subsequent modifications of these structures showed that not only was the modification of position 7 the key change in the BK sequence to produce antagonists, but also that further modifications of the antagonist structure could increase potency. However, the DPhe residue appeared to be the sole acceptable replacement for the naturally occurring Pro residue at position seven.

More recently, preliminary data have indicated that antagonists of BK can be produced by replacement of the Pro residue at position 7 with DThi, with *p*-chloro-DPhe (CDF), or with a beta-2-pyridyl-D-Ala (DPal) residue, and that simultaneous replacement of the (6-7) Ser-Pro sequence of BK with D-hydrophobic

© 1987 Pergamon Journals Ltd.
Printed in Great Britain

amino acid residues, in addition to producing potent smooth muscle-selective BK agonists (Vavrek and Stewart, 1986a), can also produce BK competetive antagonists (Vavrek and Stewart, 1986b). We report here 85 new BK analogs, many with interesting antagonistic activities.

METHODS

The BK-related peptides listed in Table 1 were synthesized by the solid-phase method (Stewart and Young, 1984), cleaved from the support with anhydrous HF, purified by countercurrent distribution, and analyzed for purity and identity by thin layer chromatography, paper electrophoresis and amino acid analysis. They were assayed _in vitro_ on isolated rat uterus (RUT) and guinea pig ileum (GPI) and _in vivo_ on blood pressure in the anesthesized rat (RBP) as described previously (Ryan and others, 1968; Vavrek and Stewart, 1985). All responses of isolated tissues were recorded isometrically except in those cases where (K) follows the entry in the RUT column in the table; in those cases BK antagonism was seen when recording was done isotonically (as with a kymograph), but not when isometric recording was used.

RESULTS

Of the analogs having a single substitution, at position 7, only the DPhe, DThi, DPal and CDF analogs antagonize the effect of BK in the GPI assay. These are D-amino acid residues having small aromatic side chains. The most potent of these on the ileum is [DThi^{7}]-BK, which had pA_2 = 6.4. None of these analogs antagonized BK on the RUT or in the RBP assay.

Replacement of the Phe residues at positions 5 and 8 of the BK sequence with the Thi residue causes a remarkable increase in BK agonist potency (Dunn and Stewart, 1971); this modification was found to increase affinity of the analogs for BK receptors (Odya and others, 1980). Similar incorporation of Thi can increase antagonist potency of some analogs (see [D-Phe^{7}]-BK), or may maintain antagonist activity without changing potency (see [D-Pal^{7}]-BK and [CDF^{7}]-BK) in the GPI assay. In the RUT assay Thi substitution can convert agonist activity to antagonist activity (see [D-Phe^{7}]-BK), or can convert antagonist activity to weak agonist activity (see [D-Thi^{7}]-BK). [Thi^{5,8}-DPhe^{7}]-BK antagonizes the depressor effect of BK in the RBP assay when administered as a bolus injection in admixture with BK either IA or IV, and antagonizes an IA or IV dose of BK when the analog alone is infused IA prior to administration of BK (Vavrek and Stewart, 1985).

Addition of a DArg residue to the N-terminus of [DPhe^{7}]-BK increases antagonist potency, but addition of the dipeptide Lys-Lys decreases antagonist potency in the ileum assay. Under isotonic recording conditions on the RUT, addition of the DArg residue converts [DPro^{3}-DPhe^{7}]-BK to a weak antagonist of BK. In the ileum assay DArg-[D-Phe^{7}]-BK is as potent an antagonist as [Thi^{5,8}-DPhe^{7}]-BK. Addition

TABLE 1. Bradykinin Analogs[a]

BRADYKININ ANALOG	RUT	GPI	RBP-IA	RBP-IV	%DESTR
BK	100	100	100	100	99
[Gly^{7}]-BK	0.2	0.07	0.6	0.7	95
[DAla^{7}]-BK	0.3	0	0.03	2	0
[DPro^{7}]-BK	1.5	-	0.5	-	0
[DPhe^{7}]-BK	4	5.9(5)	2	4	36
[DPro^{3}-DPhe^{7}]-BK	0	0	(IB)	(IB)	-
[CDF^{7}]-BK	7	5.4(5)	1	14	50
[DThi^{7}]-BK	0.9	6.4(7)	0	0	-
[DNal^{7}]-BK	0.9	0.08	0.2	14	0
[MDY^{7}]-BK	AG	0	0	0	-
[DPal^{7}]-BK	0.1	5.3(4)	0	0	-
[DPhg^{7}]-BK	AG	0			
[DHis^{7}]-BK	2	0			
[DTrp^{7}]-BK	4	0			
[DTyr^{7}]-BK	0.2	0			
[DhPhe^{7}]-BK		0.7	1	27	54
[Thi^{5,8}]-BK	1000	200	36	41	97
[Thi^{5,8}-DPro^{7}]-BK	0.6	0			
[Thi^{5,8}-DPhe^{7}]-BK	6.5(K)	6.3(10)	(IB)(II)	(II)	
[DPro^{3}-Thi^{5,8}-DPhe^{7}]-BK	0	0			
[Thi^{5,8}-DThi^{7}]-BK	4.2(K)	0	0	0	-
[Thi^{5,8}-CDF^{7}]-BK	3	5.2(5)	5	137	39
[Thi^{5,8}-DPal^{7}]-BK	MXD	5.3(5)	0	0	-
[Thi^{5,8}-DhPhe^{7}]-BK	15	0.2	0		
[Thi^{5,8}-Gly^{6}-DhPhe^{7}]-BK	56	2			
DArg-BK	143	103	66	218	85
DLys-BK	123	114	105	408	85
Lys-Lys-BK	35		750		0
DLys-Lys-BK	126	9	158	309	42
DArg-[DAla^{7}]-BK	0	0	0	0	-
DArg-[DPhe^{7}]-BK	0.1	6.3(5)	(IB)	(IB)	-
DArg-[DPro^{3}-DPhe^{7}]-BK	4.6(K)	0			
Lys-Lys-[DPhe^{7}]-BK	0.3	5.6(5)	(IB)	(IB)	-
DArg-[DPhg^{7}]-BK	0	0			
Lys-Lys-[DPhg^{7}]-BK	0	0			
DArg-[DTrp^{7}]-BK	0.3	0			
Lys-Lys-[DTrp^{7}]-BK	0.3	0			

TABLE 1. (Continued- Part 2)

BRADYKININ ANALOG	RUT	GPI	RBP-IA	RBP-IV	%DESTR
DArg-[DTyr^{7}]-BK	0.08	0			
Lys-Lys-[DTyr^{7}]-BK	0.1	0			
DArg-[DHis^{7}]-BK	0	0			
Lys-Lys-[DHis^{7}]-BK	0.1	0			
DArg-[DhPhe^{7}]-BK	11	0.2			
Lys-Lys-[DhPhe^{7}]-BK	0.3	0	AG	AG	-
DArg-[Thi^{5,8}-DPhe^{7}]-BK	5.5(K)	6.1(2)	(IB)	(IB)	-
Lys-Lys-[Thi^{5,8}-DPhe^{7}]-BK	MXD	5.6(6)	(IB)	(IB)	-
DArg-[DPro^{3}-Thi^{5,8}-DPhe^{7}]-BK	5.2(K)	0			
DArg-[Thi^{5,8}-DhPhe^{7}]-BK	4	0	AG		
DArg-[Thi^{5,8}-Gly^{6}-DhPhe^{7}]-BK	57	2			
Lys-Lys-[Thi^{5,8}-DThi^{7}]-BK	5.1(K)	0	0.3	4	70
Lys-Lys-[Thi^{5,8}-DPal^{7}]-BK	0.5	0			
[DPhe^{6}-Gly^{7}]-BK	9	6	7	36	81
[CDF^{6}-Gly^{7}]-BK	84	60	114	326	87
[DPhe^{6}-DAla^{7}]-BK	11	6	0.6	18	0
[CDF^{6}-DAla^{7}]-BK	111	120	57	1151	77
[DPhe^{6}-DPro^{7}]-BK	0.7	0			
[CDF^{6}-DPro^{7}]-BK	1	0.3			
[DPhe^{6}-DVal^{7}]-BK	1	0	(PRS/DPRS)	(PRS/DPRS)	-
[CDF^{6}-DVal^{7}]-BK	1	0	0.9	14	35
[DPhe^{6}-DPhe^{7}]-BK	0.1	0	(PRS)	(PRS)	-
[CDF^{6}-DPhe^{7}]-BK	0.3	0.01	0	0	-
[CDF^{6}-CDF^{7}]-BK	0.2	0	(PRS)	(PRS)	-
[DPhe^{6}-CDF^{7}]-BK	0.6	5.9(6)	1	9	86
[DPhe^{6}-DThi^{7}]-BK	0.06	0			
[DPhe^{6}-MDY^{7}]-BK	0.1	0			
[DPhe^{6}-DPal^{7}]-BK	0.1	0.05			
[DPhe^{6}-DNal^{7}]-BK	0.02	0			
[Hyp^{2}]-BK	200	100	90	500	90
[Hyp^{3}]-BK	100	35	85	300	94
[Hyp^{2}-DPhe^{7}]-BK	0.8	5.8(5)	(IB)	(IB)	-
[Hyp^{3}-DPhe^{7}]-BK	0.4	5.6(5)	(IB)	(IB)	-
[Hyp^{2,3}-DPhe^{7}]-BK	0.4	5.3(5)	(IB)	(IB)	-
Lys-Lys-[Hyp^{2}-DPhe^{7}]-BK	0.1	5.7(5)	(IB)	(IB)	-

TABLE 1. (Continued- Part 3)

BRADYKININ ANALOG	RUT	GPI	RBP-IA	RBP-IV	%DESTR
Lys-Lys-[Hyp^{3}-$DPhe^{7}$]-BK	0.3	5.4(5)	(IB)	(IB)	-
Lys-Lys-[$Hyp^{2,3}$-$DPhe^{7}$]-BK	0.2	5.4(5)	(IB)	(IB)	-
DArg-[Hyp^{2}-$DPhe^{7}$]-BK	0.5	6.3(6)	(IB)	(IB)	-
DArg-[Hyp^{3}-$DPhe^{7}$]-BK	6.5(5)	5.9(7)	(IB)	(IB)	-
DArg-[$Hyp^{2,3}$-$DPhe^{7}$]-BK	6.5(5)	6.3(6)	(IB)	(IB)	-
[Hyp^{2}-$Thi^{5,8}$-$DPhe^{7}$]-BK	MXD	6.0(5)	(IB)	(IB)	-
[Hyp^{3}-$Thi^{5,8}$-$DPhe^{7}$]-BK	MXD	5.7(6)	(IB)	(IB)	-
[$Hyp^{2,3}$-$Thi^{5,8}$-$DPhe^{7}$]-BK	6.0(4)	5.9(5)	(IB)	(IB)	-
Lys-Lys-[Hyp^{2}-$Thi^{5,8}$-$DPhe^{7}$]-BK	0.2	6.6(5)	(IB)	(IB)	-
Lys-Lys-[Hyp^{3}-$Thi^{5,8}$-$DPhe^{7}$]-BK	MXD	5.5(5)	(IB)	(IB)	-
Lys-Lys-[$Hyp^{2,3}$-$Thi^{5,8}$-$DPhe^{7}$]-BK	MXD	5.6(5)	0.2	0.5	82
DArg-[Hyp^{2}-$Thi^{5,8}$-$DPhe^{7}$]-BK	5.4(5)	6.5(6)	(IB)	(IB)	-
DArg-[Hyp^{3}-$Thi^{5,8}$-$DPhe^{7}$]-BK	6.9(5)	6.3(6)	(IB)	(IB)	-
DArg-[$Hyp^{2,3}$-$Thi^{5,8}$-$DPhe^{7}$]-BK	6.9(4)	6.0(6)	(IB)	(IB)	-

a. Biological activities of BK analogs on rat uterus (RUT), guinea pig ileum (GPI) and on rat blood pressure (RBP) by intraarterial (IA) or intravenous (IV) administration. Apparent pulmonary destruction (%DESTR) is given for agonists only, and is determined by the difference in IA and IV potency of the agonist. Smooth muscle and RBP agonist potency is relative to BK = 100%, and is NOT underlined. AG indicates very weak agonist activity. Values followed by (K) indicate isotonic determination on kymograph; all other smooth muscle data were obtained by isometric recording. RUT and GPI antagonist potency is given as the pA_2 value of Schild (7), underlined, with number of tissues used in parentheses. MXD indicates weak agonist activity followed by antagonist activity. RBP antagonist activity is not quantitated; (IB) indicates antagonism of an ED_{25mm} dose of BK when analog and BK are administered as a bolus mixture; (II) indicates antagonism of an ED_{25mm} dose of BK during infusion of antagonist. PRS indicates pressor effect; PRS/DPRS indicates pressor effect followed by depressor effect. Most antagonists did not change blood pressure *per se*. Abbreviations for residues: CDF = *p*-chloro-DPhe; DhPhe = D-homo-Phe; Hyp = 4-hydroxy-L-Pro; hPhe = L-homo-Phe; MDY = OMe-D-Tyr; Nal = ß-2-naphthyl-L-Ala; NMF = N-Methyl-L-Phe; Pal = ß-2-pyridyl-L-Ala; PCF = *p*-chloro-L-Phe; Phg = L-phenylglycine; Thi = ß-2-thienyl-L-Ala

of these basic residues to the N-terminus of the antagonists was done because of the dramatic increase in BK potency *in vivo* of Lys-Lys-BK; this increase is due to lack of cleavage of this analog by angiotensin converting enzyme, a major kininase *in vivo* (Ryan and others, 1968). A general reduction in the antagonist potency of analogs in the GPI assay results from the combination of the $Thi^{5,8}$ substitution and N-terminus extension with DArg or Lys-Lys (see $[DPhe^7]$-BK *vs* DArg-$[Thi^{5,8}$-$DPhe^7]$-BK). However, this combination converts $[DPro^3$-$Thi^{5,8}$-$DPhe^7]$-BK from a weak agonist to an antagonist on the uterus.

Combination of a D-aromatic amino acid residue substitution for Ser at position 6 of the BK sequence with a D-amino acid residue at position 7 generally reduces biological activity, as evidenced by the generally lower RUT and GPI agonist potencies, compared with the singly substituted position 7 analogs. However, $[DPhe^6$-$CDF^7]$-BK exhibited good antagonist potency in the GPI assay. Both antagonist potency and the overall spectrum of activity are quite similar to those of $[DPhe^7]$-BK itself. $[DPhe^6$-$CDF^7]$-BK represents a new structural class of BK antagonists, one which contains adjacent D-aromatic amino acids at positions 6 and 7 (Stewart and Vavrek, 1985).

The substitution of the Pro residues at positions 2 and 3 of BK with the 4-hydroxyproline (Hyp) residue does not significantly alter BK agonist potency. In $[DPhe^7]$-BK and its Lys-Lys-extended analog, Hyp substitution reduces BK antagonist potency slightly on the GPI and reduces RUT agonist potency by 90%. Hyp substitution in DArg-$[DPhe^7]$-BK does not alter GPI antagonist potency, but leads to potent antagonists of BK on the RUT, especially when the Hyp residue is at position 3. The Hyp^3- and $Hyp^{2,3}$-analogs of DArg-$[DPhe^7]$-BK are potent smooth muscle antagonists of BK. Hyp substitution for Pro in $Thi^{5,8}$-containing analogs of $[DPhe^7]$-BK produces effects similar to those in non-$Thi^{5,8}$-containing antagonists, especially when the Hyp residue is at position 3 of the antagonist sequence. The Hyp^3- and $Hyp^{2,3}$-analogs of DArg-$[Thi^{5,8}$-$DPhe^7]$-BK are the most potent BK antagonists in the RUT assay to date (pA_2 = 6.9). In the ileum assay, Lys-Lys-$[Hyp^2$-$Thi^{5,8}$-$DPhe^7]$-BK is the most potent antagonist (pA_2 = 6.6).

In addition to the biological results presented here, other investigators have begun to report actions of the BK antagonists in various systems. Certain of these antagonists have blocked the ability of BK to increase vascular permeability in the rabbit (Longridge and others, 1985; Stewart, Vavrek and Whalley, 1985), to bind to and stimulate phosphatidyl inositiol turnover in cultured neuroblastoma-glioma cells (Braas and others, 1985), to enhance prostaglandin E_2 release from cultured human pulmonary fibroblasts (Crecelius and others, 1986), to release norepinephrine from neurons in rat vas deferens

(Llona and others, 1987) and to evoke pain on the human blister base (JMS, personal observation). Benetos and others (1986a, 1986b) have shown that part of the blood pressure-lowering effect of the angiotensin converting enzyme inhibitor enalapril in hypertensive rats is due to protection of circulating BK, and that BK evidently makes some contribution to normal regulation of blood pressure.

It was suggested above that differential antagonist potency on GPI and RUT might indicate that there are different BK receptor types in these tissues. A more striking apparent receptor difference was seen in the work of Huidobro-Toro on rat vas deferens (Llona and others, 1987), where $[Thi^{5,8},DPhe^{7}]$ was a BK antagonist on smooth muscle but an agonist on neurons. Other more potent antagonist analogs were effective antagonists of BK on both tissues.

CONCLUSIONS

In the two years since we described the first sequence-related, competitive antagonists of BK over 30 new antagonists of BK have been produced by the classical approach to peptide molecular modification. The potency of antagonism has been increased an order of magnitude in the GPI assay (pA_2 5.9 to 6.6) and half an order of magnitude in the RUT assay (pA_2 6.5 to 6.9) from the first antagonists in each assay.

Antagonists of BK can be produced by modification of position 7 with small aromatic D-amino acid residues, or by the substitution of positions 6 and 7 with certain D-aromatic amino acid residues. These simple substitutions can provide GPI-selective antagonism. Additional substitution of the Phe residues at positions 5 and 8 with Thi can increase antagonist potency. Replacement of the Pro residues of BK with Hyp, in combination with other modifications, leads to potent antagonists having a broad spectrum of activity. The antagonism is competitive and is specific for BK and related kinin peptides.

ACKNOWLEDGMENTS

We thank Frances Shepperdson and Garima Vashistha for the bioassays and Virginia Sweeney for amino acid analyses. Supported by grants from the NHLBI-NIH and from Nova Pharmaceutical Corporation.

REFERENCES

Benetos, A., I. Gavras and H. Gavras (1986). Hypertensive effect of a bradykinin antagonist in normotensive rats. Hypertension, 8, 1089-1092.
Benetos, A., H. Gavras, J. M. Stewart, R. J. Vavrek, S. Hatinoglou and I. Gavras (1986). Vasodepressor role of endogenous bradykinin assessed by a bradykinin antagonist. Hypertension, 8, 971-974.
Braas, K. M., D. C. Manning, V. S. Wilson, D. C. Perry, J. M. Stewart, R. J. Vavrek and S. H. Snyder (1985). Characterization of bradykinin antagonists in a cultured neuronal cell line. Abstracts, Soc. Neuroscience 15th meeting, Dallas, TX 1985, p. 414.

Crecelius, D. M., J. M. Stewart, R. J. Vavrek, T. M. Balasubramanian and N. Baenziger (1986). Interaction of bradykinin with receptors on human lung fibroblasts. Federation Proc., 45, 454.

Dunn, F. W., and J. M. Stewart (1971). Analogs of bradykinin containing ß-2-thienyl-L-alanine. J. Med. Chem., 14, 779-781.

Llona, I., H. Miranda, R. J. Vavrek, J. M. Stewart and J.-P. Huidobro-Toro (1987). Identification of pre- and post-synaptic bradykinin receptor sites in the vas deferens: evidence for different structural prerequisites. J. Pharmacol. Exp. Therap., in press.

Longridge, D. J., M. Schachter, J. M. Stewart, Y. Uchida and R. J. Vavrek (1985). Synthetic competitive antagonists of bradykinin on vascular permeability in the rabbit. J. Physiol., p. 77P.

Odya, C. E., T. L. Goodfriend and C. Peña (1980). Bradykinin receptor-like binding studied with iodinated analogs. Biochem. Pharmacol., 29, 175-185.

Ryan, J. W., J. Roblero and J. M. Stewart (1968). Inactivation of bradykinin in the pulmonary circulation. Biochem. J., 110, 795-797.

Schild, H. O. (1947). pA, a new scale for the measurement of drug antagonism. Brit. J. Pharmacol., 2, 189-206.

Stewart, J. M., and Vavrek, R. J. (1985). Bradykinin competitive antagonists for classical kinin systems. Federation Proc., 44, 1242.

Stewart, J. M., and R. J. Vavrek (1986). Bradykinin competitive antagonists for classical kinin systems. In L.M. Greenbaum (Ed.), "Kinins 84", Plenum Press, New York, in press.

Stewart, J. M., R. J. Vavrek and E. T. Whalley (1985). Antagonism of the vascular permeability actions of bradykinin in the rabbit. Proc. Brit. Pharmacol. Soc. Dec. 1985, p. 167.

Stewart, J. M., and J. D. Young (1984). Solid Phase Peptide Synthesis, 2nd ed. Pierce Chemical Co, Rockford, IL.

Vavrek, R. J., and J. M. Stewart (1985). Competitive antagonists of bradykinin. Peptides, 6, 161-164.

Vavrek, R. J., and J. M. Stewart (1986a). Smooth muscle selectivity in bradykinin analogs with multiple D-amino acid substitutions. In L. M. Greenbaum (Ed.), "Kinins 84", Plenum Press, New York, in press.

Vavrek, R. J., and J. M. Stewart (1986b). Development and modification of competitive antagonists of bradykinin. In V. J. Hruby, C. M. Deber and K. D. Kopple (Eds.), "Peptides, Proceedings of the 9th American Peptide Symposium", Pierce Chemical Co., Rockford, IL, pp 655-658.

PEPTIDE TRANSPORT

Peptide Transport: The Concept and Its Development

D. M. Matthews

Department of Experimental Chemical Pathology, Westminster Medical School, London, UK
Present address: 7 First Avenue, Muswell Hill, London N10, UK

ABSTRACT

The development of the concept of peptide transport in the animal body, which resulted from investigations of protein absorption, is outlined. The part played by peptide transport in the absorption of protein digestion products is described, and the salient features of peptide transport in animal intestine and other tissues are summarised, as are the main features of peptide transport in microorganisms and higher plants.

KEYWORDS

Peptide transport; protein absorption; protein nutrition; small intestine; transport kinetics; microorganisms; germinating barley.

DEVELOPMENT OF THE CONCEPT OF PEPTIDE TRANSPORT

The concept of active transmembrane transport of small peptides in the animal body did not become firmly established until about 15 years ago, with the demonstration of active transport of di- and tripeptides by the small intestine in the early 1970s as the result of investigations of the mechanism of absorption of protein digestion products (Matthews, 1975). Yet it is no new idea, having been in the minds of physiologists about a century ago. A brief survey of the history of the study of protein absorption will throw light on the development of the concept. Where no references to history or to recent findings are given here, they will be found in Reid (1900), Cathcart (1921), Verzár and McDougall (1936), Fruton (1972), Matthews (1975), Matthews and Payne (1975), Matthews (1977), Matthews and Payne (1980) or in a recent comprehensive review prepared for the Symposium on Balanced Nutrition and Therapy held by Pfrimmer of Erlangen in 1986 (Matthews - to be published). Standard 3-letter abbreviations for amino acids are used throughout, and amino acids are in the L-form unless otherwise indicated.

In the mid-19th century, it was thought that 'protein', originally thought to be essentially only one substance and synthesised solely by plants, was absorbed with little or no chemical change, the functions of the proteolytic enzymes of the stomach and pancreas being limited to 'solubilisation' of coagulated protein. There was some experimental evidence apparently supporting this view, but as the

© 1987 Pergamon Journals Ltd.
Printed in Great Britain

heterogeneity of proteins and the extent of intra-lumen hydrolysis were appreciated, it was succeeded by the hypothesis that proteins were hydrolysed to polypeptides 'proteoses and peptones' and absorbed in this form. However, attempts to demonstrate proteoses and peptones in the blood gave unconvincing results, and this, combined with the finding by Schmidt-Mulheim in Ludwig's laboratory in 1880 that intravenous injection of polypeptides produced the potentially lethal 'peptone shock', gave rise to the 'hypothesis of resynthesis' according to which proteoses and peptones ('peptones') for short were resynthesised to proteins, probably plasma proteins, in the intestinal wall. Abderhalden became the most prominent exponent of this hypothesis. Before the end of the 19th century, both Heidenhain and Waymouth Reid had become convinced that uptake of peptones by the intestinal mucosa involved a selective mechanism which was probably active - Heidenhain used the term 'Triebkraft'.

At the turn of the century came a spectacular advance which inaugurated the modern era in the investigation of both protein absorption and metabolism. Otto Cohnheim (1873-1953), working in Kühne's laboratory in Heidelberg, showed in 1901 that the intestinal mucosa possessed enzyme activity which broke down peptones to amino acids. He named the enzyme(s) responsible 'erepsin'. At first he was uncertain of the site of action of erepsin, but by 1902 he had concluded that it was intracellular, acting on peptides which had entered the mucosal cells. Thus the germ of the modern view that peptides enter the cells by a specific mechanism and are hydrolysed within them was in existence some 85 years ago. Unfortunately, these early ideas were to be lost sight of.

Cohnheim's discovery led to a period of rapid advance, terminating only in 1914. At first, it was not possible to provide a satisfactory demonstration of amino acids in the blood, though they were found in the intestinal lumen, but in 1912 Van Slyke and Meyer, using the newly developed nitrous acid method, showed that there was a rise in the amino-N of portal blood during protein absorption and in 1913 Abel, Rowntree and Turner and Abderhalden isolated individual amino acids from the plasma. At this stage Abderhalden, who had continued to adhere to the hypothesis of resynthesis, merely substituting 'amino acids' for 'peptones' after the discovery of erepsin, abandoned this hypothesis and it became generally accepted that proteins were absorbed into the blood in the amino acid form. In 1913, Van Slyke and Meyer were able to summarise the state of knowledge as follows: 'It appears probable, therefore, that normal digestion proceeds in the intestinal lumen <u>and wall</u> until most, if not all, of the proteolytic products are reduced to the stage of free amino acids' (my underlining). This would serve quite well as a (rather cautious) summary of the position today.

Around 1920, if not before, a grave error arose. This was the baseless assumption that proteins were completely hydrolysed to amino acids in the intestinal lumen by erepsin, believed to be in the 'succus entericus', and taken up by the absorptive cells only in this form - probably as the result of 'simple diffusion'. This view, now known as the 'classical hypothesis of protein absorption', probably originated mainly with the writers of textbooks - there was never any real evidence for it. But it dominated thinking for more than 50 years, acquiring great authority as the years went by, so that in the end it was extremely difficult to overturn. Everyone who had been taught it believed implicitly that 'it had been proved' that digestion products left the lumen only as amino acids. When in 1949 Dent and Schilling showed by paper chromatography that many individual amino acids increased in the portal plasma after a protein meal, referring to the intestinal mucosa as an 'efficient barrier' to absorption of peptides, and Wiseman in 1951 showed that transmural transport of amino acids was an active process, these findings appeared to provide 'final vindication' of the classical hypothesis - and for the next 20 years nearly all research effort was concentrated on elucidating the details of amino acid absorption.

This was unfortunate, for while the uptake of amino acids from the intestinal lumen and their transfer to the blood was undeniable, a good deal of evidence had accumulated over the years showing that the classical hypothesis could not possibly provide an adequate explanation of protein absorption. This has been discussed elsewhere (Matthews, 1975; Matthews and Payne, 1975) and I shall mention only a few points here. It was known as early as 1915 that protein absorption was very rapid - nearly as rapid as the absorption of amino acids and far more rapid than the complete digestion of protein _in vitro_. Several investigators had had reported before 1914 that the absorption of peptones was faster than that of amino acids. Cajori in the 1930s and Florey and his colleagues a few years later had concluded with Cohnheim that the final hydrolysis of peptides could not take place in the lumen and must be a function of the intestinal mucosa. There was evidence that during the absorption of gelatin (which contains much Hyp, giving rise to hydrolysis-resistant peptides) peptides appeared in both blood and urine. When in 1960 Milne and his colleagues showed that in Hartnup disease the specific absorptive mechanism for most neutral amino acids was lost, there was no reasonable explanation for the relatively satisfactory protein nutrition of the patients. Finally in 1962 it was reported that during protein absorption the concentrations of several amino acids peaked in the blood _before_ they were maximal in the lumen. If all the evidence available by 1962 had been put together and considered, it is hard to see how the classical hypothesis could have survived. But inconvenient observations were ignored or explained away (sometimes by the very investigators who had made them) and it required new and startling experimental evidence before the overthrow of the classical hypothesis was initiated.

In the 1950s the results of early experiments using small intestine _in vitro_ did in fact suggest the possibility that peptides might be taken up by the absorptive cells and hydrolysed within them, and Agar, Hird and Sidhu were the first to restate this hypothesis in 1953. Then between 1957 and 1962, Newey and Smyth, stimulated by R.B. Fisher's criticisms of the classical hypothesis in his monograph 'Protein Metabolism', published in 1954, reported a series of investigations of peptide absorption carried out _in vivo_ and _in vitro_. They arrived at a similar conclusion, referring to peptide uptake followed by intracellular hydrolysis as a 'second mode' of protein absorption (they were apparently unaware of the state of opinion reached by 1913). For a variety of reasons, this work had very little impact. There was nothing in the results which could not be ascribed to hydrolysis in the brush border apart from the transmural transport of small quantities of Gly-Gly, the behaviour of which might have been anomalous. None of the distinctive phenomena of peptide absorption, such as its relative rapidity in relation to that of amino acids and the avoidance of competition for transport between amino acids when peptides are transported was demonstrated, and satisfactory evidence for the active transport of peptides could not be provided. Moreover Miller and Crane had shown in 1961 that the hydrolysis of disaccharides was a function of the brush border and it was widely assumed that the handling of peptides was likely to be analogous. The impression prevalent in 1962 was concisely expressed by Wilson in his book on Intestinal Absorption '... _in vitro_ experiments have confirmed the view that peptides are poorly absorbed'. Nothing more was published on the absorption of peptides for six years.

In 1966 my colleagues and I began some work which was to lead within a few years to the first demonstration of active transport of peptides by the small intestine. With the suggestions of Newey and Smyth in mind, we started what I thought was going to be a fairly unexceptional piece of clinical research on absorption and malabsorption of peptides of Gly. In establishing control values we were astonished to find that equivalent doses (about 10 g) of Gly were more rapidly absorbed from Gly-Gly than from the free amino acid and most rapidly from Gly-Gly-Gly. This made the hypothesis of intracellular hydrolysis seem much more credible, since intralumen or brush border hydrolysis with release of amino acid in free solution could not possibly account for the results, even if it were instantaneous. This work was published, together with confirmatory findings obtained in the rat, in 1968. In the same year Siamak Adibi in Pittsburgh published similar results obtained by jejunal perfusion in man. He was equally surprised by his findings and concluded, like me, that small peptides were likely to be taken up intact and then hydrolysed, probably within the absorptive cells. Such an interpretation provided a direct challenge to the classical hypothesis.

After these initial findings, which were at first received with some incredulity, rapid progress was made in my laboratory, Adibi's laboratory and elsewhere. The relative rapidity of peptide absorption was found to be a common phenomenon and it was shown that during peptide absorption competition for uptake between amino acids was largely avoided. Both Adibi and I realised at an early stage that absorption of small peptides was not entirely a matter of peptide transport followed by hydrolysis - in most cases there was also some brush border hydrolysis followed by uptake of free amino acids, so that a dual mechanism was involved. The independence of intestinal transport of peptides and amino acids was established between 1969 and 1971 as the result of investigations in Hartnup disease by Milne, myself and colleagues, and experiments on competition for transport by Rubino and co-workers, and it was shown that very many di- and tripeptides competed for transport between themselves, but not with amino acids.

At this stage, intestinal transport of peptides remained, strictly, no more than a working hypothesis - since most peptides were hydrolysed so rapidly in the intestinal wall that it was impossible to demonstrate their presence in it or their large-scale transport across it, and active transport of peptides could not of course be shown. In this situation, interdisciplinary cooperation came to our aid. It had been known as long ago as the 1880s that many bacteria required 'peptones' in the culture medium in order to flourish, and during the 1950s and '60s the existence of peptide transport had been recognised in bacteria. I was naturally interested in this, since there appeared to be resemblances between the handling of peptides in bacteria and in the intestine - the remarkable rapidity of peptide uptake being the most striking. In 1971, at a CIBA Symposium, Dr. J.W. Payne told me that certain bacteria could accumulate the hydrolysis-resistant dipeptide glycylsarcosine (Gly-Sar). This led directly to the demonstration, using Gly-Sar and other hydrolysis-resistant peptides, that di- and tripeptides could be actively transported by hamster small intestine _in vitro_. When this had been done I found that the scepticism which had prevailed about the whole question of intestinal peptide transport, particularly in the United States, was beginning to disappear.

By 1975 most of the salient features of intestinal peptide transport had been established, and some of the interest of investigators began to turn to the possibility of mediated peptide uptake by other cells and tissues. This was shown in kidney, liver, muscle and Ehrlich ascites carcinoma cells, confirming and extending previous reports scattered throughout the literature. It appeared that the first to demonstrate active transport of peptides in the animal body had been Abraham, Pisano and Udenfriend in 1964, working with brain slices - but the possible general implications of their work had been ignored at the time. In the same year there occurred another striking example of cross-fertilisation between disciplines. For some time it had been my ambition to see whether peptide transport was demonstrable in higher plants, and I had toyed with the idea of using carnivorous plants for this purpose. However, at this stage I learned that Professors T.M. Enari and J. Mikola in Finland suspected that small peptides might be taken up by the scutellum (absorptive organ) of germinating barley, since this contained oligopeptidases, whereas none were secreted into the food-store. As a result Dr. T. Sopanen visited my laboratory, and in 1976 we were able to report active transport of Gly-Sar and Gly-Sar-Sar in germinating barley. This was confirmed in Payne's laboratory and extended to other seeds. These findings helped to establish the global distribution of peptide transport in Nature.

In recent years, increasing interest has been taken in the intestinal absorption of biologically active peptides. A range of antibiotics and at least one tripeptide releasing factor have been shown to make us of the active transport mechanism for peptides of dietary origin. This, in conjunction with the realisation that it is more physiological to use oligopeptide-amino acid mixtures for enteral nutrition than mixtures containing only free amino acids, and extension of work showing that small peptides are well utilised when given intravenously, show that the existence of peptide transport has important therapeutic implications.

This short survey of the development of the concept of transmembrane transport of small peptides ends here, and in the remainder of the account I shall try to summarise what is known of peptide transport at the time of writing.

PEPTIDE ABSORPTION IN THE INTESTINE

Following a protein meal, the lumen of the small intestine contains a mixture of peptides and amino acids in which peptides predominate. Hydrolysis of peptides of up to about 10 amino acid residues takes place partly in the brush border. Tripeptides, dipeptides and amino acids are taken up by the absorptive cells, hydrolysis of the peptides being largely completed within the cells. Small peptides vary greatly in their susceptibility to hydrolysis, some (like Gly-Gly, carnosine and peptides of Pro and Hyp) undergoing very little brush border hydrolysis and even escaping complete intracellular hydrolysis so that a proportion enters the blood intact. The total extent of entry of protein digestion products into the blood in peptide form is uncertain, though it is unlikely to be large. It probably varies widely between one protein and another. Results from my laboratory have suggested that, during the absorption of casein by the guinea-pig, some 10% of the total entering the blood may be in peptide form. No estimate has yet been made of the proportion of protein digestion products taken up from the lumen as small peptides and as amino acids. My own guess is that about half, possibly more, of the total upake may be in the form of peptides.

The following aspects of peptide absorption are of particular interest.

1. Active transport of peptides. Both di- and tripeptides are actively transported by the small intestine. Transport appears to be limited to peptides of not more than three amino acid residues. The recent work of Leibach and of Hoshi indicates that transport is probably proton-linked, rather than sodium ion-linked, as previously believed.

2. Independence of transport of peptides and amino acids. A very large body of evidence leaves no doubt that intestinal transport of peptides and amino acids are completely separate processes.

3. Competition for peptide transport. Very many di- and tripeptides have been shown to compete for transport. Di- and tripeptides compete with each other, as do dipeptides of basic, neutral and acidic amino acids. Until recently, it seemed likely that there might be only one peptide transport system, but in the last few years work from my own laboratory and that of Hoshi has suggested that some neutral peptides at least may be transported by more than one system.

4. Structural requirements for peptide transport. The requirements for transport of peptides are in some respects similar to those for amino acids - thus substitution of the terminal groups reduces affinity for transport and transport is stereochemically specific, favouring L-forms. However, for peptide transport a peptide bond (or bonds) is essential (α-linkages are preferred) and transport is relatively indifferent to the nature of side-chains.

5. Kinetics of peptide transport. Early work on the kinetics of intestinal peptide transport was carried out under so many different experimental conditions

that the results cannot easily be compared. It was, however, soon established that though peptide transport conformed approximately to Michaelis-Menten kinetics, the kinetic characteristics of amino acid and peptide transport were quite different. Adibi suggested that transport of peptides might be characterised by higher Vmax values than that of amino acids, which he thought might account for the relatively rapid absorption of peptides. More recently (Matthews and Burston, 1984; Matthews - to be published) we have carried out an investigation of the kinetics of transport of a series of neutral dipeptides - Gly-Gly, Ala-Ala, Val-Val and Leu-Leu - and the corresponding amino acids, using hamster jejunum in vitro. It was found that though Kt and Vmax for the amino acids fell greatly with increasing bulk of side-chain, as expected from previous work, this tendency was not apparent with the peptides, Kt and Vmax varying by only a small factor between one peptide and another. According to our findings, Vmax values for the peptides were lower than for the corresponding amino acids. Other findings were that the tripeptide Gly-Sar-Sar had a higher Kt and a higher Vmax than the dipeptide Gly-Sar and that the kinetics of transport were such that influx of Gly in the form of Gly-Gly was more rapid than that of the equivalent free Gly at low concentrations, but less rapid at high concentrations. The reverse had been found to be true of Lys-Lys and Lys. With Ala-Ala and Leu-Leu, total transport of amino acids was more rapid from the peptides than from the free amino acids at all concentrations, whereas with Val-Val it was less rapid.

5. Relative rates of absorption of peptides and amino acids. To see to what extent the relatively rapid absorption of peptides was a general phenomenon, investigations were carried out with Dr. D.B.A. Silk and others on the absorption in man of individual amino acids (free plus peptide-bound) from partial hydrolysates of proteins containing mainly small peptides and the equivalent mixtures of free amino acids. Total absorption was more rapid from the partial hydrolysates, and it was also found that amino acids which were absorbed least rapidly from the mixtures were much more rapidly absorbed from the partial hydrolysates, while those which were absorbed most rapidly from the mixtures tended to be absorbed at about the same rate or more slowly from the partial hydrolysates. Thus absorption of amino acids was much more 'even' from the partial hydrolysates. The possibility that this might lead to more simultaneous presentation of amino acids to the tissues and thus to more effective protein synthesis has already been suggested (Matthews and Payne, 1975). Why is it that the absorption of peptides is often more rapid than that of the equivalent amino acids? The cause is certainly multifactorial, and possible factors include the following. Firstly, uptake of peptides which undergo some brush border hydrolysis involves both peptide and amino acid transport systems. Secondly, when peptides are absorbed, competition for transport between free amino acids is at least partly avoided. Thirdly, the kinetic characteristics of transport of peptides are such that in some cases these may be transported (at least at certain concentrations) at a rate which transfers amino acids more rapidly than when they are transported in the free form.

PEPTIDE TRANSPORT IN OTHER ANIMAL CELLS AND TISSUES

Transport of small peptides has been observed in many animal cells other than the the absorptive cells of the small intestine. It occurs in liver, kidney, brain, muscle, the mammary gland, red cells and neoplastic cells. In kidney, brain and red cells, transport is active. In other cases it may be, but few details of the mechanism are known. Adibi and others have shown that small peptides are rapidly cleared from the bloodstream, and can maintain protein nutrition quite satisfactorily.

PEPTIDE TRANSPORT IN MICROORGANISMS AND IN HIGHER PLANTS

Active transport of small peptides followed by intracellular hydrolysis occurs in many microorganisms including yeasts and fungi, and in the scutellum of the germinating seeds of barley and other plants; probably also in the leaves of carnivorous plants. In Escherichia coli and in barley scutellum, the situation is uncomplicated by surface hydrolysis, which makes the study of peptide transport easier than it is in the gut. In E.coli, which may be taken as an example of a microorganism, transport is active, stereochemically specific, independent of that of amino acids and directly dependent on phosphate-bond energy. Peptides of up to 5-6 amino acid residues can be transported. Transport of peptides may be many times more rapid than that of amino acids. Transport is little affected by the nature of amino acid side-chains. There are two transport systems - a dipeptide system and one which also accepts longer oligopeptides. In barley, scutellar transport of peptides is active and in many other respects, including its rapidity, similar to that in E.coli. However, it is probably proton-linked. It requires the presence of thiol groups (this may also apply to the intestine). There are no separate di- and oligopeptide uptake systems, and whether there is more than one peptide transport system is uncertain.

PHYSIOLOGICAL ADVANTAGES OF PEPTIDE TRANSPORT

Membrane transport of small peptides appears to have a number of special advantages. The first is that it should tend to conserve metabolic energy. In general, it should take no more energy to transport a molecule of a peptide containing two or more amino acids than is needed to transport a single amino acid molecule. Conservation of energy may also be contributed to by the entry of peptides into the cell down a steep concentration gradient maintained by rapid intracellular hydrolysis. The second advantage is nutritional. In animals, this aspect has not yet been properly explored, but in many bacteria, the nutritional importance of peptide transport is undeniable. One important factor in this is the avoidance of competition for transport between amino acids when peptides are taken up. The third advantage lies in the fact that should an amino acid transport system be lost by genetic error, the ability to transport peptides can maintain protein nutrition. This is seen in Hartnup disease and cystinuria in man, and analogous phenomena occur in bacteria.

THE HYPOTHESIS OF MEMBRANE HYDROLYSIS

The concept of transmembrane transport of peptides is now very firmly established, applying to many disparate organisms. In microorganisms and seeds, it seems an inescapable conclusion that peptide transport precedes hydrolysis. On the other hand, alternative schemes have been proposed to account for the phenomena of peptide absorption in the intestine and I cannot conclude my account without mention of them. The development of such schemes has been encouraged by the fact that in the intestine the situation is complicated by the occurrence of surface hydrolysis with uptake of amino acids, on which all are agreed. Hydrolysis with uptake of amino acids from free solution cannot, however, account for many of the features of peptide absorption, so that some other mechanism must be invoked to do so. My own opinion is that the simultaneous occurrence of peptide transport will account adequately for all the observations, while exponents of 'membrane hydrolysis', of whom Ugolev is the most distinguished, hold that hydrolase-linked transport of amino acids is the mechanism of prime importance. In the scheme proposed by Ugolev (Ugolev, Timofeeva, Smirnova and co-workers, 1977), peptides are hydrolysed by peptidases of the brush border membrane and the constituent amino acids are taken up without release in free solution by special hydrolase-related carriers which are available only to amino acids released by the peptidases. This scheme will readily account for some phenomena, but it is, in my view, open to a number of serious criticisms. Transport and hydrolysis may be completely dissociated, as shown by the active transport of hydrolysis-resistant peptides - so that Ugolev admits that some transport of intact peptides must take place. Another difficulty is that many peptides which are hydrolysed by different peptidases behave as if they share the same uptake system(s), which leads to topographical problems. Our recent observations suggest that affinity for brush border hydrolysis and apparent affinity for transport are not related, and show that the hydrolysis-resistant Gly-Sar can completely inhibit transport of Val-Val, which does undergo a good deal of hydrolysis in the brush border - these findings are not readily explained by 'membrane hydrolysis'. If the intestine does appear to contain multiple peptide transport systems, this may introduce further problems. Finally, there are formidable difficulties in the observation that partial removal of the brush border peptidases with papain does not reduce transfer of amino acids from peptides but tends to increase it. I feel that the hypothesis of membrane hydrolysis is uncomfortable with all the observations it now has to explain. On the other hand, I believe that the simple scheme of hydrolysis followed by uptake of free amino acids plus peptide transport followed by hydrolysis will account without difficulty for all the phenomena observed so far, and it is compatible with the subcellular distribution of peptidases and the very wide biological distribution of peptide transport. It seems to me that though we should keep an open mind (history emphasises the importance of this) this scheme should be adhered to unless compelling new evidence requires its modification.

ACKNOWLEDGMENT

Work on peptide transport in my laboratory was supported mainly by grants from the Medical Research Council of Great Britain. I am deeply grateful to my colleague Dr. David Burston, without whom most of the work might never have been done, to Dr. M.L.G. Gardner for criticism, fruitful suggestions and assistance in computation, and to my wife for her forbearance.

REFERENCES

Cathcart, E.P. (1921). The Physiology of Protein Metabolism. Longmans, Green, London.

Fruton, J.S. (1972). Molecules & Life. Wiley-Interscience, New York.

Liebig, J. von (1847). Researches on the Chemistry of Food. (Trans.W.Gregory). Taylor Walton, London. p.18.

Matthews, D.M. (1975). Intestinal absorption of peptides. Physiol.Rev., 55, 537-608.

Matthews, D.M. (1977). Memorial lecture: Protein absorption then and now. Gastroenterology, 73, 1267-1279.

Matthews, D.M. and D. Burston (1984). Uptake of a series of neutral dipeptides including L-alanyl-L-alanine, glycylglycine and glycylsarcosine by hamster jejunum in vitro. Clin.Sci., 67, 541-549.

Matthews, D.M. and J.W. Payne (Eds.), Peptide Transport in Protein Nutrition. North-Holland, Amsterdam.

Matthews, D.M. and J.W. Payne (1980). Transmembrane transport of small peptides. Current Topics in Membranes and Transport, 14, 331-425.

Reid, E.W. (1900). On intestinal absorption, especially on the absorption of serum, peptone and glucose. Phil.Trans.Roy.Soc.Lond.(Biol.Sci.), 192, 211-296.

Ugolev, A.M., N.M. Timofeeva, L.F. Smirnova, P. de Laey, A.A. Gruzdkov, N.N. Iezuitova, N.M. Mityusheva, G.M. Roschina, E.G. Gurman, V.M. Guzev, V.A. Tsvetkova and G.G. Shcherbakov (1977). Membrane and intracellular hydrolysis of peptides: differentiation, role and interrelations with transport. In K. Elliott and M. O'Connor (Eds.) CIBA Symposium: Peptide Transport and Hydrolysis. Elsevier, Amsterdam. pp. 221-238.

Verzár, F. and E.J. McDougall (1936). Absorption from the Intestine. (Facsimile ed.) Hafner, New York.

Properties and Physiological Functions of Mammalian Renal Peptide Transport System

Vadivel Ganapathy, Yusei Miyamoto and Frederick H. Leibach

Department of Cell and Molecular Biology, Medical College of Georgia, Augusta, GA 30912-3331, USA

ABSTRACT

The brush-border membrane of proximal tubular cells of the mammalian kidney possesses a transport system that is specific for small peptides. This peptide transport system differs from other transport systems of the renal brush-border membrane in that it is driven by a proton gradient rather than a sodium gradient. Since the peptide-proton cotransport is electrogenic, a membrane potential can also energize the transport system. Even though all segments of the proximal tubule are capable of proton-coupled peptide transport, the system is more active in the pars recta than in the pars convoluta. The powerful peptidases of the brush-border membrane are responsible for the luminal liberation of peptide substrates for the transport system from larger polypeptides and proteins that are filtered at the glomerulus. The driving force, the electrochemical proton gradient, is generated and maintained by the combined actions of the basal-lateral sodium-potassium-ATPase and the brush-border sodium-proton exchanger. The peptide transport system, together with the brush-border peptidases, may play a significant role in the recapture of peptide-bound amino acids from the tubular lumen and may also participate in the regulation of the plasma level of small peptides.

KEYWORDS

Proximal tubule; brush border; electrochemical proton gradient; peptide-proton cotransport; peptidases; segmental localization.

INTRODUCTION

The brush-border membrane of proximal tubular cells contains an array of transport systems that are responsible for the efficient reabsorption of organic solutes and inorganic ions from the glomerular filtrate. Most of these transport systems are active, and a sodium gradient serves as the driving force for these systems (Ullrich, Rumrich and Kloss, 1982). In recent years, the existence of a transport system that is specific for small peptides (2-4 amino acids) in the renal brush-border membrane has been recognized (Ganapathy and Leibach, 1986). This transport system is unique among the renal transport systems because it is energized by a proton gradient rather than a sodium gradient. Studies with

© 1987 Pergamon Journals Ltd.
Printed in Great Britain

purified brush-border membrane vesicles as well as with isolated tubules have provided a wealth of information on the localization and characteristics of this novel transport system. However, insight into the mechanistic aspects of the transport system and the nature of the transport protein itself has begun only recently. In this paper, we will attempt to describe in a concise manner what is currently known about the properties and physiological functions of the renal peptide transport system.

CHARACTERISTICS OF THE RENAL PEPTIDE TRANSPORT SYSTEM

Table 1 summarizes the salient features of the renal peptide transport system. The intestinal peptide transport system also exhibits similar characteristics. Even though the existence of a transport mechanism that is specific for small peptides was recognized in the small intestine almost a quarter of a century earlier than in the kidney, recent progress in the energetics and other molecular aspects of peptide transport has occurred simultaneously in the small intestine and in the kidney. Use of isolated brush-border membrane vesicles from the absorptive cells of these tissues for transport studies and judicial selection of hydrolysis-resistant peptide substrates have been largely responsible for this progress. Prior to the advent of the vesicle technology, investigations on peptide transport were primarily conducted in the small intestine using various intact tissue preparations and the results of these early investigations have been extensively reviewed and critically analyzed by Matthews (1975). With the discovery that the transport of peptides in the small intestine and in the kidney is energized by a proton gradient rather than a sodium gradient, the field of peptide transport has entered an exciting new phase. These interesting developments have been the subject of recent reviews by Ganapathy and Leibach (1985, 1986) and Hoshi (1985).

TABLE 1 Properties of Renal Peptide Transport System

1. Substrate Specificity
    - Specificity for small peptides (2-4 amino acids)
    - Preference for peptides containing L-amino or imino acids
    - Requirement for free carboxyl and free amino termini
2. Energetics
    - Transport against a concentration gradient
    - Sodium-independence
    - Energization by an electrochemical proton gradient
    - Electrogenicity
    - Peptide-proton cotransport
    - Tertiary active transport
3. Molecular aspects
    - Involvement of histidyl and thiol groups in the transport function
    - Resistance to papain digestion

One important question in the field of peptide transport to which we are yet to find a clear-cut answer is whether there are multiple peptide transport systems, analogous to the amino acid transport systems. It is highly unlikely that a single transport system can accept all dipeptides and tripeptides (possibly tetrapeptides) as substrates, especially when one considers the fact that the number of theoretically possible dipeptides exceeds 400 and that of tripeptides exceeds 8000. Moreover, all peptides do not exist as zwitter ions under physiological conditions and their charge content differs depending upon the nature of their constituent amino acids. At the least, we can expect the relative contribution of the proton gradient and the membrane potential to the active transport of these peptides to vary significantly. We believe that, in all probability, future investigation will reveal several distinct species of carrier systems for small peptides. The existence of multiple peptide transport systems in the small intestine and in the kidney may confer upon the organism a distinct biological advantage because, if there is a single peptide transport system, a genetic defect in the system would be highly lethal to the organism. Obviously, more work is needed in this area.

LIBERATION OF PEPTIDE SUBSTRATES IN THE PROXIMAL TUBULAR LUMEN

While the existence of peptide transport system(s) in the small intestine has been well recognized for several years, there was considerable skepticism even until very recently regarding the occurrence of similar transport systems in the proximal tubule. Since digestion and absorption of dietary proteins is one of the major physiological functions of the small intestine, most investigations on peptide transport were initially conducted, quite understandably, in this tissue. On the contrary, the concentrations of small peptides in the circulation are very low and doubts have therefore been often raised as to the need for peptide transport systems in the renal proximal tubule. Even though there is an abundance of peptidases in the renal brush-border membrane (Kenny and Booth, 1978), these enzymes were either considered functionless owing to the lack of substrates (Crane, 1980) or were believed to participate in amino acid absorption (Kenny and Booth, 1978). In fact, the very low levels of circulating small peptides and the relatively very high levels of brush-border peptidases were primarily responsible for the initial dogma that the renal proximal tubule does not contain specific peptide transport systems. It was argued that small peptides that are filtered at the glomerulus can be split into free amino acids so rapidly in the proximal tubular lumen by the brush-border peptidases that the amino acid transport systems in the proximal tubule are solely responsible for the efficient recapture of peptide-bound amino acids.

When the existence of a transport system specific for small peptides in the renal brush-border membrane was discovered in this laboratory a few years ago (Ganapathy, Mendicino and Leibach, 1981), we set out to demonstrate that the renal peptide transport system is neither a functionless, vestigial component of the brush-border membrane nor a mere biochemical curiosity, but it serves an important function in the reclamation of peptide nitrogen under physiological conditions. We hypothesized that one of the major functions of the brush-border peptidases of the proximal tubule is to generate peptide substrates for the renal peptide transport system (Ganapathy and co-workers, 1984). Although the concentration of di- and tripeptides filtered at the glomerulus is negligible, the filtrate contains larger peptides and proteins in significant quantities. These polypeptides may be hydrolyzed to liberate small peptides by the concerted actions of brush-border peptidases and thus the actual concentration of di- and tripeptides in the tubular lumen may be many-fold greater than in the circulation. Among the renal brush-border peptidases, dipeptidylpeptidase IV is unique in a number of ways. The mammalian kidney is the richest source of this enzyme which constitutes as much as 4% of the total brush-border membrane protein (Kenny and Booth, 1978). This enzyme generates dipeptides, and not amino acids like other aminopeptidases, from the amino terminus of large polypeptides in a sequential manner. We have recently presented evidence using isolated brush-border membrane vesicles that this enzyme rapidly hydrolyzes large peptides to release dipeptides into the medium and these dipeptides are then actively transported into the vesicles by the peptide transport system (Miyamoto and co-workers, 1987).

DISTRIBUTION OF THE PEPTIDE TRANSPORT SYSTEM ALONG THE PROXIMAL TUBULE

The proximal tubule of the mammalian nephron is divided into a convoluted part, pars convoluta, and a part which is less convoluted, called pars recta. The transport systems responsible for reabsorption of glucose and amino acids in different parts of the proximal tubule exhibit different kinetic properties. In general, the transport systems in pars convoluta are of the low-affinity and high-capacity type whereas the transport systems in pars recta are of high-affinity and low-capacity type. Since a large part of the filtered glucose and amino acids is absorbed in pars convoluta, the luminal concentrations of these solutes are high in pars convoluta, but relatively very low in pars recta. Therefore, the different kinetic properties of the glucose and amino acid transport systems in these two segments ensure maximal reabsorption of these solutes under physiological conditions.

We have recently undertaken studies to investigate the distribution and properties of the peptide transport system in pars convoluta and in pars recta. There were two compelling reasons for this endeavor. Firstly, the glucose and amino acid transport systems whose characteristics have so far been studied in different segments of the proximal tubule are sodium-dependent transport mechanisms. The peptide transport system is unique among the renal transport systems because it is energized not by a sodium gradient but by a proton gradient. Nothing is known about the distribution and properties of this novel transport system along the proximal tubule. Secondly, the concentration profile of the substrates of the peptide transport system in different parts of the proximal tubule is quite opposite to that of glucose and amino acids. The concentration of small peptides at the beginning of the proximal tubule is negligible. However, the filtrate in this region contains larger polypeptides and proteins in significant quantities. As these polypeptides pass down the proximal tubule, they are acted upon by the peptidases of the brush-border membrane to release smaller peptides. Therefore, the concentration of small peptides (2-4 amino acids) gradually increases along the proximal tubule and reaches significantly higher levels in pars recta than in pars convoluta.

For these reasons, we investigated the characteristics of the peptide transport system in brush-border membrane vesicles derived from pars convoluta and pars recta. The transport system is active, electrogenic and is driven by a proton gradient in both membrane preparations. Interestingly, however, the peptide transport system is more active in the pars recta than in the pars convoluta. This is quite different from the properties of the glucose and amino acid transport systems in these segments. The existence of a low-capacity peptide transport system in the pars convoluta and a high-capacity peptide transport system in the pars recta ensures maximal reabsorption of small peptides from the tubular lumen under physiological conditions. The affinity of the transport system to its peptide substrates is however, similar in both the segments and so is the number of carriers. It therefore appears that the difference in the activities of the peptide transport system in these segments may be due to the difference in the affinity of the transport system for protons and/or the difference in the proton:peptide stoichiometry.

We have also recently studied the distribution of the peptide transport system in the mammalian nephron using a microperfusion technique (Silbernagl and co-workers, 1986). In this investigation, we measured the absorption of a hydrolysis-resistant dipeptide, glycylsarcosine, from the tubular lumen when the dipeptide was perfused into different parts of rat superficial nephrons. More than 65% of the dipeptide was absorbed if microperfusion started at the early or late sections of the proximal convoluted tubule, but the absorption was reduced to 18% when early distal tubular sections were perfused. These data suggest that an effective peptide transport mechanism exists in the nephron section between the last superficial loop of the convoluted part of the proximal tubule and the earliest superficial loop of the distal tubule. The proximal straight tubule (pars recta) appears to be the most probable site of peptide transport though the other parts of the Henle's loop cannot be totally excluded. We then studied peptide transport in the isolated proximal straight tubules of the rabbit using glycylsarcosine as the peptide substrate (Barfuss, Ganapathy, and Leibach, unpublished data). The absorptive cells of this segment of the proximal tubule were indeed found to be capable of uphill transport of the dipeptide, the cellular glycylsarcosine concentration reaching a value that was many-fold greater than the luminal glycylsarcosine concentration.

GENERATION OF THE DRIVING FORCE FOR PEPTIDE TRANSPORT

Coupling of the movement of an organic solute against its concentration gradient to the movement of a monovalent cation down its concentration gradient is one of the fundamental principles in biology. Two ionic species, proton and sodium, play the role of the coupling ions. The pathways involved in the generation of transmembrane proton and sodium gradients and the coupling of the energy stored in these ion gradients to the active transport of organic solutes in various organisms are described in Fig. 1. There is reasonable evidence to assume that proton rather than sodium was the initial coupling ion in evolution. In micro-organisms, membrane-linked respiratory chain and ATP hydrolysis directly produce an electrochemical proton gradient across the cytoplasmic membrane. According to the chemiosmotic hypothesis of Mitchell (1963), this proton motive force is the immediate driving force for many processes including uphill solute transport in bacterial cells. In animal cells, however, owing to the presence of the sodium pump in the plasma membrane, ATP hydrolysis results in the generation of an electrochemical sodium gradient across this membrane. This sodium motive force appears to be the primary driving force for the uphill solute transport in these cells. The substitution of sodium for proton may have some advantages in the overall cellular metabolism. The most important of these advantages is the fact that many intracellular enzymes are more sensitive to changes in the proton concentration than in the sodium concentration. Thus, a cell with the membrane mechanisms to produce the electrochemical sodium gradient across the membrane and to utilize this gradient in the energization of active solute transport appears to be more progressive in evolutionary terms than a cell which utilizes proton instead of sodium in these processes. Since the sodium pump and the proton pump directly link the metabolic energy to the uphill movement of these ions, these pumps are traditionally classified as primary active transport systems. These ion gradients in turn provide energy for the uphill transport of various organic solutes and, therefore, the proton-coupled amino acid- and sugar transport systems of microorganisms and the sodium-coupled amino acid- and sugar transport systems of animal cells are called secondary active transport systems.

Although most bacterial transport systems are driven by a proton motive force, several examples of sodium-coupled transport systems in microorganisms have appeared in the recent literature (Skulachev, 1985). Similarly, though it has long been believed that proton-coupled transport systems do not exist in the plasma membrane of animal cells, recent studies have shown this belief to be incorrect. The proton gradient-driven transport of small peptides described in this paper is only one of several examples of proton-coupled transport systems in animal cells. While the transport of some amino acids in animal cells has been unequivocally shown to be dependent upon an electrochemical sodium gradient, several other amino acids such as leucine, valine, isoleucine, and phenylalanine are transported by a sodium-independent system. Most recent studies by Mitsumoto and co-workers (1986) have demonstrated that this system (conventionally designated as system l) is driven by a proton motive force in Chang liver cells. Lysine transport, also sodium-independent in many cells, has been reported to be energized by a proton gradient in Mullet kidney (Lee and Pritchard, 1983). Furthermore, the transport of glutamate appears to be dependent on both sodium and proton (Nelson and co-workers, 1983; Berteloot, 1986). Most recently, Hoshi and co-workers (1986) have demonstrated that the sodium-independent glucose transport in the presence of an inside-negative membrane potential can be stimulated by a proton gradient.

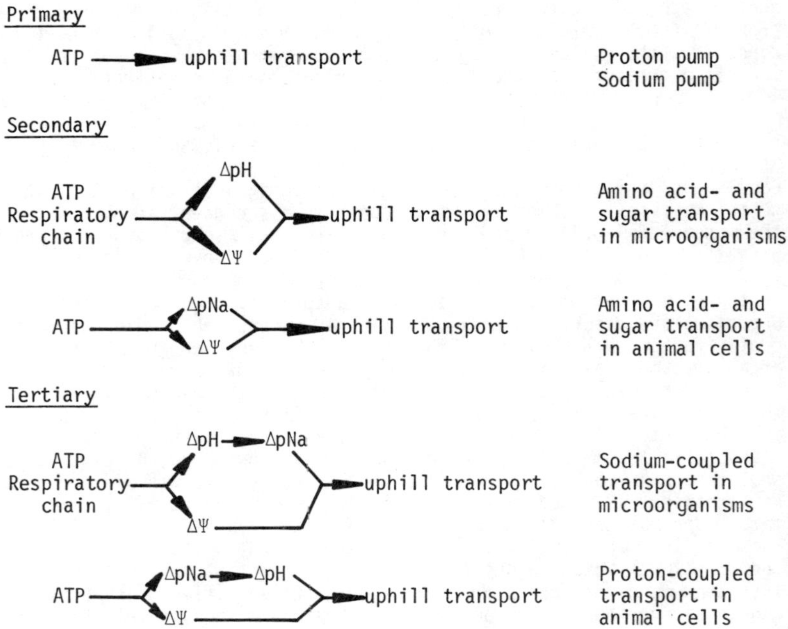

Fig. 1. Classification of active transport systems

These discoveries of sodium-coupled transport systems in microorganisms and proton-coupled transport systems in animal cells raise an important question. What are the mechanisms involved in the generation of a sodium gradient across bacterial cell membranes and generation of a proton gradient across animal cell membranes? This question now appears to have been satisfactorily answered with the demonstration of the presence of a sodium-proton exchanger in the plasma membrane of all types of cells (Seifter and Aronson, 1986). This exchanger executes successfully the interconversion of the chemical gradients of proton and sodium across these membranes. This plasma membrane protein has been the focus of numerous investigations in recent years because it plays a crucial role in the regulation of intracellular pH and in pH-dependent metabolic processes such as those involved in cell growth and cell proliferation (Seifter and Aronson, 1986). In polarized cells such as the epithelial cells of the small intestine and renal proximal tubule and the syncytiotrophoblast of the human placenta, the sodium-proton exchanger is predominantly localized in the brush-border membrane (Murer, Hopfer and Kinne, 1976; Balkovetz and co-workers, 1986). In addition to its role in the regulation of intracellular pH, this exchanger is responsible for the maintenance of an acid microclimate pH on the luminal surface of the intestinal mucosal epithelium and for the acidification of urine as well as for the generation of a proton gradient across the brush-border membranes of these cells. Maintenance of this proton gradient is achieved by the constant presence of a sodium gradient across the brush-border membrane which in turn is generated and maintained by the action of sodium-potassium ATPase located in the basal-lateral membrane. In other animal cells that are unpolarized, the sodium-potassium ATPase and the sodium-proton exchanger coexist in the plasma membrane, and a proton gradient can be generated across this membrane by the concerted actions of these two transport systems. Thus, metabolic energy is utilized by the sodium-potassium ATPase to produce a sodium gradient across the plasma membrane of the animal cells and the sodium-proton exchanger then generates a proton gradient in response to this sodium gradient. This sequence of events is reversed in bacterial cells in which a proton gradient is initially generated by the hydrolysis of ATP and the membrane-bound respiratory chain and this proton gradient subsequently drives the sodium-proton exchanger to produce a sodium gradient. This means, under physiological conditions, the sodium-proton exchanger of animal cells extrudes protons whereas that of the bacterial cells extrudes sodium. Based on the above mechanisms of the origin of a sodium gradient in bacterial cells and a proton gradient in animal cells, the sodium-coupled transport in microorganisms and the proton-coupled transport in animal cells can be classified as tertiary active transport systems.

FUNCTIONS OF THE RENAL PEPTIDE TRANSPORT SYSTEM

The proximal tubule of the mammalian kidney plays an important role in the determination of concentrations and half-lives of circulating proteins and polypeptides of sizes that are smaller than that of albumin. Endocytosis of filtered polypeptides followed by intralysosomal hydrolysis and luminal breakdown of polypeptides by brush-border peptidases followed by absorption of the breakdown products are responsible for this important physiological function. It was originally believed that free amino acids are the only absorbable breakdown products. The recent discovery that proximal tubular cells are also capable of absorbing small peptides adds a new dimension to the role of the kidney in the regulation of circulating levels of small peptides as well as in the reabsorption of peptide-bound amino acids. It also appears that the renal peptide transport system may accept as substrates some non-physiological, peptide-like compounds, many of which may possess pharmacological actions (e.g. aminocephalosporins) (Inui and co-workers, 1984) and thus the peptide transport system may participate in the renal handling of these substances. Furthermore, the existence of two different energy sources for reabsorption of two related groups of organic solutes, namely a sodium gradient for amino acid reabsorption and a proton gradient for peptide reabsorption, may be advantageous to the organism, because such on arrangement would eliminate or reduce any competition between these solutes for the same driving force.

ACKNOWLEDGEMENTS

We thank Ida O. Thomas for her expert secretarial assistance. This study was supported by a NIH grant DK-28389 and a grant from the National Kidney Foundation of Georgia. This is contribution number 1035 from the Department of Cell and Molecular Biology, Medical College of Georgia.

REFERENCES

Balkovetz, D. F., F. H. Leibach, V. B. Mahesh, L. D. Devoe, E. J. Cragoe, Jr., and V. Ganapathy (1986). The Na^+-H^+ exchanger of human placental brush-border membrane. Identification and characterization. Am. J. Physiol., 251, C852-C860.

Berteloot, A. (1986). Membrane potential dependency of glutamic acid transport in rabbit jejunal brush-border membrane vesicles: K^+ and H^+ effects. Biochim. Biophys. Acta, 861, 447-456.

Crane, R. K. (1980). Brush-border membrane disease. Biochem. Soc. Trans., 8, 688-690.

Ganapathy, V., and F. H. Leibach (1985). Is intestinal peptide transport energized by a proton gradient? Am. J. Physiol., 249, G153-G160.

Ganapathy, V., and F. H. Leibach (1986). Carrier-mediated reabsorption of small peptides in renal proximal tubule. Am. J. Physiol., 251, F945-F953.

Ganapathy, V., J. F. Mendicino, and F. H. Leibach (1981). Evidence for a dipeptide transport system in renal brush border membranes from rabbit. Biochim. Biophys. Acta, 642, 381-391.

Ganapathy, V., D. H. Pashley, M. C. Fonteles, and F. H. Leibach (1984). Peptidases and peptide transport in mammalian kidney. Contr. Nephrol., 42, 10-18.

Hoshi, T. (1985). Proton-coupled transport of organic solutes in animal cell membranes and its relation to Na^+ transport. Jpn. J. Physiol., 35, 179-191.

Hoshi, T., N. Takuwa, M. Abe, and A. Tajima (1986). Hydrogen ion-coupled transport of D-glucose by phlorizin-sensitive sugar carrier in intestinal brush-border membranes. Biochim. Biophys. Acta, 861, 483-488.

Inui, K-I., T. Okano, M. Takano, H. Saito, and R. Hori (1984). Carrier-mediated transport of cephalexin via the dipeptide transport system in rat renal brush-border membrane vesicles. Biochim. Biophys. Acta, 769, 449-454.

Kenny, A. J., and A. G. Booth (1978). Microvilli: their ultrastructure, enzymology, and molecular organization. Essays Biochem., 14, 1-44.

Lee, S. H., and J. B. Pritchard (1983). Proton-coupled L-lysine uptake by renal brush border membrane vesicles from Mullet (Mugilcephalus). J. Membr. Biol., 75, 171-178.

Matthews, D. M. (1975). Intestinal absorption of peptides. Physiol. Rev., 55, 537-608.

Mitsumoto, Y., K. Sato, T. Ohyashiki, and T. Mohri (1986). Leucine-proton cotransport system in Chang liver cell. J. Biol. Chem., 261, 4549-4554.

Miyamoto, Y., V. Ganapathy, A. Barlas, K. Neubert, A. Barth, and F. H. Leibach (1987). Role of dipeptidylpeptidase IV in the uptake of peptide nitrogen from β-casomorphin in rabbit renal brush-border membrane vesicles. Am. J. Physiol., in press.

Murer, H., U. Hopfer, and R. Kinne (1976). Sodium/proton antiport in brush-border membrane vesicles isolated from rat small intestine and kidney. Biochem. J., 154, 597-604.

Nelson, P. J., G. E. Dean, P. S. Aronson, and G. Rudnick (1983). Hydrogen ion cotransport by the renal brush border glutamate transporter. Biochemistry, 22, 5459-5463.

Seifter, J. L., and P. S. Aronson (1986). Properties and physiologic roles of the plasma membrane sodium-hydrogen exchanger. J. Clin. Invest., 78, 859-864.

Silbernagl, S., V. Ganapathy, F. H. Leibach, and K. Volker (1986). Renal transport and metabolism of dipeptides (Abstract). Kidney Int., 29, 1252.

Skulachev, V. P. (1985). Membrane-linked energy transductions. Bioenergetic functions of sodium: H^+ is not unique as a coupling ion. Eur. J. Biochem., 151, 199-208.

Ullrich, K. J., G. Rumrich, and S. Kloss (1982). Transport of inorganic and organic substances in the renal proximal tubule. Klin. Wochenschr., 60, 1165-1172.

Passage of Intact Peptides Across the Intestine

Michael L. G. Gardner

School of Biomedical Sciences, University of Bradford,
Bradford, West Yorkshire, BD7 1DP, UK

ABSTRACT

This paper reviews evidence, obtained by several independent approaches, which shows convincingly that small amounts of some intact peptides can cross animal and human small intestine. Although apparently not of nutritional significance, this process has potential physiological and pathological implications since biologically or pharmacologically active peptides can arise during enzymic digestion of dietary protein, and since some active peptides can exert systemic activity even after oral administration. Also, intestinal permeability is known to be increased in numerous diseases. It is proposed that the extent of entry of intact peptides into the circulation depends upon a balance between three factors: (a) protein or peptide indigestibility (or protection from enzymic hydrolysis), (b) the intestinal luminal and mucosal proteolytic and peptidolytic activity, and (c) intestinal permeability (or 'leakiness') especially that via the paracellular route. The first of these factors probably dominates. It is argued that critical quantitative assessment of the significance of assimilation of intact peptides should now be undertaken.

KEYWORDS

Peptides; small intestine; intestinal absorption; biologically active peptides; peptide transport; digestion; intestinal permeability.

INTRODUCTION

Two aspects of peptide biology in which substantial advances have been made in the last decade are: (i) the characterisation of specific transport mechanisms for small peptides in the brush-border of the small intestine responsible for peptide transport (followed by hydrolysis within intestinal mucosal cytoplasm) which comprise a major route of protein absorption, and (ii) the conclusion that numerous small peptides are biologically active, many possessing neurotransmitter or 'neuromodulator' activity. However in spite of this, the possibility that biologically (or pharmacologically) significant quantities of small peptides arising from partial digestion of dietary protein might escape further digestion and might enter the circulation in such a form as to be active in peripheral

© 1987 Pergamon Journals Ltd.
Printed in Great Britain

tissues, including central nervous system, has received remarkably little attention. Although the conventional 'dogma' of protein absorption is that only free amino acids enter the circulation, there is much evidence scattered throughout the literature suggesting that absorption of intact peptides does occur, but probably on a small scale, i.e. one too small to be of nutritional relevance although even this is not unequivocal: selected evidence for absorption of intact peptides is reviewed here, fuller discussion and references having recently been given elsewhere (Gardner, 1984 - see also Gardner, 1985, 1986). Further, it is now recognised that the healthy small intestine in man and animals is 'leaky', i.e. the mucosal epithelium is an imperfect barrier between lumen and bloodstream; additionally, recent studies show that this 'leakiness' or permeability can be increased in a substantial number of gastrointestinal diseases and in various other common disorders ranging from alcoholism to schizophrenia. In this context the term 'absorption' is used to refer to transmucosal passage (whether by mediated, active or passive mechanisms; also whether by transcellular or paracellular routes - see below) into sub-mucosal tissue fluid, blood, and/or lymph. In contrast, the term 'absorption' is used in most of the literature on intestinal peptide transport to refer specifically to transport across the brush-border membrane into the mucosal cytosol. For further details of the latter process, the key reviews by Matthews (1975a, 1975b, 1977, 1987), Matthews & Payne (1980) and Grimble & Silk (1986) may be consulted.

Technical Difficulties in the Study of Absorption of Intact Peptides

Outsiders to this particular field of investigation will be surprised by the inability of current knowledge to define the significance and extent of absorption of intact peptides: this issue has received remarkably little serious investigation and, apart from being connected with a general failure to recognise the potential biological and medical significance of such a phenomenon, this has probably arisen largely due to serious technical and analytical difficulties including, for example, the following:

1. Many active peptides are so potent that entry to the circulation of very small amounts, possibly below the limits of detection by conventional methods (especially chemical ones), may be biologically or pharmacologically significant;

2. Although there are methods, albeit tedious ones (e.g. the historical Van Slyke method), for quantitation of free amino acids as a class, and automated ion-exchange amino acid analysis is available for estimation of all individual amino acids in mixtures, there is no comparable method available for quantitation of peptides as a general class in a mixture; the resolution or 'mapping' and quantitation of all individual peptides can be seen to be an unrealistic task when one realises that there are 400 possible dipeptides, 8,000 tripeptides, 160,000 tetrapeptides .... etc.

3. Acid hydrolysis of physiological fluids such as plasma (unlike the hydrolysis of relatively pure proteins) leads to large lossess of amino acids, so that 'the difference method' of amino acid analysis before and after total hydrolysis (see Gardner, 1981) of deproteinised plasma is an unreliable and certainly insensitive method for measurement of trace amounts of peptides in plasma even though this approach works well with less complex materials;

4. Enzymic digestion of proteins by the array of proteases and peptidases present in the gastrointestinal tract (including intracellular enzymes) normally produces very heterogeneous mixtures of peptides together with free amino acids, with individual peptide components each being present in very small amounts. Attempts to predict the structures of individual peptides present in the intestinal lumen during protein digestion *in vivo* are likely to be fruitless;

5. Sensitive and specific methods for detection of peptides, e.g.

radioimmunoassay or pharmacological bio-assay, are applicable only if one already knows what the nature of the absorbed peptides may be; so these methods are not generally applicable, and they may lead to an underestimate of the magnitude of the phenomenon;

6. Too little is known about the binding of small peptides to plasma proteins;

7. Studies on absorption _in vivo_ in animals or in man ideally require direct access to portal blood and to intestinal lymph;

8. Studies on absorption in animal intestine _in vitro_ require the provision of evidence that the tissue is being maintained in a physiologically and structurally viable state, and these requirements are often not stringently met (Gardner & Plumb, 1979; and unpublished observations). Intestinal tissue _in vitro_ can readily become more 'leaky', giving an unrealistically large impression of trans-mucosal passage of molecules that physiologically would be non-absorbable; conversely, leakage of cellular peptidases can result in extra-cellular hydrolysis of transported peptides giving a false impression that no intact peptide has crossed the tissue.

EVIDENCE FOR ABSORPTION OF INTACT PEPTIDES

The technical difficulties mentioned above dictate that no single experimental approach can provide unequivocal evidence. Fortunately, five independent lines of enquiry all point to the same conclusion, that significant amounts of some peptides, including some biologically active ones, can cross the small intestine in intact form although the quantitative and medical significance of the process remains unclear.

1. _Oral Protein Tolerance Tests in Man_. The early experiments of Folin & Berglund (1922) showed that ingestion of gelatin was followed by a substantial increase in the non-protein non-amino acid non-urea nitrogen content of plasma; Matthews (1975a) interpreted this as showing the absorption of a substantial quantity (_sic_) of intact peptides especially relatively indigestible ones containing hydroxyproline in which gelatin is particularly rich. More recent experiments confirm a large increase in conjugated hydroxyproline (presumed to be peptide-bound) in an ultrafiltrate of serum and in urine after a gelatin meal (Bronstein and others, 1966), and the major peptide appearing in urine appeared to be Pro-Hyp (Meilman and others, 1963; Huekel & Rogers, 1970). However, in contrast my own experiments on ingestion of culinary-grade gelatin do _not_ show a major increase in bound hydroxyproline in peripheral plasma; it seems likely that the extent of 'refinement' of the gelatin may be a critical factor influencing its digestibility and hence the amount of peptide absorbed without complete hydrolysis.

Other protein-loading experiments with more readily digestible proteins (e.g. Stein & Moore, 1954; Frame, 1958) have yielded equivocal results but, although they have generally been interpreted as showing no absorption of intact peptide, close inspection of the data casts doubt upon this interpretation. It is only when one attempts to repeat these experiments that one fully realises the analytical difficulties and the frailties of interpretation mentioned above.

2. _Absorption of Biologically Active Peptides in vivo_. These experiments, although difficult to interpret quantitatively, provide among the most cogent evidence of the potential significance of absorption of intact peptides.

Various peptide hormones are active when administered orally. Good examples are provided by LHRH (a decapeptide) and TRH (a tripeptide), both of which are active on oral administration, albeit at a much higher dosage than would be required to elicit a pharmacological response after intravenous administration (e.g. Ormiston, 1972). Like many endogeneous peptides, these have blocked N-termini (pyroglutamate) and C-termini (amidated), and these features are undoubtedly important in permitting intestinal absorption without complete intestinal hydrolysis and/or permitting expression of systemic biological activity before clearance from the circulation. Thus, resistance to enzymic hydrolysis, rather than specificity for some transport mechanism, appears to be a prime factor in determining whether a peptide may enter the body in intact form. Saffran and co-workers (1979) reported that vasopressin (an amidated nonapeptide), given orally, retains significant antidiuretic activity in rats, and its efficacy is enhanced by co-administration of a protease inhibitor.

Insulin (mol. wt. 6,000) is normally not significantly absorbed across the intestinal tract; hence the need for other routes of administration to diabetics. However, in animals, hypoglycaemic activity can be elicited by oral insulin if especially large doses are given, and this effect can be enhanced by presentation of the insulin together with soy-bean trypsin inhibitor or, especially in diabetic animals, if the insulin is entrapped within liposomes. These experiments emphasise the dependence of intact peptide transport upon a fine balance between transport and digestion. Absorption of insulin in active form is enhanced also in the presence of hypertonic luminal solutions, this emphasising the relationship between increased intestinal permeability (probably paracellular) and enhanced absorption of intact peptides. For further references, please see Gardner (1984).

3. Other Animal Experiments in vivo. Dent & Schilling's (1949) classical experiments on paper chromatographic analysis of dogs' portal blood during protein absorption raised several still unanswered controversies: e.g. why did they fail to find any increase in the free amino acid content of portal plasma during assimilation of homologous plasma proteins and were homologous proteins really being absorbed in intact form? Their analyses (see Christensen's appendix to their paper) did show a definite increase in conjugated amino-N in the portal plasma during protein absorption, but it is not known how much of this might have been peptide. Their cautious conclusion that 'at most a minor part of the protein was absorbed in peptide form' epitomises the uncertainty in their results and their failure to realise that absorption of even minute amounts of peptide might be biologically significant.

Bouillin and co-workers (1973) showed that various dipeptides instilled at high doses in tied loops of rat intestine appeared in mesenteric venous plasma; numerous other studies have shown the absorption of structurally atypical peptides such as carnosine (β-ala-his), β-aspartyl peptides etc. Experiments on absorption of partial (enzymic) digests of proteins in vivo have also shown passage of bound amino-N into mesenteric blood, but the analytical difficulties make it difficult to assess the quantitative contribution of this - up to 10% of the amino-N could have been absorbed as peptides (Gardner and colleagues, 1983).

4. Experiments on Absorption in vitro. Corresponding experiments on perfused segments of animal intestine and on everted sacs of intestine have also shown trans-mucosal passage of some, but not all, peptides in intact form, although undoubtedly most peptides are extensively hydrolysed during absorption. These experiments could be open to question since the integrity of the tissue may have been broached; however, many of these experiments yield results closely corroborating those obtained in vivo. Gardner (1982) noted that the extent of absorption of intact peptide from various protein digests depended upon the source of the protein; up to 30% of the transported amino-N could have been peptide-bound in some experiments. See also Gardner (1978).

5. Other Relevant Observations. These include:

(i) Absorption of intact protein - it is now clear that immunologically significant quantities of protein cross the intestine, so that healthy individuals have circulating antibodies to food proteins. It is thought that the 'M-cells' overlying Peyer's patches may provide a specialised route for access of dietary proteins to lymphocytes. The consequences of absorption of toxins such as Botulinum toxin (a fragment of about 10^6 daltons enters the circulation) are well known!

(ii) Studies on intestinal permeability - recent clinical studies on intestinal 'leakiness' have led to the introduction of non-invasive tests using probe molecules, generally of mol. wt. below 400 but sometimes up to 4,000; these show that traces of many substances, previously regarded as 'non-absorbable', are excreted in urine after oral administration and that intestinal permeability to them is increased in many common diseases. These probes include sugar alcohols (e.g. mannitol), oligosaccharides (e.g. cellobiose, lactulose), polyethylene glycols and Cr-EDTA complex. See Menzies (1984) and Peters & Bjarnason (1985).

ROUTES FOR PASSAGE OF PEPTIDES ACROSS INTESTINE

These and mechanistic aspects have been discussed fully elsewhere (Gardner, 1984). It seems clear that both transcellular and paracellular routes can be utilised. In the former case, both carrier-mediated mechanisms and, especially in 'M-cells', pinocytosis appear to be involved although it is impossible at present to define the specificities of the mechanisms. It is likely that they are fairly broad: for example, amino-β-lactam antibiotics interact competitively with the gly-gly carrier in rat jejunum (Nakashima and others, 1984). Although receptor-binding is generally regarded as a prerequisite for pinocytosis, this mechanism is also implicated in the absorption of relatively inert particles, e.g. carbon and iron particles, in experimental situations. While the paracellular route (including tight junctions and extrusion zones) is probably important mainly for hydrophilic molecules with a radius of over about 0.5 nm, this route (the major route of fluid transport) is likely also to be influenced by solvent drag; also the permeability of tight junctions can be increased reversibly in the presence of hypertonic luminal contents, and the effect of hypertonic solutions is accentuated in the presence of gastrointestinal disease.

PRODUCTION OF ACTIVE PEPTIDES DURING DIGESTION

A particularly dramatic series of findings was initiated by Zioudrou, Streaty & Klee (1979) and Brantl & Teschemacher (1979): they showed that enzymic digestion of gluten and of casein, respectively, can yield peptides with opioid activity. These were named 'exorphins' (i.e. exogeneous counterparts of the endorphins) and 'casomorphins'. The active peptides from α-casein and from β-casein have been sequenced (see Loukas and others, 1983; also references cited by Gardner, 1985), but it is not yet clear whether these arise during physiological digestion of casein or whether they can be (or are) absorbed in intact form. Such peptides, if absorbed and not cleared rapidly from the circulation, could elicit important effects in central nervous system; also, Morley and his colleagues (1983) have shown that peptic hydrolysates of gluten can produce a naloxone-reversible increase in plasma somatostatin-like activity in man and a reversible increase in intestinal transit-time. Hence, such peptides could have a physiological regulatory role, and so Morley has invented the term 'formones' for such 'food-derived hormones'. Alternatively or additionally, they could interfere pathologically with the peptide-modulated integration of neural and hormonal regulation.

The original account by Zioudrou, Streaty & Klee (1979) implied that other, not strictly opioid, active peptides in addition to 'exorphins' were present in the enzymic digests of gluten: furthermore, theoretical considerations arising from a search through a computerised database of known primary structures of proteins suggest that the occurrence of pharmacologically active peptide sequences in proteins should be regarded as 'likely' rather than 'rare' (Gardner, 1985).

Wheat gluten is the protein to which patients with coeliac disease are sensitive. While there undoubtedly is increased intestinal permeability and an increased susceptibility to hypertonic luminal stimuli in untreated patients, it is not certain whether these patients have peptidaemia and, if so, whether this has any pathophysiological consequences. Also of great interest are the suggested relevance of gluten toxicity to a sub-set of schizophrenic patients (Vlissides, Venulet & Jenner, 1986) and the observation that intestinal permeability is increased in some such patients (Prof. R.H.S. Mindham, Department of Psychiatry, University of Leeds, Personal Communication, 1986). However it is premature to propose that biologically active peptides may have been absorbed in these patients: this provocative suggestion nevertheless merits further appraisal. For fuller references see Dohan (1980) and Gardner (1984).

CONCLUSIONS

We can conclude beyond reasonable doubt that:

1. Peptides predominate over free amino acids in the intestinal lumen during protein digestion, both as time progresses and as distance down the alimentary tract progresses;

2. Biologically active peptides can be produced by enzymic digestion, certainly *in vitro* but probably also *in vivo*, of dietary proteins;

3. Some peptides can and do cross the intestine in intact and active form in health, and it is likely that this can be augmented in a variety of pathological circumstances including those known to cause enhanced intestinal 'permeability' and those associated with impaired digestive capacity.

4. Protein or peptide indigestibility (or protection from hydrolysis) tends to promote absorption of intact peptides. Thus, the extent of this absorption appears to depend upon a fine balance between digestibility, permeability, and mucosal digestive capacity.

However, in spite of the substantial *potential* implications, no firm conclusions can yet be drawn about the real consequences of absorption of intact peptides. Assessment of whether or not there are implications of clinical or physiological significance should now be regarded as a matter for urgent investigation. Despite the limited sensitivity of chemical methods, modern chromatographic techniques (preferably in conjunction with pharmacological ones) may allow a more conclusive assessment of the potential significance of peptidaemia and peptiduria than has yet been achieved.

ACKNOWLEDGEMENTS

I am indebted to The Royal Society and to the University of Bradford for Travel Grants enabling me to attend the Meeting in Halle, G.D.R. at which this paper was presented.

REFERENCES

Boullin, D.J., R.F. Crampton, C.E. Heading and D. Pelling (1973). Intestinal absorption of dipeptides containing glycine, phenylalanine, proline, β-alanine or histidine in the rat. Clin.Sci.molec.Med., 45, 849-858.
Brantl, V. and H. Teschemacher (1979). A material with opioid activity in bovine milk and milk products. Naunyn-Schmiedebergs Arch.exp.Path.Pharmak., 306, 301-304.
Bronstein, H.D., L.J. Haeffner and O.D. Kowlessar (1966). The significance of gelatin tolerance in malabsorptive states. Gastroenterology, 50, 621-630.
Dent, C.E. and J.A. Schilling (1949). Studies on the absorption of proteins: the amino acid pattern in the portal blood. Biochem.J., 44, 318-335.
Dohan, F.C. (1980). Hypothesis: genes & neuroactive peptides from food as cause of schizophrenia. Adv.biochem.Psychopharmacol., 22, 535-548.
Folin, O. and H. Berglund (1922). The retention and distribution of amino-acids with especial reference to the urea formation. J.biol.Chem., 51, 395-418.
Frame, E.G. (1958). The levels of individual free amino acids in the plasma of normal man at various intervals after a high-protein meal. J.clin.Invest., 37, 1710-1723.
Gardner, M.L.G. (1978). Amino acid and peptide absorption from partial digests of proteins in isolated rat small intestine. J.Physiol., 284, 83-104.
Gardner, M.L.G. (1981). Amino acid analysis in the study of protein digestion and absorption. In J.M. Rattenbury (Ed.), Amino Acid Analysis. Ellis-Horwood, Chichester. pp. 158-187.
Gardner, M.L.G. (1982). Absorption of intact peptides: studies on transport of protein digests and dipeptides across rat small intestine in vitro. Q.J. Physiol. 67, 629-637.
Gardner, M.L.G. (1984). Intestinal assimilation of intact peptides and proteins from the diet - a neglected field? Biol.Rev., 59, 289-331.
Gardner, M.L.G. (1985). Production of pharmacologically active peptides from foods in the gut. In J.O. Hunter and V. Alun Jones (Ed.), Food and the Gut. Bailliere Tindall, London. pp. 121-134.
Gardner, M.L.G. (1986). Intestinal absorption of peptides. In J.E. Morley, M.B. Sterman and J.H. Walsh (Ed.), Nutritional Modulation of Neural Function. In the Press.
Gardner, M.L.G., B.S. Lindblad, D. Burston and D.M. Matthews (1983). Transmucosal passage of intact peptides in the guinea-pig intestine in vivo: a re-appraisal. Clin.Sci., 64, 433-439.
Gardner, M.L.G. and J.A. Plumb (1979). Release of dipeptide hydrolase activities from rat small intestine perfused in vitro and in vivo. Clin.Sci., 57, 529-534.
Grimble, G.K. and D.B.A. Silk (1986). Physiology and clinical applications of peptide absorption. B.J.clin.Practice. In the Press.
Hueckel, H.J. and Q.R. Rogers (1972). Prolylhydroxyproline absorption in hamsters. Can.J.Biochem., 50, 782-790.
Loukas, S., D. Varoucha, C. Zioudrou, R.A. Streaty and W.A. Klee (1983). Opioid activities and structures of α-casein-derived exorphins. Biochemistry, 22, 4567-4573.
Matthews, D.M. (1975a). Absorption of peptides by mammalian intestine. In D.M. Matthews and J.W. Payne (Ed.), Peptide Transport in Protein Nutrition. North-Holland, Amsterdam. pp. 61-146.
Matthews, D.M. (1975b). Intestinal absorption of peptides. Physiol.Rev., 55, 537-608.
Matthews, D.M. (1977). Memorial lecture: Protein absorption, then and now. Gastroenterology, 73, 1267-1279.
Matthews, D.M. (1987). Development of the concept of peptide transport (This volume).
Matthews, D.M. and J.W. Payne (1980). Transmembrane transport of small peptides. Current Topics in Membranes and Transport, 14, 331-425.
Meilman, E., M.M. Urivetzky and C.M. Rapoport (1963). Urinary hydroxyproline peptides. J.clin.Invest, 42, 40-50.
Menzies, I.S. (1984). Transmucosal passage of inert molecules in health and

disease. In E. Skadhauge and K. Heintze (Ed.), Intestinal Absorption and Secretion. MTP Press, Lancaster. pp. 527-543.

Morley, J.E., A.S. Levine, T. Yamada, R.L. Gebhard, W.F. Prigge, R.B. Shafer, F.C. Goetz and S.E. Silvis (1983). Effect of exorphins on gastrointestinal function, hormonal release and appetite. Gastroenterology, 84, 1517-1523.

Nakashima, E., A. Tsuji, H. Mizuo and T. Yamana (1984). Kinetics and mechanism of in vitro uptake of amino-β-lactam antibiotics by rat small intestine and relation to the intact-peptide transport system. Biochem.Pharmac., 33, 3345-3352.

Ormiston, B.J. (1972). Clinical effects of TRH on TSH after i.v. and oral administration in normal volunteers and patients with thyroid disease. In (R. Hall, I. Werner and H. Holgate (Ed.), Thyrotropin Releasing Hormone (Frontiers of Hormone Research, Vol. 1) Karger, Basle. pp. 45-52.

Peters, T.J. and I. Bjarnason (1985). Intestinal Permeability. In J.O. Hunter and V. Alun Jones (Ed.), Food and the Gut . Bailliere Tindall, London. pp. 30-44.

Saffran, M., R. Franco-Saenz, A. Kong, D. Papahadjopoulos and F. Szoka (1979). A model for the study of the oral administration of peptide hormones. Can.J.Biochem., 57, 548-553.

Stein, W.H. and S. Moore (1954). The free amino acids of human blood plasma. J.biol.Chem., 211, 915-926.

Vlissides, D.N., A. Venulet and F.A. Jenner (1986). A double-blind gluten-free/gluten-load controlled trial in a secure ward population. B.J.Psychiatry, 148, 447-452.

Zioudrou, C., R.A. Streaty and W.A. Klee (1979). Opioid peptides derived from food proteins. The exorphins. J.biol.Chem., 254, 2446-2449.

Transport of Glutathione in Intestine, Kidney, and Lung

Tory M. Hagen and Dean P. Jones

Department of Biochemistry, Emory University of Medicine, Atlanta, Georgia 30322, USA

ABSTRACT

Isolated enterocytes, proximal tubule cells and alveolar type II cells from rat take up GSH by Na^+-dependent mechanisms. Characterization of uptake by basal-lateral membrane vesicles from kidney and small intestine show that a Na^+-dependent, electrogenic system is present that is specific for compounds with a γ-glutamyl moiety. The uptake system allows cells to utilize exogenous GSH for protection against oxidative injury. Experiments with isolated, vascularly perfused small intestinal segments demonstrated that luminal GSH can be transported intact into the vasculature. Thus, GSH from either the diet or from the tissues that efflux GSH is available to epithelial cells for use in detoxication processes.

KEYWORDS

t-butylhydroperoxide; menadione; paraquat; toxicity, in epithelia

GLUTATHIONE TRANSPORT IN EPITHELIAL CELLS

Transport of GSH has not been studied extensively until recently because early studies showed that the enzymes required for GSH synthesis are widespread in mammalian tissues and little GSH is present in blood plasma (Tietz, 1969). Strivastava and Beutler (1969) demonstrated GSSG efflux from red blood cells, and Sies *et al* (1972) showed that it is transported out of liver. Bartoli and Sies (1978) found that GSH is also transported out of the liver at a rate compatable with the turnover rate for GSH in liver. Subsequent studies *in vivo* confirmed these results and showed that GSH efflux occurs both on the sinusoidal (12-15 nmol/min/g liver) and canalicular surfaces (1-3 nmol/min/g liver) (Sies, 1983). In studies of isolated rat liver sinusoidal membrane vesicles, Inoue and co-workers (1984) found that GSH efflux was due to 2 carrier-mediated transport systems with apparent K_m values for GSH of 0.34 and 3.3 mM, respectively. Subsequent research with perfused liver by Ookhtens et al (1985) also demonstrated saturation kinetics, (K_m, 3.2 μmol/g; V_{max}, 20 nmol/min per g liver). This rate corresponds with the low affinity carrier in the vesicle studies and shows that transport is near saturation at normal concentrations of hepatic GSH. Generation

© 1987 Pergamon Journals Ltd.
Printed in Great Britain

of the bromosulfophthalein-GSH conjugate intracellularly inhibited efflux and also indicated a carrier mechanism. Studies by Aw *et al* (1986) confirmed these findings in isolated hepatocytes. Recent evidence (Sies and Graf, 1985) shows that vasopressin, phenylephrine, and adrenaline cause a rapid increase in thiol (60-70% GSH) release by perfused rat liver, specifically into the plasma.

Efflux of glutathione in the bile was initially detected only as GSSG. However, Eberle *et al* (1981) demonstrated that GSH rapidly autooxidizes in the bile. Biliary efflux does not appear to be saturable under *in vivo* conditions, (Kaplowitz *et al*, 1983) indicating that GSH transport into the bile could be a diffusional process. However, Inoue *et al* (1983) using canalicular membrane vesicles obtained evidence that a carrier-mediated transport of GSH occurs with a K_m of transport of 0.33 mM. GSH transport was inhibited by S-benzylglutathione and GSSG. Whether this translocase functions physiologically or whether a diffusional component as well as carrier-mediated GSH efflux exists into the bile remains to be assessed.

Whereas the efflux of GSH into the plasma and bile occurs physiologically, neither biliary nor sinusoidal GSH transport into the liver was demonstrated (Lauterberg, *et al* 1984). High hepatic versus plasma concentrations of GSH as well as an unfavorable cellular membrane potential combine to make the carrier-derived transport of GSH essentially an irreversible process.

Since hepatic GSH efflux accounts for nearly all the GSH present in the plasma (Lauterberg *et al*, 1984), investigations into the functions of released GSH have been ongoing. Initial studies indicated that plasma GSH was degraded to the amino acid components by γ-glutamyltransferase (γ-GT) in kidney and other organs. These amino acids (especially Cys) are available for synthesis of GSH or proteins. Hence, plasma GSH was thought to be a storage form, mainly for Cys (Tateishi and Sakamoto 1983). However, plasma Cys and cystine are substantially higher than plasma GSH and GSSG (Lash and Jones, 1985b), and recent studies show that transport of intact GSH also occurs in a number of epithelial tissues. Moreover, this transported GSH can protect epithelial cells from various forms of chemical injury. Thus, the liver may release GSH as a mechanism to supply GSH to epithelial cells which are susceptible to injury from oxidative agents and electrophiles (Fig. 1).

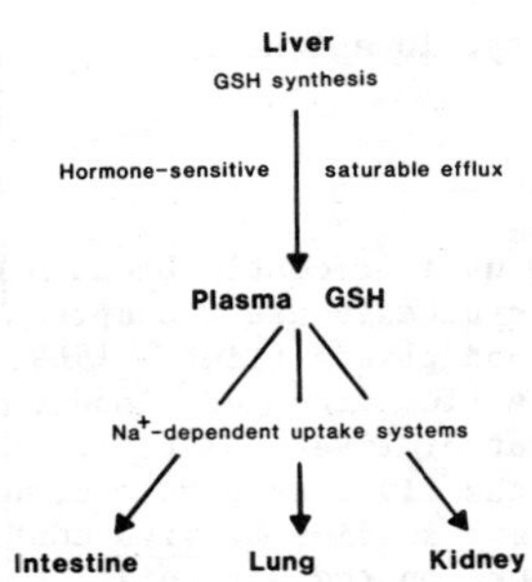

Fig. 1. Origin and function of plasma GSH.

The kidney primarily accounts for the clearance of plasma GSH, but the glomerular filtration of plasma does not account for all GSH extracted (Rankin and Curthoys, 1982). While a portion of the filtered GSH is degraded by γ-GT and dipeptidases in the lumen of renal peritubules, fate of the remaining GSH is not accounted for solely by degradation. Rankin and Curthoys (1982), using a single pass, *in situ* perfusion technique, suggested that a transport system for GSH exists on the basal-lateral membrane of renal tubular cells. This was established by vesicle studies of Lash and Jones (1984). Transport of GSH into isolated basal-lateral membrane vesicles was found to be Na^+-dependent and electrogenic. Maximal uptake was 19.5 nmol/min per mg protein with a stoichiometry of 2 Na^+/GSH. At physiological concentrations of plasma and cellular Na^+, the membrane potential and Na^+ gradient are sufficient to drive GSH uptake even against the substantial GSH concentration gradient that occurs. Transport of GSH was inhibited by γ-Glu-Glu and ophthalmic acid, but little effect on GSH transport was observed with Gly, Cys, Glu, or Cys-Gly. Thus, specificity for GSH transport is in the γ-glutamyl moiety. GSSG is also transported, but at only 20% of that of the reduced species, i.e., 2.1 nmol/min/mg protein. The glutathione conjugate S-(1,2-dichlorovinyl)GSH is also transported by this system with the same properties as observed for GSH (Lash and Jones, 1985a).

More recent studies in our laboratory have established that this transport mechanism also exists in isolated renal epithelial cells. Uptake of GSH was demonstrated by incubation of [Gly-2-^{3}H]GSH with cells. Intact GSH transport was confirmed by HPLC and, again, transport was Na^+-dependent and saturable. The γ-glutamyl analog, γ-Glu-Glu, inhibited GSH transport by 67% (Fig. 2), and the organic anion inhibitor, probenecid, provided a similar extent of inhibition.

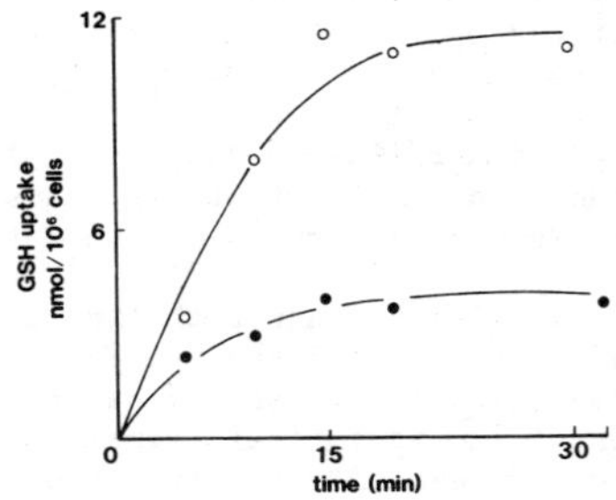

Fig. 2. GSH uptake (o) by kidney cells and its inhibition by 5 mM γ-glutamylglutamate (•).

Recently, Scott and Curthoys (1986) have described the secretion of GSH into the tubular lumen in rats. By inhibiting γ-GT with acivicin, arterial and urinary GSH increased from 5 μM to 23 μM. In other experiments, they infused acivicin into rats and later added purified γ-GT to the blood to degrade plasma GSH. Results showed that plasma GSH levels decreased to 0.3 μM while urinary GSH remained at 1.2-1.8 μM/hr, or 10-20 times the filtered load. These studies show that GSH is secreted into the lumen and this secretion may account for a significant amount of GSH catabolism by the kidney. Other epithelial cells from different organs also take up intact GSH, but the extent of transcellular transport and degradation in other cells is not known.

Dawson et al (1984) used isolated lung cells and whole perfused lung to show that enzymatic conjugation of GSH to chlorodinitrobenzene was enhanced only when GSH was added. These results indicated that transport of GSH may occur. Further investigation showed that cells in the presence of 100 μM GSH contained 180% of the control amount after 5 min. Subsequent work by Berggren et al (1984) showed that inhibition of γ-GT only partially blocked the GSH-dependent replenishment of GSH depleted cells. Thus, these studies indicated that uptake of intact GSH may occur in lung.

In recent studies with freshly isolated alveolar type II cells from rat, we found that intact GSH is taken up by cells by a mechanism similar or identical to that in the kidney basal-lateral membrane (Hagen et al, 1986). Cells were pre-treated with acivicin and buthionine sulfoximine (BSO) to inhibit γ-GT and GSH synthesis, respectively. [Gly-2-^{3}H]GSH (1 mM) was added to cell suspensions, and aliquots were taken over a 2 hr time course and filtered through Percoll to remove any unincorporated radiolabel. Radiolabel increased in cell lysates over time and reached a steady state after 30 min. The initial rate of radiolabel uptake was 1.2 nmol/min/10^6 cells with the radiolabel species confirmed as intact GSH by HPLC analysis. Transport was found to be dependent on Na^+ as in the kidney. Saito et al have also found that intact GSH is taken up by neonatal rat alveolar type II cells (1986). Also, HPLC analysis of the radiolabel species transported proved that intact GSH was accumulated by these cells. These authors also found that 97% of GSH synthesized de novo from added Cys is excreted into the culture medium. Thus, GSH transport may occur in both directions in this cell type.

Transport of intact GSH has also been found in intestine. Linder et al (1984) demonstrated transport of GSH into osmotically sensitive brush border membrane vesicles of porcine jejunum. Hydrolysis was shown to be minimal during the time course of their experiments and further analysis revealed saturation kinetics with net transport against a concentration gradient. The presence of Na^+ stimulated GSH transport into the vesicles.

We recently found that GSH uptake also occurs in the basal-lateral membrane of rat small intestine (Lash et al, 1986) with characteristics identical to that of the rat kidney basal-lateral membrane. Uptake is Na^+-dependent, electrogenic, and inhibited by γ-glutamyl analogs and probenecid. Furthermore, GSH was shown to be transported into isolated enterocytes. Transport occurred against a concentration gradient and was Na^+-dependent. Inhibition of cellular GSH synthesis and degradation of GSH by BSO and γ-GT did not markedly affect GSH transport. These studies establish that GSH transport systems exist on both the brush border and basal-lateral membranes of the enterocyte.

To determine if GSH transepithelial transport can occur in intestine as earlier suggested (Hunjan and Evered, 1985; see also Yoshimura et al, 1982) we used a single pass, in situ, vascular perfusion technique to monitor GSH appearance from the lumen of the intestinal segment into the mesenteric vasculature by analyzing perfusate samples collected from the portal vein (Hagen and Jones, 1986). Using a 1 mM ^{3}H-GSH solution introduced into the luminal segment, GSH appearance in the perfusate increased to 24.1 nmol/ml after 10 min (Fig. 3). Transport of intact

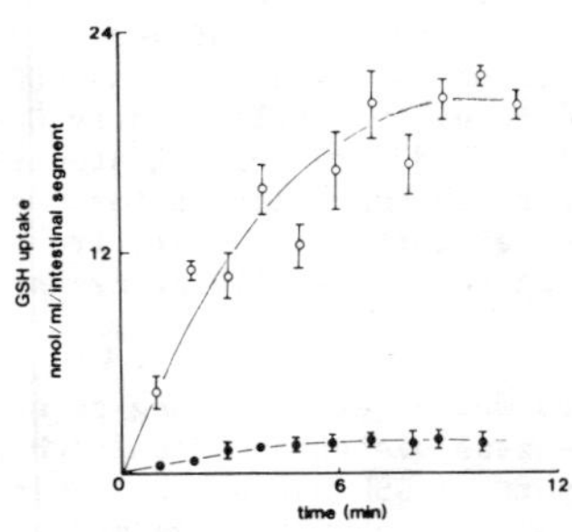

Fig. 3. Transepithelial transport of intact GSH from the lumen into the vasculature (o) and inhibition (•) in Na^+-free medium.

GSH was confirmed by HPLC analysis. Appearance of GSH in the perfusate was minimally affected by inhibitors of γ-GT. Replacement of NaCl by choline·Cl in the lumen resulted in rates less than 50% of that for the Na^+-containing rate, indicating that the process is Na^+-dependent. Other studies revealed that GSH transport is not affected by luminal pH, in contrast to other peptides transported by the intestine (Vadivel and Leibach, 1985). Transport rates at different concentrations of GSH indicated that the overall process is carrier-mediated, with saturation occurring around 3 mM. Transport is inhibited by the γ-glutamyl analogs, γ-Glu-Glu and opthalmic acid, by greater than 60%. This indicates that specificity for transport involves the γ-glutamyl moiety.

To determine whether GSH could be taken up *in vivo*, rats were given 90 μmol GSH in 0.9% (w/v) saline by gavage and plasma GSH levels were monitored. GSH levels in plasma increased from control values of 10-12 nmol/ml to 30 nmol/ml plasma after 90 min. Administration of equivalent molar amounts of cysteinyl residues as Cys or cystine did not increase plasma GSH. Thus, the results show that GSH is transported intact from the lumen into the vascular perfusate and can increase plasma GSH *in vivo*. These results suggest that dietary GSH may be directly utilized by the body, that biliary GSH may be reabsorbed in the intestine, and that orally administered GSH may be useful as a therapeutic measure to increase plasma GSH. Because the kidney, lung, and intestinal epithelial cells can take up exogenous GSH, these results indicate that transport of GSH may serve to provide a reserve supply of GSH for detoxication processes in epithelial cells.

FUNCTIONS OF TRANSPORTED GSH

GSH plays a critical role in protecting cells against xenobiotics and other compounds that can produce chemical injury. The presence of high concentrations of reactive electrophiles or peroxides can overwhelm the capacity of cells to maintain GSH and result in GSH depletion. Thus, cells able to transport intact GSH may be able to augment their GSH pools and protect themselves against chemical damage. It is interesting to note that all cells known to take up GSH are epithelial cells that can encounter high xenobiotic concentrations. Therefore, transported GSH may be vital in protecting these cells against such injury. Our laboratory has tested this hypothesis using isolated kidney, intestinal, and pulmonary alveolar type II cells (Lash and Jones, 1984; Lash *et al*, 1986; Hagen *et al*, 1986).

Exogenous GSH substantially protects alveolar type II cells against the redox cycling agent, paraquat. In control incubations where only 250 μM paraquat was added to these cells, the cell viability declined from over 90% initially to approximately 25% after a 2 hr incubation. However, cells with 1 mM GSH added to the medium along with 250 μM paraquat exhibited only a 10% drop in viability. Similar results were obtained with or without inhibition of γ-GT and γ-Glu-Cys synthetase by acivicin and BSO, respectively. Further studies confirmed that physiological plasma concentrations of GSH (20 μM) also showed this protection. To test whether exogenous GSH must be taken up by these cells to provide protection, we preincubated γ-Glu-Glu (1 mM) and probenecid (0.25 mM) with type II cells prior to addition with GSH and paraquat. Results showed minimal protection of type II cells with substantial cell death in these incubations. Therefore, exogenous GSH protects type II cells, and uptake of GSH is required for protection.

The redox cycling agent, menadione, as well as *t*-butylhydroperoxide was used to examine protection of enterocytes by GSH (Lash *et al*, 1986). Substantial protection was observed against these xenobiotics when both 1 mM and 20 μM GSH was added to cell suspensions. Protection was lost upon preincubation of cells with γ-Glu-Glu or probenecid. Moreover, preincubation of cells with 0.1 mM probenecid caused loss of protection at 20 μM GSH but only partial loss at 1 mM GSH. Since probenecid inhibits only basal-lateral transport, protection appears to be mainly derived from GSH transport from the blood at physiological concentrations. However, this does not preclude protection by luminally derived GSH as well.

Protection against oxidative injury due to *t*-butylhydroperoxide has also been found in experiments where exogenous GSH (1 mM or 20 μM) was supplied to isolated

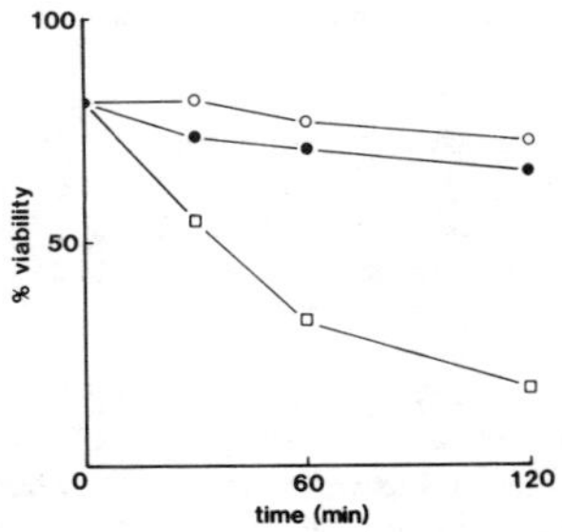

Fig. 4. Injury induced by t-butylhydroperoxide (□) is prevented by addition of 1 mM GSH (•) in isolated kidney cells. Control (o).

kidney cells (Fig. 4). Again, protection was eliminated when γ-glutamyl compounds were preincubated with cells prior to t-butylhydroperoxide and GSH addition. Thus, the results of studies with three types of epithelial cells show that the cells take up exogenous GSH and can utilize this extracellular GSH to protect against chemically induced injury.

CONCLUDING COMMENTS

Substantial evidence now exists that transport of GSH is important in determining GSH distribution and availability for detoxication processes. These functions can be outlined as follows:

The liver has a relatively high synthetic capacity both because the synthesis rate is high and because the liver is a relatively large organ. This high synthetic capacity serves to maintain hepatic GSH supply to protect against injury from the numerous oxidative processes that occur in liver and to supply GSH to the plasma as a supplemental source of GSH for detoxication processes in epithelial cells.

GSH can be absorbed intact from the small intestine. Thus, dietary GSH, which may occur in the range of 100-150 mg/day, is potentially usable in detoxication processes. Furthermore, if biliary GSH is not degraded or consumed in luminal reactions, it may be reabsorbed into the vasculature.

A final possibility, based upon the uptake of GSH by the basal-lateral membrane and GSH release on the brush border membrane in kidney, is that epithelial cells may secrete GSH to protect against extracellular reactive species. For instance, the renal tubule may secrete GSH into the tubular lumen to detoxify any reactive species which could be concentrated in urine and cause collecting duct, ureter, or bladder injury. The marked glutathionuria due to inhibition of γ-glutamyltransferase indicates that GSH secretion in kidney is substantial. The lung could secrete GSH into the alveolar exudate to protect against reactive species that are inhaled. High concentrations of GSH in this exudate have been reported (Sutherland et al, 1985). Thus, a potentially important aspect of GSH transport is the supply of GSH to extracellular fluids to detoxify reactive chemicals before they reach the cells.

Finally, although earlier studies have indicated that administration of GSH is not of therapeutic benefit (Cook and Sherlock, 1956), this possibility should not be summarily dismissed. We are only beginning to understand which cell types can utilize GSH and the conditions under which administered GSH is useful. With the recent advances in our understanding of the critical roles of thiols in protection against various types of cell injury, it may be possible to select specific disease states which can benefit from exogenously administered GSH.

ACKNOWLEDGEMENTS

Research upon which this manuscript is based was supported by NIH Grants HL 30286 and GM 36538.

REFERENCES

AW, T.Y., M. OOKHTENS, C. REN, and N. KAPLOWITZ (1986). Kinetics of glutathione efflux from isolated rat hepatocytes. Am. J. Physiol. 250, G236-G243.

BARTOLI, G.M., and H. SIES (1978). Reduced and oxidized glutathione efflux from liver. FEBS Lett. 86, 89-91.

BERGGREN, M., J.R. DAWSON, and P. MOLDEUS (1984). Glutathione biosynthesis in the isolated perfused rat lung: utilization of extracellular glutathione. FEBS Lett. 176, 189-192.

COOK, G.C., and S. SHERLOCK (1965). Results of a controlled clinical trial of glutathione in cases of hepatic cirrhosis. Gut 6, 472-476.

DAWSON, J.R., K. VAHAKANGAS, B. JERNSTROM, and P. MOLDEUS (1984). Glutathione conjugation by isolated lung cells and the isolated perfused lung. Effect of extracellular glutathione. Eur. J. Biochem. 138, 439-443.

EBERLE, D., R. CLARKE, and N. KAPLOWITZ (1981). Rapid oxidation in vitro of endogenous and exogenous glutathione in bile of rats. J. Biol. Chem. 256, 2115-2117.

HAGEN, T.M., and D.P. JONES (1986). Transepithelial transport of glutathione in vascularly perfused small intestine of rat. Am. J. Physiol., in press.

HAGEN, T.M., L.A. BROWN, and D.P. JONES (1986). Protection against paraquat-induced injury by exogenous GSH in pulmonary alveolar type II cells. Biochem. Pharmacol. in press.

HUNJAN, M., and D.F. EVERED (1985). Absorption of glutathione from the gastro-intestinal tract. Biochim. Biophys. Acta 815, 184-188.

INOUE, M., R. KINNE, T. TRAN, and I.M. ARIAS (1984). Glutathione transport across hepatocyte plasma membranes. Analysis using isolated rat liver sinusoidal membrane vesicles. Eur. J. Biochem. 138, 491-495.

INOUE, M., R. KINNE, T. TRAN, and I.M. ARIAS (1983). The mechanism of biliary secretion of reduced glutathione. Analysis of transport process in isolated rat liver canulicular membrane vesicles. Eur. J. Biochem. 134, 467-471.

LASH, L.H., and D.P. JONES (1985a). Uptake of the glutathione conjugate S-(1,2-dichlorovinyl)glutathione by renal basal-lateral membrane vesicles and isolated kidney cells. Molec. Pharm. 28, 278-282.

LASH, L.H., and D.P. JONES (1985b). Distribution of oxidized and reduced forms of glutathione and cysteine in rat plasma. Arch. Biochem. Biophys. 240, 583-592.

LASH, L.H., T.M. HAGEN, and D.P. JONES (1986). Exogenous glutathione protects intestinal epithelial cells from oxidative injury. Proc. Natl. Acad. Sci., USA. 83, 4641-4645.

LASH, L.H., and D.P. JONES (1984). Renal glutathione transport. Characteristics of the sodium-dependent system in the basal-lateral membrane. J. Biol. Chem. 259, 14508-14514.

LAUTERBURG, B.H., J.D. ADAMS, and J.R. MITCHELL (1984). Hepatic glutathione homeostasis in the rat: Efflux accounts for glutathione turnover. Hepatol. 4, 586-590.

LINDER, M., G. DE BURLET, and P. SUDAKA (1984). Transport of glutathione by intestinal brush border membrane vesicles. Biochem. Biophys. Res. Comm. 123, 929-936.

OOKHTENS, M., K. HOBDY, M.C. CORVASCE, T.Y. AW, and N. KAPLOWITZ (1985). Sinusodial efflux of glutathione in the perfused rat liver. Evidence for a carrier-mediated process. J. Clin. Invest. 75, 259-265.

RANKIN, B.B., and N.P. CURTHOYS (1982). Evidence for renal paratubular transport of glutathione. FEBS Lett. 147, 193-196.

SAITO, K., J.B. WARSHAW, and R.A. PROUGH (1986). Glutathione homeostasis in isolated type II alveolar cells. Fed. Proc. 45, 908.

SCOTT, R.D. and N.P. CURTHOYS (1986). Renal tubular secretion of glutathione (GSH). Fed. Proc. 45, 1789.

SIES, H. (1983). Reduced and oxidized glutathione efflux from liver. Y. Sakamoto, T. Higashi, and N. Tateishi (Ed.), In Glutathione: Transport and Turnover in Mammals. Japan Sci. Soc. Press. Tokyo/VNU Science Press, Utrecht., 63-88.

SIES, H., CH. GERSTENECKER, H. MENZEL, and L. FLOHE (1972). Oxidation in the NADP system and release of GSSG from hemoglobin-free perfused rat liver during peroxidatic oxidation of glutathione by hydroperoxides. FEBS Lett. 27, 171-175.

SIES, H., and P. GRAF (1985). Hepatic thiol and glutathione efflux under the influence of vassopressin, phenylephrine, and adrenaline. Biochem. J. 226, 545-549.

SRIVASTAVA, S.K., and E. BEUTLER (1969). The transport of oxidized glutathione from the erythrocytes of various species in the presence of chromate. Biochem. J. 114, 833-837.

SUTHERLAND, M.W., M. GLASS, J. NELSON, Y. LYEN, and H.J. FORMAN (1985). Oxygen toxicity: Loss of lung macrophage function without metabolite depletion. J. Free Radicals Biol. Med. 1, 209-214.

TATEISHI, N., and Y. SAKAMOTO (1983). Nutritional significance of glutathione in rat liver. Y. Sakamoto, T. Higashi, and N. Tateishi (Ed.) In Glutathione: Transport and Turnover in Mammals. Japan Sci. Soc. Press Tokyo/VNU Science Press, Utrecht., pp 13-38.

TIETZ, F (1969). Enzymatic method for quantitative determination of nanogram amounts of total and oxidized glutathione: Applications to mammalian blood and other tissues. Analyt. Biochem. 27, 502-522.

VADIVEL, G., and F.H. LEIBACH (1985). Is intestinal peptide transport energized by a proton gradient? Am. J. Physiol. 249, G153-G160.

YOSHIMURA, K., Y. IWAUCHI, S. SUGUJAMA, T. KUSOAMURA, Y. ODAKA, T. SATOH and H. KATAGAWA (1982). Transport of L-cysteine and reduced glutathione through biological membranes. Res. Comm. Chem. Path. Pharm. 37, 171-186.

Stoichiometry and Specificity of the Proton/Peptide Cotransport Across Intestinal Brush Border Membranes

T. Hoshi, M. Abe and A. Tajima

Department of Physiology, Faculty of Medicine,
University of Tokyo, Tokyo 113, Japan

ABSTRACT

In the complete absence of Na^+, various di- and tripeptides evoked saturable increments of the transmural potential (ΔPD_t) of isolated everted small intestine. Na^+-independent peptide-evoked potentials were dependent on both extracellular and intracellular pH. Simultaneous measurements of Gly-Gly-induced increases in short-circuit current and Gly-Gly influxes in the absence of Na^+ suggested that the stoichiometry of the H^+/peptide cotransport was 2 : 1. Comparison of kinetic parameters of the ΔPD_t's induced by various dipeptides and Gly-Gly indicated that a variety of glycyl- and alanyl-dipeptides shared a common carrier, while some other dipeptides, such as Leu-Leu, Phe-Phe and charged dipeptides, appeared to be transported by separate carriers.

KEYWORDS

H^+/peptide cotransport; stoichiometry; specificity; brush border membrane; small intestine; peptide-evoked potential; transmural potential.

INTRODUCTION

Recent studies on transport of intact dipeptides across renal and intestinal brush border membrane vesicles have disclosed that the vesicular membranes possess a H^+/dipeptide cotransport system (Ganapathy and Leibach, 1983; Ganapathy, Burckhardt and Leibach, 1984; Takuwa and others, 1985a, 1985b,). Namely, these studies have shown that transport of neutral dipeptides is entirely independent of Na^+, markedly stimulated by a pH gradient, and electrogenic. Furthermore, a marked overshoot uptake of dipeptides was demonstrated to occur when an inward-directed H^+ gradient was imposed across the vesicular membranes (Takuwa and others, 1985a, 1985b; Miyamoto, Ganapathy and Leibach, 1985). To understand further details of the H^+/peptide cotransport system(s), we need information concerning the stoichiometry and specificity of the cotransport. In this communication, the results of our recent experiments which aimed at investigation of these aspects of the H^+/peptide cotransport in the small intestine are to be summarized.

© 1987 Pergamon Journals Ltd.
Printed in Great Britain

MATERIALS AND METHODS

All experiments were performed in everted preparations of the upper intestine of toads (Bufo vulgaris). The preparations were found to be extremely resistant to Na^+-free conditions and to exhibit quite reproducible peptide-induced changes in the transmural potential (PD_t) for more than several hours in Na^+-free media. After anesthetizing animals by intracisternal injection of 0.3 ml of 25% urethane, the upper intestine, about 3 cm long from the hepatopancreatic duct, was excised, everted and rinsed with a Na^+-free Ringer's solution several times. The everted intestine was fixed over a fenestrated polyethylene tube, and PD_t was led out by means of thin polyethylene bridges filled with 2% agar in 1 M KCl. The bridges were connected to calomel electrodes and PD_t was directly recorded on a high-sensitivity DC recorder. In experiments under Na^+-free conditions, the following incubation medium was employed (in mM): D-mannitol, 204; K_2HPO_4, 1.26; KH_2PO_4, 0.42; $CaSO_4$, 1.5; $MgSO_4$, 1.0. pH of the solutions was adjusted by adding 10 mM Tris/Hepes (pH 7.4) or 10 mM Hepes/Mes (pH 6.5 or 5.5). In experiments in the presence of Na^+, 100 mM mannitol was replaced with 50 mM Na_2SO_4. To measure short-circuit current (I_{sc}), polyethylene bridges filled with 2% agar in 1M Tris-sulfate connected to Zn-$ZnSO_4$ electrodes were used for passing electric current. The tips of the inner and outer bridges were fixed remote from the intestinal wall. Determination of the flux ratio was carried out by measuring glycylglycine(Gly-Gly)-induced increase in I_{sc} and Gly-Gly influx across the everted intestines simultaneously as described in a previous paper (Hoshi and others, 1976). Gly-Gly influx was measured with [1-^{14}C]Gly-Gly (RCC Amersham, custom made). The uptake of Gly-Gly during the initial 2 min was measured, while H^+ flow was determined from the mean value of the increased I_{sc} during the same period.

RESULTS

Na^+-independent and pH-dependent Nature of the Peptide-induced PD_t changes

In contrast to amino acids, all naturally occurring di- and tripeptides tested were found to induce reproducible increments of PD_t (ΔPD_t), when added to the mucosal bathing solution, regardless of the presence of Na^+ in the medium. In the present study, 18 dipeptides, including glycyl- and alanyl-dipeptides and some of other dipeptides, and 4 different tripeptides, including glycyl- and alanyl-tripeptides, were examined. The characteristics of the PD_t changes induced by these peptides in the absence of Na^+ were essentially the same. The PD_t changes were saturable increases in the mucosal negativity which conformed to simple Michaelis-Menten kinetics as first observed by Kohn, Smyth and Wright (1968) in the presence of Na^+. There was no difference in the time course between ΔPD_t's recorded in the presence and absence of Na^+, but the amplitude was usually greater in the absence of Na^+. The pattern of the ΔPD_t recorded in the absence of Na^+ was unaffected by addition of Na^+ or K^+ in the medium to a concentration of 10-30 mM during its time course. This indicates that the electrogenic process does not require these cations as demonstrated in other electrophysiological studies with intracellular microelectrodes (Boyd and Ward, 1982; Shimada and Hoshi, 1986). Also, the time course and amplitude of the ΔPD_t were independent of Ca^{2+}, Mg^{2+} in the medium.

The amplitude of the Na^+-independent dipeptide-induced PD_t changes were strongly influenced by both extracellular and intracellular pH. Lowering external pH from 7.4 to 6.5 and 5.5 caused stepwise increases in amplitude. A typical example of such an effect of external pH on the Gly-Gly-induced ΔPD_t is shown in Fig. 1a. Kinetically, the effect of lowering external pH was to decrease the K_t value for peptide without changing the maximum ΔPD_t (ΔPD_tmax) (Hoshi, 1986a). Similar K-type (affinity-type) kinetic effect was also seen in Gly-Gly influx across intestinal brush border membrane vesicles (Takuwa and others, 1985b).

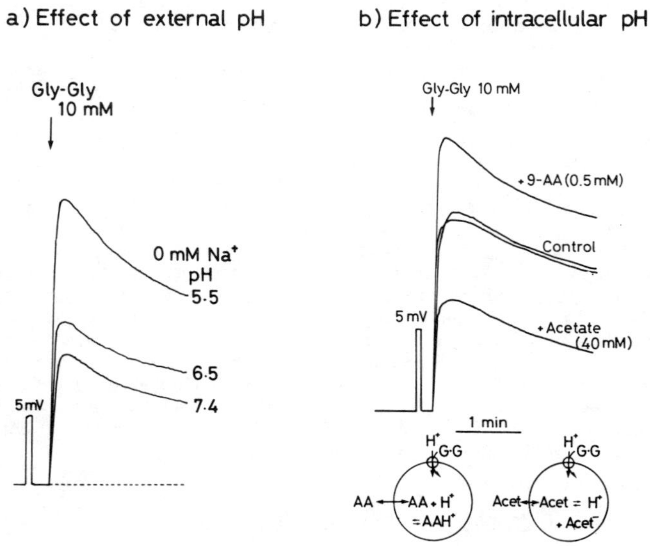

Fig. 1. Effects of extracellular pH(a) and modulation of intracellular pH(b) on the Gly-Gly-evoked potential in the absence of Na^+.

Addition of 9-aminoacridine (9-AA), a weak base which can enter the cells by the mechanism of non-ionic diffusion and act as a H^+ sinker, caused a marked increase in amplitude. Acetate, which is known to enter the cells by non-ionic diffusion and dissociate to release H^+ within the cells, strongly depressed the amplitude of Gly-Gly-induced ΔPD_t (Fig. 1b). These findings together with other findings described above support that the electrogenic property of the peptide transport is directly related to H^+ current generated by the H^+/peptide cotransport.

Stoichiometry of the H^+/peptide cotransport

Simultaneous measurements of Gly-Gly influx and Gly-Gly-induced increase in I_{sc} (ΔI_{sc}) were performed in the same samples at different Gly-Gly concentrations and different pH. From measured ΔI_{sc} the amount of monovalent cation transferred was calculated and given as $nmoles \cdot min^{-1} \cdot g^{-1}$ tissue wet weight. As the measurements were performed in the complete absence of Na^+, and in view of clear dependence of ΔPD_t on external and internal pH, it may be safe to consider that the monovalent cation transferred is H^+. The average ratio of the H^+ flow to Gly-Gly influx thus obtained from 9 different preparations was 1.78 ± 0.12 (S.E.). A linear relationship between H^+ flow and Gly-Gly influx was seen as reported elsewhere preliminarily (Hoshi, 1986a). The regression line indicates that monovalent cation/Gly-Gly flux ratio is approximately 2. Similar measurements by the same techniques for Na^+ and D-galactose fluxes in guinea pig small intestine gave the ratio of approximately 1 : 1 (Hoshi and others, 1976).

To obtain the stoichiometry from the above data, a special analysis of the relationship between the cotransport-associated membrane current across the apical membrane (i_m) and ΔI_{sc} is needed. In the case of the H^+-coupled cotransport, hydrogen ions entered the cells are partly buffered, trapped by intracelular organellae and extruded by undefined mechanisms. The current flows across the basolateral membrane is carried by other ions. However, even such a case, i_m and ΔI_{sc} is closely related. An analysis using an equivalent electric circuit for the intestinal epithelial cell revealed that the relationship between i_m and ΔI_{sc} could be given as a function of the apical and basolateral membrane resistances (R_a and R_b), paracellular shunt resistance (R_l) and transcellular resistance (R_{tc}) (Hoshi, 1986c). In that analysis, cotransport-related apical membrane current was assumed to be induced by a gate-opening mechanism in the cotransporter and driven by the driving force for the cotransported ion across the membrane (Hoshi, 1986b). The result of analysis indicated that $\Delta I_{sc}/i_m$ ratio was dependent on the membrane resistance divider ratio ($R_a/(R_a+R_b)$) and R_l/R_{tc}. Only when $R_a/(R_a+R_b)$ and R_l are sufficiently high, the current ratio becomes unity, otherwise the ratio is less than 1. The value of R_l was relatively high (1,210 $\pm$ 100 Ωcm^2) in the present experimental conditions. Although the value of $R_a/(R_a+R_b)$ has been determined to be 0.63 in Necturus small intestine(Gunter-Smith, Grasset and Schultz, 1982) exact value of this ratio has not yet been determined for the present preparation. However, these data and the result of our analysis suggest that true flow coupling ratio may be slightly higher than 1.78, suggesting the stoichiometry of the H^+/peptide cotransport being 2 : 1. Also, it can be said that as twice as many cations are cotransported with one substrate in the case of the H^+/peptide cotransport as compared to the case of the Na^+/sugar cotransport.

Specificity of the H^+/peptide Cotransport System(s)

All dipeptides and tripeptides tested thus far were found to evoke similar saturable changes in PD_t in the absence of Na^+. In the absence of Na^+, involvement of Na^+/amino acid cotransport-related electrical changes were completely eliminated since amino acid-evoked potentials were absolutely dependent on Na^+. Therefore, the recording of electrical responses to stepwise increases in peptide concentration gave the information concerning kinetic property of H^+-coupled transport of intact peptides purely. Moreover, in this series of experiments, PD_t changes caused by cumulative additions of Gly-Gly were recorded before and after the recording of PD_t changes induced by another (test) peptide, and Lineweaver-Burk plot of the latter was compared with that of means of Gly-Gly-evoked ΔPD_t's recorded immediately before and after the test peptide. This seems of particular importance since the presence of an inhibitor in the medium, as usually seen in uptake inhibition experiments, may cause alteration of the driving force for H^+, i.e. the transmembrane potentials and intracellular pH, which frequently leads to an equivocal conclusion concerning the identity of the carrier.

Glycyl-dipeptides, such as Gly-Pro, Gly-Phe, Gly-Leu and Gly-Sar, had exactly the same ΔPD_tmax as that of Gly-Gly (Hoshi, 1986a), indicating that these dipeptides are transported by the same carrier and at the same stoichiometry in regard to coupling with H^+. Other dipeptides, including Ala-Ala, Ala-Phe and Phe-Ala, and some tripeptides, such as Gly-Gly-Gly and Ala-Ala-Ala, also had the same ΔPD_tmax values as that of Gly-Gly (Fig. 2). These findings are well in agreement with the current concept that a large number of dipeptides and tripeptides share a common carrier (Matthews and Payne, 1980; Matthews, 1984). However, some dipeptides and tripeptides, such as Leu-Gly, Leu-Gly-Gly, carnosine, Phe-Phe, Phe-Leu and Leu-Leu were found to have ΔPD_tmax values definitely different from that of Gly-Gly. Some of these peptides, Leu-Gly and Leu-Gly-Gly, did not cause clear additive increments of PD_t when added after generation of almost saturated PD_t change by Gly-Gly. At the present time, it is not certain whether these peptides are transported by the same carrier as that for Gly-Gly but at a different H^+/peptide stoichiometry or mediated by separate carrier system(s). Among these peptides, carnosine seems of particular interest since flux measurements indicated that this dipeptide and Gly-Gly were purely competitive (Himukai, 1985) but PD_t measurements show that ΔPD_tmax for

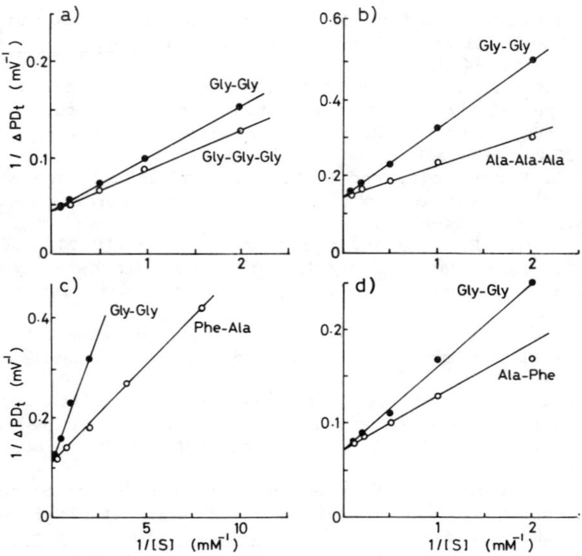

Fig. 2. Comparison of Lineweaver-Burk type plot of ΔPD_t's induced by 4 different peptides with that induced by Gly-Gly.

carnosine is about half that for Gly-Gly. This seems to indicate that, in the case of H^+/carnosine cotransport, the stoichiometry may be 1 : 1 exceptionally. It is known that Leu-Leu uptake is not inhibited by Gly-Gly or Gly-Sar (Matthews, 1984), suggesting that the hydrophobic dipeptides may be transported by a separate carrier. But further analysis seem to be needed until we can solve this question more clearly. Glu-Glu and Gly-Lys exhibited quite different kinetics as compared to Gly-Gly. Also, the same question remains for these charged dipeptides.

In summary, naturally occurring dipeptides and tripeptides were found to generate quite large and reproducible increments of PD_t in the small intestine in the absence of Na^+. The present study proved that the recording and quantitative analyses of the peptide-evoked potentials were quite useful for studies of kinetic properties, stoichiometry and specificity of the H^+-coupled transport of peptides. The PD_t changes, which obeyed Michaelis-Menten kinetics, were independent of Na^+, K^+, Ca^{2+} and Mg^{2+}, but greatly influenced by both external and intracellular pH, indicating that the evoked PD_t change is related to H^+ current associated with the peptide transport. Simultaneous measurements of Gly-Gly-induced increase in I_{sc} and Gly-Gly influx revealed that H^+ current/Gly-Gly flux ratio was 1.78 ± 0.12. Analysis using equivalent electric circuit suggests that the stoichiometry of H^+/peptide transport is 2 : 1. Glycyl-dipeptides and some of other dipeptides and tripeptides were ascertained to be transported by the same carrier at the same stoichiometry. However, some of other dipeptides including Leu-Leu, Phe-Phe, Glu-Glu, and Gly-Lys, had a ΔPD_tmax value different from that for Gly-Gly, suggesting that these peptides are transported by a separated carrier or carriers.

REFERENCES

Boyd, C. A. R., and M. R. Ward (1982). A micro-electrode study of oligopeptide absorption by the small intestinal epithelium of Necturus maculosus. J. Physiol., **324**, 411-428.

Ganapathy, V., and F. H. Leibach (1983). Role of pH gradient and membrane potential in dipeptide transport in intestinal and renal brush-border membrane vesicles from the rabbit. Studies with L-carnosine and glycyl-L-proline. J. biol. Chem., **258**, 14189-14192.

Ganapathy, V., G. Burckhardt, and F. H. Leibach (1984). Characteristics of glysylsarcosine transport in rabbit intestinal brush-border membrane vesicles. J. biol. Chem., **259**, 8954-8959.

Gunter-Smith, P. J., E. Grasset, and S. G. Schultz (1982). Sodium-coupled amino acid and sugar transport by Necturus small intestine. An equivalent electrical circuit analysis of a rheogenic cotransport system. J. Membrane Biol., **66**, 25-39.

Himukai, M. (1985). The characteristics of carnosine transport and carnosine-induced electrical phenomena by the everted intestine of guinea pig. Jpn. J. Physiol., **35**, 945-952.

Hoshi, T. (1986a). Proton-coupled transport of dipeptides across renal and intestinal brush border membranes. In F.Alvarado and C.H. van Os (Ed.), Ion Gradient-Coupled Transport, INSERM Symposium No.26, Elsevier, Amsterdam/New York/Oxford. pp. 183-191.

Hoshi, T. (1986b). Electrophysiological aspects of Na^+-coupled cotransport and short-circuit current increase across transporting epithelial cell membranes. In F. Bronner, and A. Kleinzeller (Ed.), Current Topics in Membranes and Transport, Vol. 14, Academic Press, New York. pp. 331-425.

Hoshi, T. (1986c). Relationship between apical membrane current induced by cation-coupled cotransport and short-circuit current increase across transporting epithelia. A theoretical consideration. Biomed. Res., **7**, suppl. 2, in press.

Hoshi, T., Y. Suzuki, T. Kusachi, and Y. Igarashi (1976). Interrelationship between sugar-evoked increases in transmural potential difference and sugar influxes across the mucosal border in the small intestine. Tohoku J. exp. Med., **119**, 201-209.

Kohn, P. G., D. H. Smyth, and E. M. Wright (1968). Effects of amino acids, dipeptides and disaccharides on the electric potential across rat small intestine. J. Physiol., **196**, 723-746.

Matthews, D. M. (1984). Absorption of peptides, amino acids and their methylated derivatives. In L. D. Stegink and L. J. Filer Jr. (Ed.), Aspartame, Physiology and Biochemistry, Marcel Dakker, New York/Basel, pp. 28-46.

Matthews, D. M. and J. W. Payne (1980). Transmembrane transport of small peptides. In F. Bronner, and A. Kleinzeller (Ed.) Current Topics in Membranes and Transport, Vol. 14, Academic Press, New York. pp. 331-425.

Miyamoto, Y., V. Ganapathy and F. H. Leibach (1985). Proton gradient-coupled uphill transport of glycylsarcosine in rabbit renal brush-border membrane vesicles. Biochem. biophys. Res. Commun., **132**, 946-953.

Shimada, T. and T. Hoshi (1986). Sodium-independent, hydrogen ion-dependent changes in membrane potential and conductance induced by dipeptides in Triturus enterocytes. Jpn. J. Physiol., **36**, 451-465.

Takuwa, N., T. Shimada, H. Matsumoto, and T. Hoshi (1985a). Proton-coupled transport of glycylglycine in rabbit renal brush border membrane vesicles. Biochim. biophys. Acta, **814**, 186-190.

Takuwa,N., T. Shimada, H. Matsumoto, M. Himukai and T. Hoshi (1985b). Effects of hydrogen ion-gradient on carrier-mediated transport of glycylglycine across brush border membrane vesicles from rabbit small intestine. Jpn. J. Physiol., **35**, 629-642.

Renal Handling of Hormonal Peptides

Darryl R. Peterson

Department of Physiology and Biophysics, University of Health Sciences, The Chicago Medical School, North Chicago, Illinois, USA

ABSTRACT

The mechanisms responsible for renal transport and hydrolysis of peptide hormones were investigated. *In vitro* microperfusion of tritiated angiotensin I through rabbit proximal straight nephron segments revealed that ^{3}H-leucine is liberated in the luminal compartment and reabsorbed across the tubular epithelium into the bathing medium. Incubations of angiotensin directly in the presence of isolated renal cortical microvilli confirmed that enzymes associated with the luminal brush border are responsible for hydrolysis of the peptide. In similar studies, luteinizing hormone releasing hormone was cleaved by brush border hydrolases, and the N-terminal di-, tri-, and tetrapeptide metabolites were subsequently reabsorbed intact. Analytical cell fractionation studies in concert with tubular infusions and perfusions of glucagon indicated that cytosolic and lysosomal peptidases may contribute to renal hydrolysis of the peptide, which is initiated at the brush border of the proximal nephron. A similar approach to characterize renal handling of insulin provided additional evidence that this peptide hormone is reabsorbed in the renal proximal tubule primarily by a mechanism involving endocytosis and lysosomal digestion. Collectively, these studies indicate that circulating peptides are degraded in the kidney during reabsorption in the proximal nephron by processes involving: 1) endocytosis and lysosomal digestion, or 2) initial cleavage by hydrolases associated with the luminal brush border, after which further intracellular hydrolysis by cytosolic and/or lysosomal enzymes may occur.

KEYWORDS

Renal transport; renal catabolism; renal brush border; isolated perfused nephron segments; analytical cell fractionation

INTRODUCTION

The kidneys play an important role in removing peptide hormones from the blood (Maack and co-workers, 1979). This occurs mainly by their renal filtration, followed by hydrolysis accompanying tubular reabsorption (Carone and co-workers, 1979, 1980a, 1980b, 1982, 1984; Maack and co-workers, 1979). Thus, the kidneys serve to inactivate peptide hormones by degrading them, while conserving the constituent amino acids. The site of hydrolysis and reabsorption of metabolites

© 1987 Pergamon Journals Ltd.
Printed in Great Britain

has been shown to be the proximal tubule (Maack and co-workers, 1979). In an effort to define the cellular mechanisms involved, we examined renal tubular transport and hydrolysis of four peptide hormones: 1) angiotensin I, 2) luteinizing hormone releasing hormone, 3) glucagon, and 4) insulin.

MATERIALS AND METHODS

Renal tubular transport and hydrolysis of several peptide hormones were studied by employing a variety of techniques, including: 1) microperfusion of nephron segments *in vitro* (Burg and co-workers, 1966; Peterson and co-workers, 1979), 2) incubation of peptides directly in the presence of proximal tubular homogenates (Hjelle, Oparil, and Peterson, 1984) or isolated renal cortical microvilli (Peterson and co-workers, 1979), and 3) analytical cell fractionation methodology (Hjelle, Morin, and Trouet, 1981; Peterson and co-workers, 1984). Radiolabeled peptides and standard metabolites were either prepared in the laboratory or purchased, as previously described (Hjelle, Oparil, and Peterson, 1984; Peterson and co-workers, 1979, 1986; Stetler-Stevenson, Flouret, and Peterson, 1981). Products of hydrolysis were detected by high voltage paper electrophoresis (Peterson and co-workers, 1977), paper chromatography (Peterson and co-workers, 1982), or high pressure liquid chromatography (Stetler-Stevenson and co-workers, 1981).

RESULTS AND DISCUSSION

Angiotensin I

^{3}H-angiotensin I (^{3}H-AI) was microperfused (23.01 ± 0.24 nl/min, SEM; 3.48 pg/nl) through rabbit proximal straight nephron segments *in vitro*. Linear reabsorption of radiolabeled material into the bathing medium was observed as a function of perfusion time over 30 minutes. High voltage paper electrophoresis of the collected luminal fluids and bathing media revealed that ^{3}H-leucine was liberated into the luminal fluid and transported across the tubular epithelium into the surrounding bathing medium. Incubation of ^{3}H-AI directly in the presence of rabbit renal cortical microvilli derived from the luminal surface of proximal tubular cells resulted in cleavage of the peptide and release of ^{3}H-leucine into the incubation medium. The data indicate that ^{3}H-AI is degraded upon renal filtration by enzymes associated with the luminal brush border of the proximal nephron, followed by rapid reabsorption of metabolites.

Luteinizing Hormone Releasing Hormone

^{3}H-luteinizing hormone releasing hormone (^{3}H-LHRH) and all 10 metabolites containing tritium on the N-terminal pyroglutamic acid were prepared in the laboratory. During microperfusion of ^{3}H-LHRH through rabbit proximal straight nephron segments *in vitro*, ^{3}H-label appeared in the bathing medium linearly with respect to perfusion time. Analysis of the collected luminal fluids and bathing media by high pressure liquid chromatography (HPLC) revealed the presence of the N-terminal di-, tri-, and tetrapeptide metabolites. In addition, ^{3}H-pyroglutamic acid was found in the bath. Incubation of the radiolabeled peptide with renal cortical microvilli resulted in the formation of the same di-, tri-, and tetrapeptide metabolites observed in the luminal fluids of the micropefused nephron segments. The data suggest that filtered LHRH is degraded by enzymes associated with the brush border of the proximal tubule, and that short-chain oligopeptide metabolites may be reabsorbed intact. Furthermore, one or more of these peptide metabolites apparently undegoes some additional intracellular hydrolysis, yielding pyroglutamic acid in the peritubular fluid.

Glucagon

Although renal extraction of glucagon occurs primarily by glomerular filtration and hydrolysis accompanying proximal tubular reabsorption, significant degradation of glucagon is observed in the non-filtering kidney (Emmanouel and co-workers, 1978). This implies that a peritubular mechanism exists for the renal removal of glucagon from the plasma. A recent study has proposed that a significant amount of glucagon may be cleaved by enzymes associated with the basolateral membranes of proximal tubular cells, thus providing a possible explanation for peritubular extraction of the hormone (Talor, Emmanouel, and Katz, 1983). In approaching the present study, we proposed to characterize transport and hydrolysis of glucagon during its reabsoprtion by the proximal nephron, and to investigate the degradative role of enzymes associated with the basolateral (i.e. contraluminal) membrane of proximal tubular cells. Once again, microperfusion of rabbit proximal straight nephron segments with iodinated glucagon resulted in reabsorption of radiolabeled material, primarily ^{125}I-tyrosine. To investigate its hydrolysis further during the reabsorptive process, ^{125}I-glucagon was incubated with homogenates derived from rabbit proximal nephron segments, and hydrolysis was quantified by assaying for trichloroacetic acid soluble radioactivity as a function of pH of the incubation medium. A minor peak of activity was observed at pH 3.0, and a major peak was seen at pH 7.0-7.4. Analytical cell fraction studies of post-nuclear homogenates prepared from isolated proximal nephron segments revealed that glucagon-degrading activity is associated with the brush border and cytosol at pH 7.4, and develops a peak of activity corresponding with lysosomal marker enzymes as the pH is lowered. Hydrolases associated with the basolateral membrane contributed little to glucagon degradation. The data support the concept that renal extraction of glucagon involves its hydrolysis during reabsorption in the proximal tubule, and that the mechanisms include cleavage by brush border-associated hydrolases, and probably cytosolic and lysosomal proteases, as well. Some dipeptide metabolites might enter the cytosol during carrier-mediated transport described for the luminal membrane (Ganapathy and Leibach, 1982), and be subject to cytosolic proteases. Larger metabolites could be taken up endocytically and reach lysosomes for further cleavage. By contrast, it is unlikely that enzymes associated with the basolateral membrane of proximal tubular cells play an important role in glucagon degradation.

Insulin

Renal extraction of insulin has been shown to involve its endocytic uptake at the luminal membrane of proximal tubular cells, followed by lysosomal hydrolysis (Bourdeau, Chen, and Carone, 1973; Peterson and co-workers, 1982). To examine more closely the nature of insulin degrading proteases in the proximal nephron, the iodinated hormone was incubated in the presence of rabbit proximal tubular homogenates. As observed for glucagon, acidic and neutral peaks of hydrolytic activity were seen. Using analytical cell fractionation techniques, insulin degrading activity was found associated with the cytosol and mitochondria of proximal tubular cells at pH 7.4. Cleavage of insulin occurred primarily at lysosomes at pH 3.5. The data are consistent with the interpretation that the more complex structure of insulin makes it a better candidate for the endocytic-lysosomal degradative pathway in proximal tubular cells, rather than the mechanism involving surface digestion by brush border - associated proteases. The presence of insulin-degrading activity in the cytosol and mitochondria remains to be explained, since endocytosed insulin would not be expected to reach these sub-cellular compartments (Hjelle, Oparil, and Peterson, 1984).

CONCLUSIONS

These studies provide evidence that renal extraction of peptide hormones involves their hydrolysis during reabsorption in the proximal tubule by mechanisms including: 1) endocytosis and lysosomal degradation, and 2) surface digestion at the brush border, followed by further hydrolysis in the cytosol or at lysosomes.

REFERENCES

Bourdeau, J., E. Chen, and F. Carone (1973). Insulin uptake in the renal proximal tubule. Am. J. Physiol., 225, 1399-1404.

Burg, M., J. Grantham, M. Abramow, and J. Orloff (1966). Preparation and study of fragments of single rabbit nephrons. Am. J. Physiol., 210, 1293-1298.

Carone, F.A., D.R. Peterson, S. Oparil, and T.N. Pullman (1979). Renal tubular transport and catabolism of proteins and peptides. Kidney Int., 16, 271-278.

Carone, F.A. and D.R. Peterson (1980). Hydrolysis and transport of small peptides by the proximal tubule. Am. J. Physiol., 238, F151-F156.

Carone, F.A., D.R. Peterson, S. Oparil, and T.N. Pullman (1980). Renal tubular transport and catabolism of small peptides. In A.B. Maunsbach, T.S. Olsen, and E.I. Christensen (Eds.), Functional Ultrastructure of the Kidney, Academic Press, New York. pp. 327-340.

Carone, F.A., D.R. Peterson, and G. Flouret (1982). Renal tubular processing of small peptide hormones. J. Lab. Clin. Med., 100, 1-14.

Carone, F.A., D.R. Peterson, and G. Flouret (1984). Proximal tubular hydrolysis and transport of small peptides. In C. Bianchi, A. Bertelli, and C.G. Duarte (Eds.), Contributions to Nephrology, Vol. 42. Kidney, Small Proteins and Drugs. S. Karger, Basel. pp. 1-9.

Emmanouel, D.S., J.B. Jaspan, A.H. Rubenstein, A. H.-J. Huen, E. Fink, and A.I. Katz (1978). Glucagon metabolism in the rat. Contribution of the kidney to the metabolic clearance rate of the hormone. J. Clin. Invest., 62, 6-13.

Ganapathy, V., and F.H. Leibach (1982). Peptide transport in rabbit kidney. Biochim. Biophys. Acta, 691, 362-366.

Hjelle, J.T., J.P. Morin, and A. Trouet (1981). Analytical cell fractionation of isolated rabbit renal proximal tubules. Kidney Int., 20, 71-77.

Hjelle, J.T., S. Oparil, and D.R. Peterson (1984). Sub-cellular sites of insulin hydrolysis in renal proximal tubules. Am. J. Physiol., 246, F409- F416.

Maack, T., V. Johnson, S.T. Kau, J. Figueredo, and D. Sigulem (1979). Renal filtration, transport, and metabolism of low- molecular-weight proteins: a review. Kidney. Int., 16, 251-270.

Peterson, D.R., G. Chrabaszcz, W.R. Peterson, and S. Oparil (1979). Mechanisms for renal tubular handling of angiotensin. Am. J. Physiol., 236, F365-F372.

Peterson, D.R., F.A. Carone, S. Oparil, and E.I. Christensen (1982). Differences between renal tubular processing of glucagon and insulin. Am. J. Physiol., 242, F112-F118.

Peterson, D.R., S. Oparil, G. Flouret, and F.A. Carone (1977). Handling of angiotensin II and oxytocin by renal tubular segments perfused in vitro. Am J. Physiol., 232, F319-F324.

Peterson, D.R., J.T. Hjelle, F.A. Carone, and P.A. Moore (1984). Renal transport and hydrolysis of plasma high density lipoprotein. Kidney Int., 26, 411-421.

Peterson, D.R., E.A. Green, S. Oparil, and J.T. Hjelle (1986). Transport and hydrolysis of glucagon in the proximal nephron. Am. J. Physiol., 251, F460-F467.

Stetler-Stevenson, M.A., G. Flouret, and D.R. Peterson (1981). Handling of luteinizing hormone-releasing hormone by renal proximal tubular segments in vitro. Am. J. Physiol., 241, F117-F122.

Stetler-Stevenson, M.A., D.C. Yang, A. Lipowski, L. McCartney, D. Peterson, and G. Flouret (1981). An approach to the elucidation of metabolic breakdown of the luteinizing hormone releasing hormone, J. Med. Chem., 24, 688-692.

Talor, Z., D.S. Emmanouel, and A.I. Katz (1983). Glucagon degradation by luminal and basolateral rabbit tubular membranes. Am. J. Physiol., 244, F297-F303.

Intestinal Uptake of β-Lactam Antibiotics and Its Relationship to Peptide Transport

Akira Tsuji

Department of Pharmaceutics, Faculty of Pharmaceutical Sciences, Kanazawa University, Takara-machi, Kanazawa 920,Japan

ABSTRACT

In this kinetic experimets on the uptake of cefixime (FK027), a new oral cephalosporin, by the rat intestinal brush-border membrane vesicles, it was elucidated that all types of beta-lactam antibiotics can be absorbed from the small intestine by utilizing the peptide carrier system(s) in a co-coupled manner with proton(s).

KEYWORDS

Beta-lactam antibiotics; cefixime; cephalosporin; intestinal transport; brush-border membrane vesicles; peptide transport; carrier-mediated transport

INTRODUCTION

Over the past 10 years, by extensive studies in our laboratory using either intact animals or the in vitro intestinal everted sac technique, evidence has been accumulated to show the mechanism of transport of beta-lactam antibiotics across the small intestine (Tsuji and Yamana, 1981; Nakashima and others, 1984a, 1984b). The antibiotics without an alpha-amino group in the 6-position of the penicillin nucleus can across the intestinal brush-border membrane by rate-limiting diffusion through the lipoidal membrane barrier and the unstirred water layer existing in front of the mucosal surface. This transport process is available for highly lipid soluble penicillins such as propicillin and dicloxacillin (Tsuji and Yamana, 1981). In contrast, amino-beta-lactam antibiotics (Fig. 1) such as amoxicillin, cyclacillin, cephalexin, cephradine and cefadroxil, which have very low lipid solubility and a zwitterionic structure, have been shown to be absorbed through the dipeptide carrier systems in the intestinal brush-border membrane (Nakashima and others, 1984a, 1984b; Okano and others, 1986).

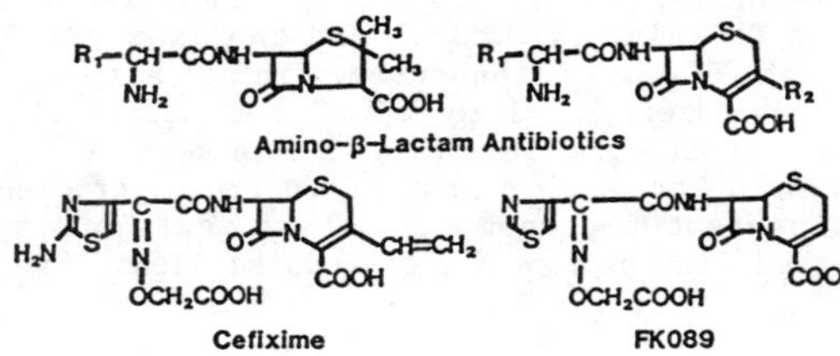

Fig. 1. Structure of beta-lactam antibiotics.

© 1987 Pergamon Journals Ltd.
Printed in Great Britain

Until now, such as a carrier-mediated transport of beta-lactam antibiotics has been believed to be sensitive only for beta-lactam antibiotics having an alpha-amino group in the side chain at the 6-position in penicillins and the 7-position in cephalosporins. Recently, cefixime (FK027), a new cephalosporin antibiotic analogous to poorly absorptive ceftizoxime, was developed for oral use against Gram-negative bacilli. The structure of cefixime is shown in Fig. 1 and completely different from those of cephalexin-analogues from the viewpoint of the lack of an alpha-amino group and anticipated to have low lipophylicity at neutral pH. However, little is known of cefixime transport across the intestinal brush-border membrane. It is of great interest to elucidate whether cefixime, an alpha-amino group deficient cephalosporin, is transported via passive diffusion system, a carrier-mediated transport system being the same to that of amino-beta-lactam antibiotics or a previously unknown system. Furthermore, we should like to answer a question what kind of beta-lactam antibiotics have the potential ability to be transported across the brush-border membrane via dipeptide carrier system(s).

EXPERIMENTAL PROCEDURES

Intestinal brush-border membrane vesicles (BBMVs) were prepared following the method of Kessler and others (1978). Uptake of cefixime into BBMVs was measured by the rapid filtration method. The conditions for the uptake experiments are described in the figure legends and table footnote.

Cefixime trapped on a Millipore filter was extracted with 500 μl of distilled water and analyzed by high-performance liquid chromatography (HPLC). The HPLC system was eqipped with a constant flow pump and UV detector (Japan Spectroscopic Co., Tokyo, Japan). The anlytical column used was Cosmosil 5-C_{18} (4.6 mm x 15 cm, Nakarai Chemicals Ltd., Kyoto, Japan). The mobile phase, a mixture of acetonitrile and water (27.5/75.5, v/v) containing 0.01 M ammonium acetate and 0.01 M tetra-N-butylammonium bromide, was used at a flow rate of 1.0 ml/min and was monitored at 290 nm.

RESULTS AND DISCUSSION

The characteristics of the intestinal transport of cefixime were studied by using the everted sac and purified BBMVs from the rat small intestine.

Cefixime was transported into osmotically responsive intravesicular space rather than binding to the membranes. The initial rate of uptake of cefixime was essentially the same in mannitol medium as that in the presence of an inward gradient of either Na^+ or other monovalent cations (K^+, Li^+, $Choline^+$). With an intravesicular pHi of 7.5, changes in the extravesicular pHo significantly affected the transport of cefixime. Optimal cefixime uptake occurred at a pHo of 5.0, with about 6-fold acceleration compared with that in the absence of an inward proton gradient (pHi = pHo = 7.5). The concentration-dependent uptake shows that cefixime can be taken up by a saturable mechanism. The uptake rates follows Michaelis-Menten kinetics. The non-linear least squares computer fitting provided the Kt-value of 0.83 mM and the Jmax value of 2.67 nmol/mg protein/30 sec. Though FCCP, a protonophore, did not alter cefixime uptake in the absence of proton gradient, addition of FCCP significantly decreased in the presence of an inward proton gradient (Fig. 2). As seen in Fig. 3, time-course of cefixime uptake in the presence of an inward proton gradient (pHi = 7.5, pHo = 5.0) gave a clear overshoot-phenomenon. These facts strongly suggest that cefixime is transported with proton by a co-coupled mechanism.

Another important aspect of proton-coupled cotransport of cefixime is in the fact that an interior-negative membrane potential produced by valinomycin significantly reduced cefixime uptake in the absence of Na^+ and in the presence of a sufficient proton gradient (pHi = 7.5, pHo = 5.0). Since this antibiotic exists predominantly as dianion in the pH range employed in this study, the uptake reduction of cefixime by valinomycin effect demonstrates that cefixime is transported in a transfer of negative charge across the membrane

From these results, we can say that cefixime uptake by the rat intestinal BBMVs is Na^+-independent and is coupled with proton. The effect of valinomycin implies that one cefixime molecule is cocoupled with one proton.

Does cefixime transport via what kind of transport system existing in the intestinal brush-border membrane? Table I summarizes the inhibitory effects of amino acids and oligopeptides and several beta-lactam antibiotics on cefixime uptake into BBMVs, in the absence of sodium ion but in the presence of an inward proton-gradient (pHi = 7.5, pHo = 5.0). There was no significant inhibition by any amino acids. It is of interest that cefixime uptake was inhibited to various extents by the peptides and beta-lactam antibiotics, since the results suggest the existence of a common transport system between cefixime and the peptides or the other beta-lactam antibiotics. L-cephalexin, a stereoisomer of clinically used D-cephalexin inhibited the uptake of cefixime 2.5 times more than did D-cephalexin, indicating that the transport system responsible for cefixime is highly stereospecific.

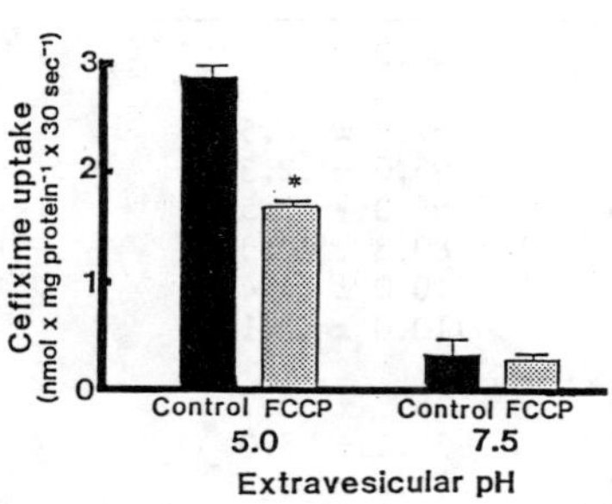

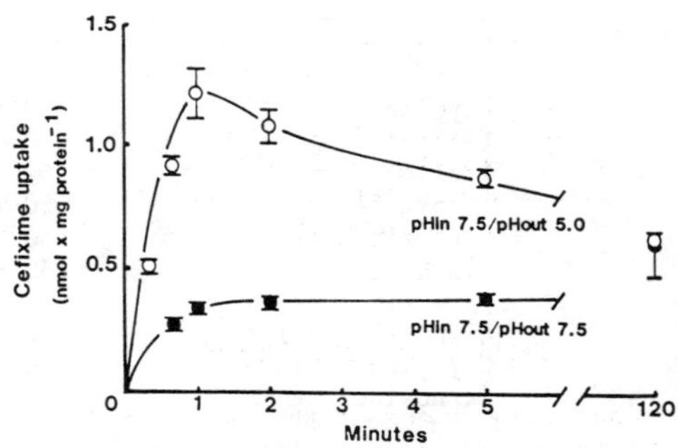

Fig. 2. Effect of cabonyl-cyanide-4-trifluoromethoxy-phenylhydrazone (FCCP) on cefixime uptake by intestinal BBMVs. (Left figure)

Membrane vesicles were preloaded with 10 mM Tris-HEPES buffer (pH 7.5) containing 270 mM mannitol (buffer A). Uptake of 2 mM cefixime was measured at 37°C for 30 sec by incubating the membrane vesicles in buffer A or 10 mM Tris-citrate buffer (pH 5.0) containing 270 mM mannitol (buffer B) with simultaneous addition of FCCP dissolved in ethanol. In the absence of FCCP (control), final concentration of ethanol was 0.5%. Final concentrations of FCCP and ethanol were 50 µM and 0.5%, respectively. Each bar represents the mean ± S.E. of 3 to 5 experiments.

Fig. 3. Time course of cefixime uptake by intestinal BBMVs in the presence (○) and absence (●) of H^+-gradient. (Right figure)

The preloaded condition of the membrane vesicles was the same as described in the legend in Fig. 2. Uptake of 1 mM cefixime was measured at 37°C by incubating the membrane vesicles in buffer A or uffer B. Each point represents the mean ± S.E. of 3 to 5 experiments.

Figure 4 shows the inhibitory effects of glycyl-L-proline, FK089 which is a structural analogue of cefixime and is under clinical trail for oral use (see Fig. 1), benzylpenicillin and cyclacillin on the transport of cefixime. The peptide and these antibiotics inhibited the uptake of cefixime competitively and the estimated inhibitor constants, Ki, were 5.29, 1.90, 1.51 and 1.28 mM, respectively. The observed competitive inhibition kinetics and the intestinal transport characteristics established previously for cyclacillin (NaKashima and others, 1984b) and glycyl-L-proline (Ganapathy and Leibach, 1985) suggest that cefixime has a common affinity site with these beta-lactam antibiotics on the dipeptide carrier protein.

To confirm the existence of a common carrier system between cefixime and the peptides or the other beta-lactam antibiotics in the brush-border membrane, some of the inhibitors were examined for a countertransport effect. The results are shown in Fig. 5, expressed as the ratio, in percent, of uptake after 10 min to that in the control experiment performed simultaneously.

TABLE 1 Inhibitory effect of amino acids, peptides and beta-lactam antibiotics on cefixime uptake by intestinal BBM vesicles.[a)]

Inhibitor		Relative uptake (% of control)
Control		100.0
Glycine	(50 mM)	87.9 ± 6.5
Leucine	(50 mM)	95.0 ± 2.3
Proline	(50 mM)	96.8 ± 16.5
Phenylalanine	(50 mM)	89.8 ± 20.3
Glutamic acid	(50 mM)	110.0 ± 4.4
Lysine	(50 mM)	110.0 ± 12.1
Glycyl-L-proline	(50 mM)	16.7 ± 3.7 *
Glycylsarcosine	(50 mM)	58.1 ± 5.0 *
Carnosine	(50 mM)	74.3 ± 3.4 *
Benzylpenicillin	(20 mM)	15.9 ± 1.3 *
Carbenicillin	(20 mM)	46.5 ± 5.5 *
Dicloxacillin	(20 mM)	26.9 ± 5.0 *
Phenoxymethylpenicillin	(20 mM)	30.0 ± 14.5 *
Propicillin	(20 mM)	53.3 ± 16.4 *
Cefazolin	(20 mM)	74.0 ± 7.9 *
Ceftizoxime	(20 mM)	67.1 ± 12.8 *
FK089	(4 mM)	36.7 ± 10.0 *
Cyclacillin	(20 mM)	53.6 ± 12.0 *
Cephradine	(20 mM)	61.7 ± 1.8 *
D-Cephalexin	(25 mM)	68.4 ± 1.7 *
L-Cephalexin	(25 mM)	24.4 ± 1.4 *

[a)]Membrane vesicles were preloaded with 10 mM Tris-HEPES buffer (pH 7.5) containing 270 mM mannitol. Uptake of 2 mM cefixime was measured at 37° for 30 sec by the addition of the membrane vesicles to 10 mM Tris-citrate buffer (pH 5.0) containing mannitol alone as a control or containing mannitol and inhibitors. Each value is expressed as the percentage of the control as the mean ± S.E. of 3 to 5 experiments. The inhibitor concentrations used are shown in parentheses.

[*)]The level of significance by Student t-test was set at $p < 0.05$.

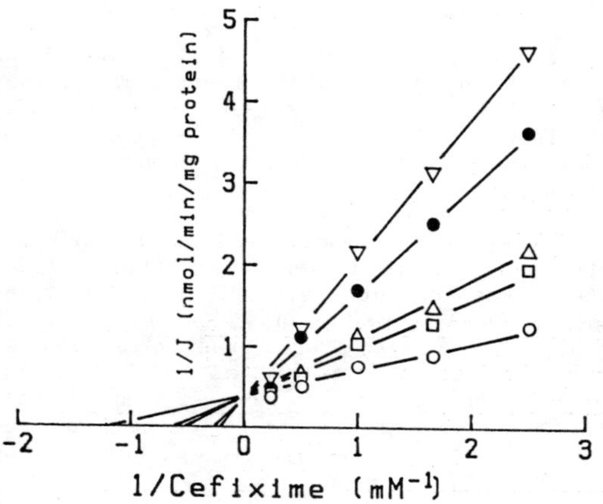

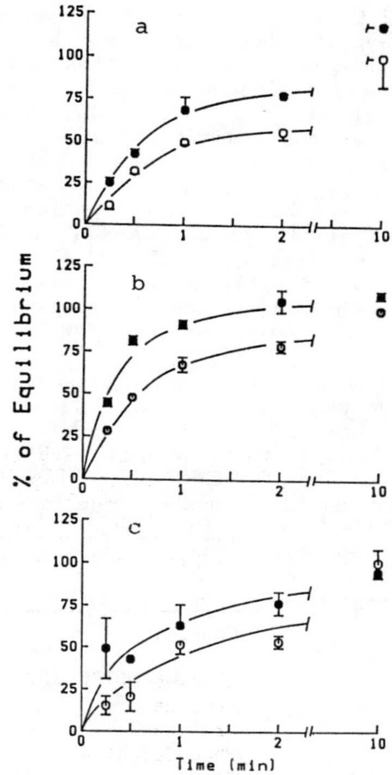

Fig. 4. Lineweaver-Burk plots of cefixime uptake into BBMVs showing inhibition by glycyl-L-proline and beta-lactam antibiotics. (Left figure)

A 20-μl sample of membrane vesicles was preloaded with 10 mM Tris-HEPS buffer (pH 7.5) containing 270 mM mannitol (buffer A). Uptake of cefixime (0.4-4 mM) was measured at 37° for 30 sec with the addition of a 20-μl sample of the membrane vesicles to 80 μl of 10 mM Tris-citrate buffer (pH 5.0) containing 270 mM mannitol (buffer B) in the presence of 5 mM glycyl-L-proline (□), 2 mM FK089 (△), 5mM benzylpenicillin (●) and 5 mM cyclacillin (▽) or in the absence of inhibitors (○). Each pont represents the mean of 3 to 5 experiments.

Fig. 5. Countertransport effect of (a) glycyl-L-proline, (b) FK089 and (c) cyclacillin on cefixime uptake into BBMVs. (Right figure)

A 20-μl sample of membrane vesicles dispersed in buffer B was preloaded at 37°C for 10 min with the addition of 20-μl of (a) 8 mM of glycyl-L-proline (b) 8 mM FK089 or (c) 10 mM of cyclacillin dissolved in buffer B (●) or 20-μl of buffer B alone as a control (○). Uptake of 2 mM cefixime into BBMVs was started by adding 360 μl of a solution of cefixime dissolved in buffer B. The mixtures were ncubated at 37°C for 10 min. The ordinate is expressed as the ratio,in percent, of uptake after 10 min to that of the control. Each point represents the mean ± S.E. of 3 to 5 experiments.

The initial uptake of cefixime was stimulated by preloading of glycyl-L-proline and both beta-lactam antibiotics with (cyclacillin) and without (FK089) an alpha-amino group in the side chain, while uptake at the steady state (10 min) was not altered. The clearly obtained competitive inhibitory effect and the countertransport effect indicate that all types of beta-lactam antibiotics share a common transport system with dipeptides in the intestinal brush-border membrane.

The additional experiment using the rat everted sac technique revealed the interesting findings that the uptake of cefixime into the everted intestinal tissue was apparently Na^+-dependent. It is also true, according to some studies using similar in vitro intestinal tissue preparations, that the transport of several dipeptides and amino-beta-lactam antibiotics across the intact tissue is reduced by replacement of Na^+ with choline.

Based on the results obtained for cefixime in the present study and the evidence accumulated for amino-beta-lactam antibiotics (Kimura and others, 1985; Nakashima and others, 1984a, 1984b; Okano and others, 1986) and dipeptides (Ganapathy and Leibach, 1986; Hoshi, 1986), a model shown in Fig. 6 can be proposed for their intestinal transport. Inward proton gradient drives the transport of amino-beta-lactam antibiotics, cefixime as well as dipeptides. The Na^+-H^+-sodium exchanger, in conjunction with the Na^+,K^+-ATPase, is most probably responsible for the apparent discrepancy observed for Na^+-effect on the intestinal transport of beta-lactam antibiotics and dipeptides in the experiments using in vitro everted tissue and BBMVs.

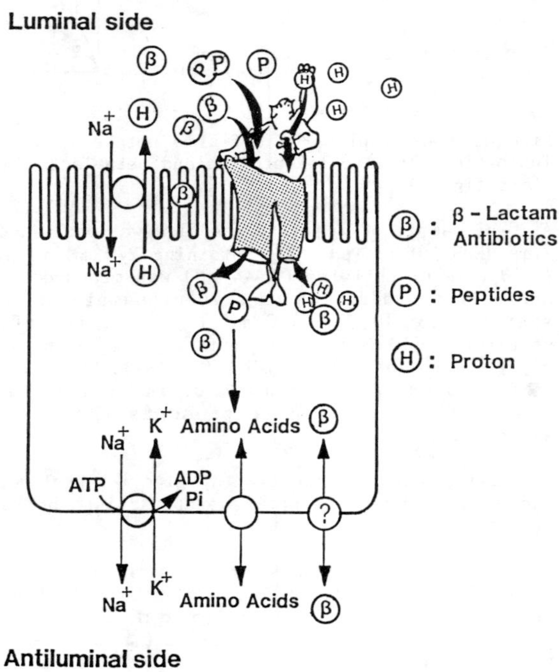

Fig. 6. A model for intestinal transport of beta-lactam antibiotics.

CONCLUSIONS

In conclusion, all types of beta-lactam antibiotics share a common transport system with dipeptides in the intestinal brush-border membrane. Although the chemical structure essential for the translocation of beta-lactam antibiotics via the peptide transport system is unclear from the present study, it can be safely said that beta-lactam antibiotics themselves have various affinity constants to peptide carrier systems depending on the structural difference in the side chains at the 3- and 7-positions in cephems and at the 6-position in penicillins. The translocation of beta-lactam antibiotics possessing sufficient lipid solubility such as benzylpenicillin, phenoxymethylpenicillin, propicillin and dicloxacillin may occur via the peptide transport carrier and/or via the lipid membrane barrier.

ACKNOWLEDGEMENTS

I am deeply indebted to my colleagues, Dr. E. Nakashima, Dr. T. Terasaki, and Mr. I. Tamai who have collaborated in this work. I would like to express my grateful thanks to Professor T. Yamana for his continuous encouragement. Many thanks are also due to Mr. S. Kagatani, Mr. H. Mizuo and Mr. H. Hirooka.

REFERENCES

Ganapathy, V., and F. H. Leibach (1985). Am. J. Physiol., 29, G153-G160.

Hoshi, T. (1985). Japn. J. Physiol., 35, 179-191.

Kessler, M., O. Acuto, C. Storelli, H. Murer, M. Muller, and G. Semenza (1978). Biochim. Biophys. Acta, 506, 136-154.

Kimura, T., T. Yamamoto, R. Ishizuka, and H. Sezaki (1985). Biochem. Pharmacol., 34, 81-84.

Nakashima, E., A. Tsuji, S. Kagatani, and T. Yamana (1984a). J. Pharmacobio-Dyn., 7, 452-464.

Nakashima, E., A. Tsuji, H. Mizuo, and T. Yamana (1984b). Biochem. Pharmacol., 33, 3345-3352.

Okano, T., K. Inui, M. Takano, and R. Hori (1986). Biochem. Pharmacol., 35, 1781- 1786.

Tsuji, A., and T. Yamana (1981). in Beta-lactam antibiotics (ed. S. Mitsuhashi), pp. 235-258. Japan Science Societies Press, Tokyo.

PEPTIDE SYNTHESIS USING PROTEASES

Protease-catalyzed Peptide Synthesis

W. Kullmann

University of Hamburg, UKE, Institut für Zellbiochemie
und klinische Neurobiologie, Martinistr. 52,
2000 Hamburg 20, Federal Republic of Germany

ABSTRACT

Under ordinary conditions the equilibrium point in protease-catalyzed reactions is near to complete hydrolysis. Therefore, proteases are commonly known for their proteolytic rather than for their proteosynthetic activities. Nevertheless, the proteases have proved to be excellent catalysts in preparative peptide synthetic chemistry. The present work describes the applicability of proteases to the synthesis of several opioid peptides. It concludes with a critical assessment of the current state-of-the-art and the future prospects of the enzymatic approach to peptide synthesis.

KEYWORDS

Protease-catalyzed peptide bond synthesis; enzymatic syntheses of opioid peptides; Leu- and Met-enkephalin; dynorphin$_{1-8}$.

INTRODUCTION

One may ask, why do enzymes enjoy a wide and still growing popularity as catalysts for organic syntheses? Certainly, enzymes achieve impressive rate enhancements for the chemical processes they promote. However, the most compelling attraction is based on the unique specificity that they display. From the organic chemist's point of view therefore, the most notable advantage of enzymes, is their capacity to combine, in a concerted fashion, different specificities such as stereo- and regiospecificity thereby opening up new possibilities for degrees of control presently unattainable in any other way. Consequently, the efficient production of chiral synthons with the targeted asymmetric centres and functional groups may be best achieved enzymatically.

Superficially, it may appear incongrous to attempt the synthesis of peptides using proteolytic enzymes. The equilibrium position in a protease-catalyzed reaction is usually far over in the direction of peptide bond hydrolysis and as a consequence, the reversal of this reaction i.e. the formation of peptide bonds - represents merely a 'quantité négligeable' under physiological conditions. However, the proteases per se cannot be held responsible for this state of affairs. Like other enzymes, they simply accelerate the attainment of equilibrium in a chemical process, whereas the equilibrium point itself is determined exclusively by thermodynamic factors. Indeed, the proteases do possess

Advances in the Biosciences Vol. 65

© 1987 Pergamon Journals Ltd.
Printed in Great Britain

the inherent capacity to catalyze both the synthesis and the hydrolysis of a peptide bond, and it is therefore the equilibrium position that actually decides upon the making or breaking of peptide bonds. Consequently, if suitable expedients could be found to shift the equilibrium point of a protease-controlled reaction in favor of the 'reverse reaction', the synthesis of peptide bonds may likely become a 'quantité considérable'.

ENZYMATIC SYNTHESES

The enzymatic syntheses of the opioid peptides Leu- and Met-enkephalin may serve as examples of the peptide synthetic potential of the proteases. Two alternative pathways finally led to the assembly of the opioid peptides each peptide bond of which was formed either by papain or α-chymotrypsin catalysis (Kullmann 1979, 1980)(Fig. 1).

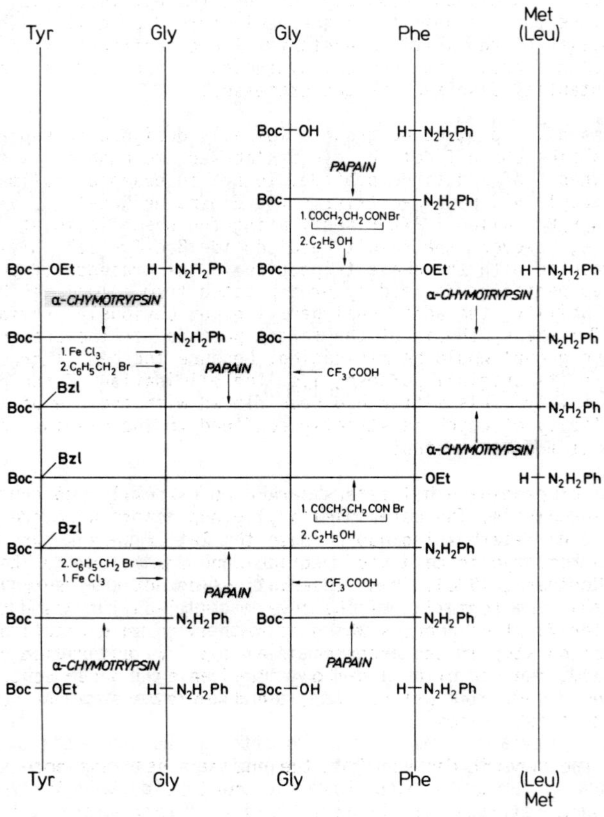

Fig. 1. Enzymatic synthesis of Leu- and Met-enkephalin

Throughout this study all the peptide bond forming steps were solubility-controlled. Thus, two birds could be killed with one stone: the poor solubility of the resulting products induced a favorable shift of the chemical equilibrium and reduced the purification procedures. All but one of these enzymatic coupling steps were performed in monophasic aqueaous-organic systems both to provoke a favorable shift of the ionic equilibria and to enhance the solubility of the initial reactants in the respective incubation mixture (Fig. 1).

To ensure the synthesis of well-defined peptides, the application of temporary and permanent protecting groups is an indispensable 'must' in chemical peptide synthesis. In enzymatic peptide syntheses, this condition proves also right as fas as temporary protection groups are concerned. Beyond that, their introduction is an essential requirement both for thermodynamic reasons. As the case may be, protecting groups serve a further purpose; they provide for a suitable solubility differential between educts and products.

During the syntheses of the enkephalins, N^{α}- and C^{α}-protection was achieved by t-butyloxycarbonyl - and phenylhydrazide groups, respectively (Fig. 1). They contributed to the poor solubility of the newly-formed products without endangering the solubility of the initial reactants. The esterification of α-carboxyl groups - a popular approach in chemical syntheses - is not advisable in view of the esterase potential displayed by the proteases.

As occasion demands, a protecting group originally designed to suppress the reactivity of a side-chain function, can be 'abused' to enable the solubility-controlled synthesis of a desired peptide. To set an example (Kullmann, 1980): an attempt to couple via papain-catalysis the dipeptide Boc-Tyr-Gly-OH with H-Gly-Phe-Leu-N_2H_2Ph failed.[1] After benzylating the phenolic moiety of the tyrosine residue, however, the resulting dipeptide Boc-Tyr(Bzl)-Gly-OH could be successfully reacted with the above tripeptide in the presence of papain (Fig. 1). The newly formed pentapeptide readily precipitated from solutions. By reducing the product-solubility, the additional benzyl group obviously provided the driving force for the synthesis of the target peptide. In the present case, the term 'protecting group' would be misleading, because the benzyl group does not serve any longer its original purpose, i.e. the elimination of the phenolic function's reactivity. This matter had been closed with the use of a protease, the regiospecificity of which is strictly confined to the α-amino- or the α-carboxyl group of the substrates.

Boc-amino acid- or peptide ethyl esters served as carboxyl components during α-chymotrypsin-catalysis. The esterified acyl group donors were preferred to those having a free α-carboxyl group because the latter were known to be less efficient in α-chymotrypsin-catalyzed peptide bond formation (Morihara and Oka, 1977; Oka and Morihara, 1978). These observations do not come surprising it one bears in mind that the formation of the covalent acyl-chymotrypsin complex represents an esterification process which is reduced to an energetically favored transesterification step in the presence of an a priori esterified acyl group donor. Conversely, the use of an α-carboxylate-free acyl group donor would require a 'de novo' esterification. I.e., the donor esters may be considered as 'pre-activated' substrates.

In addition to the benefit derived from the enzyme's esterase potential, the use of donor esters allows the reaction to proceed at a more alkaline environment (pH 10), where the proteolytic activity of α-chymotrypsin is largely

[1] Abbreviations used are: Boc, t-butyloxycarbonyl; Bzl, benzyl; Ph, phenyl; EOt, ethyl ester; OMe, methyl ester.

reduced and the esterolytic potential is still high. On the other hand, the chymotrypsin-catalyzed synthesis in the presence of α-carboxylate-free acyl group donors proceeds most efficiently near pH 7 (Oka and Morihara, 1978), where the proteolytic activity of the protease is optimally expressed as well.

Protease-inherent esterase activities also played a decisive role during the synthesis of dynorphin$_{1-8}$ (Kullmann, 1982). The sequence of the N-terminal pentapeptide of dynorphin$_{1-8}$ is identical with the amino acid sequence of Leu-enkephalin. Therefore, the synthetic design (Fig. 2) could start from the previous synthesis of Leu-enkephalin.

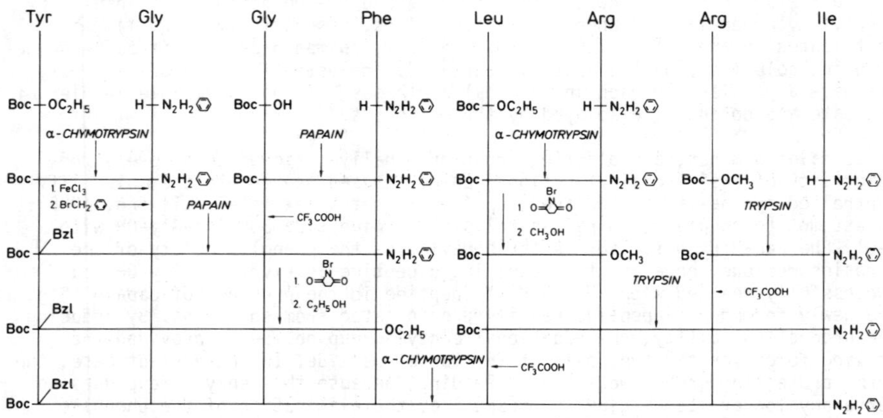

Fig. 2. Protease-catalyzed synthesis of dynorphin$_{1-8}$

The protected N-terminal tetrapeptide fragment of the dynorphin octapeptide coincided with an enkephalin subunit(cf. Fig. 1) and its assembly will not be a subject to further discussion. The synthesis of the C-terminal tetrapeptide, however, is of special interest in view of the above mentioned features. The presence of two arginine residues within the primary structure of this tetrapeptide suggested a synthetic pathway that could profit from the preference of trypsin for basic amino acid residues. Apart from their substrate specificity, chymotrypsin and trypsin display striking similarities with respect to their structural aspects and mode of action. That is, among other things, both proteases form an intermediary acyl-enzyme complex inclusive. Hence, it can hardly surprise that trypsin, too, proved to be an excellent catalyst for the synthesis (Oka and Morihara, 1977). During the preparation of dynorphin$_{1-8}$ the tryptic peptide bond forming steps were performed at an alkaline pH (10.5) in the presence of acyl donors the α-carboxylate functions of which were methylated. Thus, in analogy to the aforementioned chymotryptic syntheses, the tryptic syntheses, too, could profit from the esterase potential of the enzyme. Trypsin's proteolytic activities, however, were largely suppressed, as could be demonstrated by the survival of the pre-existing trypsin-sensitive Arg-Ile bond during the tryptic condensation of Boc-Leu-Arg-OMe and H-Arg-Ile-N_2H_2Ph.

Despite of the analogies between chymotryptic and tryptic peptide syntheses, the substrate specificity of trypsin causes a crucial difference. Opposite to the above chymotryptic syntheses, the trypsin-catalyzed syntheses were not solubility-controlled, because the resulting arginyl-peptides were quite soluble in the aqueous environment. Therefore, the peptide bond formation via trypsin-catalysis had to be performed under kinetic control. The reactions could be terminated well before secondary hydrolysis of the products came into being, because optimal yields were obtained after incubation periods of 2 and 5 minutes, respectively. First traces of secondary hydrolysis were observed only when the reaction was allowed to proceed for 50 minutes. In addition to the kinetic effects, the tryptic syntheses were enhanced by mass action; i.e. by adding a molar excess of the respective nucleophile acceptor.

CONCLUSION

The most prominent advantage of enzymatic syntheses relies on the enzymes' capacity to catalyze chemical processes with an otherwise unattainable specificity. As a consequence, the use of proteases as catalysts for the peptide bond forming step offers favorable opportunities to suppress most of the undesired side-reactions often encountered during the relatively unspecific chemical syntheses. In particular, the stereospecific action of the proteases provides for an outstanding way of maintaining the chiral integrity of the reactants. This aspect can be hardly overestimated, because racemization has been of major concern in the field of peptide synthesis throughout its history. Therefore, the following statement hits the nail on the head: 'The preservation of chiral integrity is a cardinal criterion for judging the relative merits of synthetic methods' (Gross and Meienhofer, 1979).

The hitherto existing studies on enzymatic syntheses of both model peptides and naturally occurring peptides, as well as the semisyntheses of peptides and proteins (for reviews see: Chaiken et al., 1981; Fruton, 1982; Jakubke and Kuhl, 1982, Jakubke et al., 1985; Kullmann, 1985) have proved the excellent qualification of the proteases as proteosynthetic catalysts. Notwithstanding this favorable balance account, the protease-inherent deficiencies should not be ignored. Although the enzymatic method may be well described as being rapidly developing, it is still in its infancy. Though for the time being, the enzymatic technique has not yet reached the versatility of a standard procedure, this shortcoming may be continuously moderated by systematically evaluating the proteosynthetic capacities of already known proteases and by a methodic search for new proteases with suitable catalytic properties. Insofar, impetus for further research may be provided by the statement of Laskowski, Jr. (1978): 'The continual stream of discoveries of new proteinases with very strict specificity gives me a great deal of hope'.

At the present 'state-of-the-art' the author considers the protease-mediated peptide synthesis to be a supplement rather than an alternative to chemical procedures. In the near future, the benefits of the enzymatic method will be exploited most favorably for special synthesis, for synthetic strategies which combine enzymatic as well as chemical techniques, and for semisynthetic purposes. Whether, in the far future, the enzymatic syntheses will approach or even arrive at the level of general applicability remains to be seen.

REFERENCES

Chaiken, I.M., A. Komoriya, M. Ohno, and F. Widmer (1982). Use of enzymes in peptide synthesis. Appl. Biochem. Biotechnol. 7, 385.

Fruton, J.S. (1982). Proteinase-catalyzed synthesis of peptide bonds. Adv. Enzymol. Relat. Areas Mol. Biol. 53, 239.

Gross, E., and J. Meienhofer (1979). The Peptide Bond. In Gross, E., and Meienhofer, J. (Eds). The Peptides, Vol.1, Academic Press, New York, pp. 1-64.

Jakubke, H.-D., and P. Kuhl (1982). Proteasen als Biokatalysatoren für die Peptidsynthese. Pharmazie, 37, 89.

Jakubke, H.-D., P. Kuhl, and A. Könnecke (1985). Grundprinzipien der protease-katalysierten Knüpfung der Peptidbindung. Angew. Chem. 97, 79.

Kullmann, W. (1979). Enzymatic synthesis of Leu- and Met-enkephalin. Biochem. Biophys. Res. Comm. 91, 693.

Kullmann, W. (1980). Proteases as catalysts of enzymic syntheses of opioid peptides. J. Biol. Chem. 255, 8234.

Kullmann, W. (1982). Enzymatic synthesis of dynorphin(1-8). J. Org. Chem. 47, 5300.

Kullmann, W. (1985). Proteases as catalysts in peptide synthetic chemistry. Shifting the extent of peptide bond synthesis from a 'quantité négligeable' to a 'quantité considérable'. J. Protein Chem. 4, 1.

Laskowski, M. Jr. (1978). The Use of Proteolytic Enzymes for the Synthesis of Specific Peptide Bonds in Globular Proteins. In Offord, R.E., and DiBello, C., (Eds.) Semisynthetic Peptides and Proteins. Academic Press, New York, pp. 255-261.

Morihara, K., and T. Oka (1977). α-Chymotrypsin as the catalyst for peptide synthesis. Biochem. J. 163, 531.

Oka, T., and K. Morihara (1977). Trypsin as a catalyst for peptide synthesis. J. Biochem. (Tokyo), 82, 1055.

Oka, T., and M. Morihara (1978). Peptide bond synthesis catalyzed by α-chymotrypsin. J. Biochem. (Tokyo), 84, 1277.

Enzymatic Synthesis of Processing and the Formation of Biologically Active Peptides and Proteins

Irwin M. Chaiken, Shoji Ando and Giorgio Fassina

National Institute of Diabetes and Digestive and Kidney Diseases, National Institutes of Health, Bethesda, Maryland 20892, USA

ABSTRACT

Proteases and other polypeptide modifying enzymes have become useful tools in research and technology for producing biologically active peptides and proteins by both bond formation and cleavage. In peptide and protein engineering, we have used general-specificity proteases to form semisynthetic protein mutants from polypeptide fragments. The selectivity for such reactions is provided by both local sequence and conformational organization at the peptide bond to be formed. Yet, limitations in enzymatic fragment condensation suggest the increasing need for enzymes of more restricted specificities. Ongoing investigations of biosynthetic precursors and the formation of active polypeptides from them suggest synthetic uses of precursor processing enzymes. As with general proteases, an understanding of the molecular features which regulate enzymatic protein breakdown would be helpful to define conditions for the controlled use of processing enzymes to synthesize polypeptides *in vitro*.

KEYWORDS

Enzymatic peptide synthesis; limited proteolysis; processing enzymes; semisynthesis; nucleases; hormone biosynthetic precursors; protein engineering.

INTRODUCTION: SEQUENCE VERSUS CONFORMATION IN REGULATION OF ENZYMATIC PEPTIDE SYNTHESIS

Peptides and proteins are formed and degraded in biological systems by enzymes. Increasingly, the same enzymes, especially proteases and other polypeptide modifying enzymes, have been used as synthetic tools in biomedical research and biotechnology to produce active peptides and proteins *in vitro* (Fruton, 1982; Chaiken and colleagues, 1982; Zhang, 1983; Kullmann, 1985; Jakubke, Kuhl and Konnecke, 1985). The use of enzymes in peptide synthesis has received attention largely because of the power of biospecificity inherent in enzymatic reactions but not always available with fully chemical reactions used more classically in peptide synthesis. This feature helps circumvent such problems as side reactions, racemization, and size limits when making polypeptides by chemical means alone. As a consequence, enzymatic peptide synthesis has been used effectively for both stepwise elongation and fragment condensation. Successful synthetic applications of enzymes have included total synthesis of such polypeptides as

© 1987 Pergamon Journals Ltd.
Printed in Great Britain

aspartame, enkephalins, and fragments of epidermal growth factor (Isowa and colleagues, 1979; Widmer and colleagues, 1981, 1984; Kullmann, 1984); semisynthesis of insulins, including amino acid switching in the conversion of porcine to human insulin (Inouye and colleagues, 1979; Tager and colleagues, 1980; Gattner and colleagues, 1981; Rose, Gladstone, and Offord, 1984; Obermeier and Seipke, 1984); and reconversion of noncovalent and disulfide-linked polypeptide complexes into intact proteins (Sealock and Laskowski, 1969; Homandberg and Laskowski, 1979; Homandberg and colleagues, 1979; Homandberg and Chaiken, 1980; Komoriya, Homandberg and Chaiken, 1980; Westerhuis, Tesser and Nivard, 1980; Juillerat and Homandberg, 1981; Graf and Li, 1981; Homandberg, Komoriya and Chaiken, 1982; Proudfoot and Wallace, 1983).

The regulation of proteolysis by substrate structure, both conformation and amino acid sequence, has been an important guide for effective use of enzymes in synthesis. In our own laboratory, we have used proteases in fragment condensation, to build semisynthetic proteins by reforming peptide bonds between native protein fragments and those synthesized chemically. As others, we have found that general, classical proteases such as trypsin and subtilisin often can be encouraged to synthesize (restitch) the specific peptide bonds that they most readily hydrolyze in folded proteins by limited proteolysis. In such cases, selectivity in synthesis mimics that in proteolysis and is driven by both local amino acid sequence at the peptide bond being synthesized and the conformation of the polypeptide substrates (the folded organization that juxtaposes the amino and carboxyl groups to be covalently restitched). The facilitative contribution of conformation is a powerful control of specificity in enzymatic synthesis, but it also implies an important limitation, that peptide bonds containing a local sequence predicted to be hydrolyzed by a particular protease will not necessarily be synthesized by that protease at a significant rate, even under solvent conditions favoring the synthesis reaction. Thus, while a limited set of enzymatic fragment condensation systems have been defined successfully, the more pervasive use of enzymes to build a wide range of proteins likely will require a library of proteases more expanded than that of general proteases alone.

In the discussion here, the successful use of proteases in semisynthetic fragment condensation is first reviewed. The benefits as well as some experimental limitations which have been encountered when using general proteases are cited. Finally, the occurrence of specific proteases and peptide converting enzymes in precursor processing pathways is discussed and their potential use considered as enzymatic synthesis and processing tools for the *in vitro* production of biologically active peptides and proteins.

ENZYMATIC FRAGMENT CONDENSATION IN SYSTEMS OF LIMITED PROTEOLYSIS

General proteases can catalyze the covalent attachment of protein fragments by reversal of hydrolysis,

$$R_1\text{-}COO^{\ominus} + {}^{\oplus}NH_3\text{-}R_2 \rightleftharpoons R_1\text{-}CONH\text{-}R_2 + H_2O.$$

Most often, this has been achieved for systems in which fragments form compact noncovalent complexes (Table 1) and the synthesis reaction is catalyzed by the same enzyme which catalyzes the hydrolysis reaction. In such cases, the extent of synthesis and therefore the yield of protein product is increased by carrying out the reaction with organic cosolvents which do not destroy the noncovalent fragment interaction but shift the equilibrium to favor synthesis versus hydrolysis by shifting the pK_a of the α-carboxyl group (Homandberg, Mattis and Laskowski, 1978). Examples of such enzymatic resynthesis reactions are given in Table 1A.

While the use of enzymatic fragment condensation has been rather limited so far, the general application of this approach for a diversity of proteins seems

TABLE 1 Examples of Enzymatic Fragment Condensation for Preparing Biologically Active Peptides and Proteins[a]

General case	Protein or peptide	Fragments condensed yielding active species	Enzyme used; solvent	Yield %	Reference for process shown
A: Fragments forming noncovalent complexes	Bovine pancreatic ribonuclease	(1-20) + (21-124)	Subtilisin BPN'; 90% glycerol	50	Homandberg and Laskowski, 1979
	Staphylococcal nuclease	[Gly 48](6-49) + (50-149)	Trypsin; 90% glycerol	20	Komoriya, Homandberg and Chaiken, 1980
	Horse heart cytochrome c	(1-38,heme)+(39-104)	Clostripain; 80% glycerol	25-30	Juillerat and Homandberg, 1981
	Human somatotropin	(1-134)+(135-191)	Thrombin; 90% glycerol	20	Westerhuis, Tesser and Nivard, 1980
B: Noncomplex-forming fragments	Human insulin	[des B23-B30]insulin + (B23-B30)	Trypsin; 50% dimethylformamide	58	Inouye and colleagues, 1979
	Ribonuclease S-peptide	(1-10)+(11-15) [product trapped by RNase-S-(21-124)]	Clostripain; water	15	Homandberg, Komoriya and Chaiken, 1982

[a]Adapted from Chaiken and colleagues, 1982.

optimistic given the widespread occurrence of limited enzymatic proteolysis (Bennett and Huber, 1984). This latter phenomenon appears to be rooted in macromolecular organization, with sites of most rapid proteolysis limited not only by inaccessibility of many parts of the sequence involved in ordered structural elements (α-helices, pleated sheets, etc.) but also by the nature of loops at which limited proteolysis often occurs. Current experimental (Fontana and colleagues, 1986) and theoretical (Rose, Gierasch and Smith, 1985) evaluations of loop dynamics and proteolysis sites should lead to an improved understanding and ultimate predictability of limited proteolysis and consequently of enzymatic fragment condensation.

USE IN SEMISYNTHESIS AND PROTEIN ENGINEERING

We have used the enzymatic fragment condensation approach to produce semisynthetic proteins, including site-directed mutants, from two-fragment complexes containing a combination of synthetically-derived pieces and those derived from proteolysis of native proteins. The formation of semisynthetic bovine pancreatic ribonuclease A (Homandberg and colleagues, 1979) from the semisynthetic SPeptide-SProtein complex (SRNase-S) was based on the pioneering study of Homandberg and Laskowski (1979) with the fully native bovine pancreatic protein. The SRNase-S to A conversion, by synthesis of the Ala 20-Ser 21 bond, was effected (Fig. 1) in high concentrations of glycerol by subtilisin, the protease which rapidly hydrolyzes the 20-21 bond under normal aqueous conditions. Even though many peptide bonds in both the SPeptide and SProtein components are expected to be susceptible to subtilisin proteolysis based on sequence alone, hydrolytic destruction of fragments and product is slow compared to fragment restitching (Fig. 1), due to the conformational compactness of the substrates and juxtaposition of residues 20 and 21 by fragment complexation.

Semisynthetic staphylococcal nuclease also has been formed by enzymatic fragment restitching of the tryptic complex, Nase-T. Here (Fig. 2), the trypsin synthetic

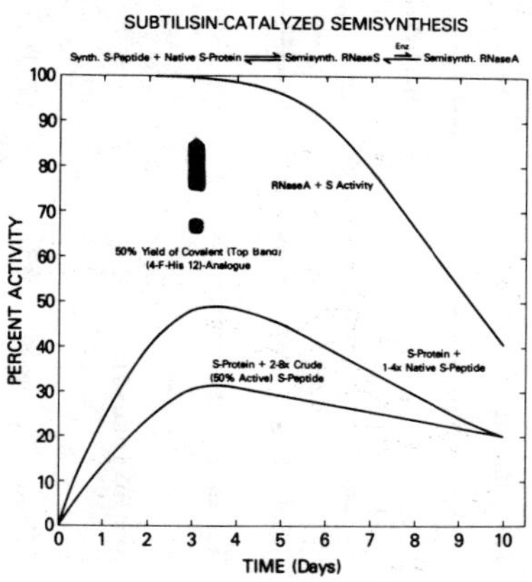

Fig 1. Conversion of RNase-S complexes [native and semisynthetic (with crude synthetic SPeptide)] to covalent forms, in 90% glycerol, catalyzed by subtilisin BPN'. Lower curves: activity of covalent (RNase-A) forms only. Upper curves: activity of total RNase-A and S. Adapted from Homandberg and colleagues, 1979.

reaction with the two native-fragment complexes [(6-48):(49-149)] and [(6-48):(50-149)] yielded (Homandberg and Chaiken, 1980) the inactive product [des Lys 49]Nase(6-149). The resynthesis reaction was successful (Fig.2) but the presence of the Lys 48-Lys 49-Gly 50 sequence caused eventual deletion of Lys 49 by recycling between synthesis and hydrolysis at the Lys-X bonds. Distortion of folded structure, including the catalytically important Glu 43 (Chaiken and Anfinsen, 1971; Chaiken and Sanchez, 1972; Cotton, Hazen and Legg, 1979; Hibler and colleagues, 1986), likely is the major cause of loss of enzymatic activity in the trypsin resynthesis product (see Fig. 3). Nonetheless, an active resynthesis product eventually was obtained (Komoriya, Homandberg and Chaiken, 1980) by trypsin-catalyzed condensation of the complex containing (50-149) and the synthetic mutant fragment, [Gly 48](6-49). The mutant protein, [Gly 48]Nase-(6-149), is resistant to enzymatic loss of Lys 48, since no Gly 48-Lys 49 hydrolysis occurs with trypsin.

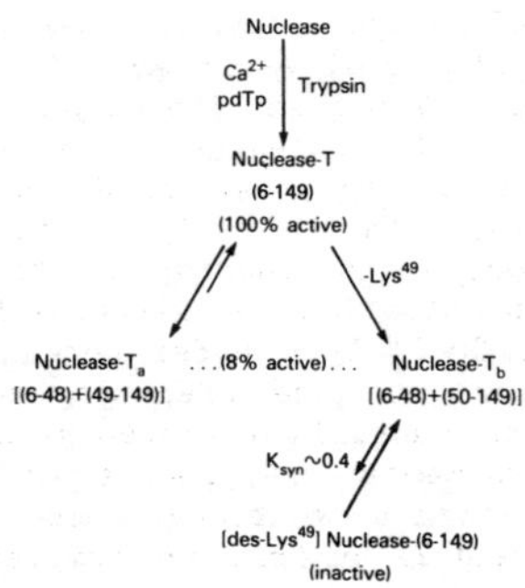

Fig.2. Enzymatic synthesis and hydrolysis reactions leading to enzymatically inactive [des Lys 49]nuclease-(6-149). From Homandberg and Chaiken, 1980.

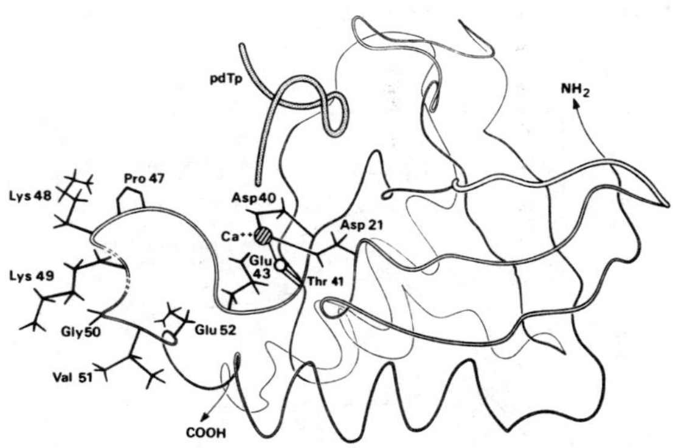

Fig. 3. Scheme of nuclease crystal structure (Cotton, Hazen and Legg, 1979). Glu 43 interacts with the Ca^{++} through one of two Ca^{++}-liganded water molecules. Trypsin-sensitive peptide bonds carbonyl to residues 49 and 50 are indicated by broken lines. Adapted from Homandberg and Chaiken, 1980.

As shown in the nuclease case, conversion of semisynthetic fragment complexes to covalent proteins offers a useful general methodology to form protein mutants, by synthesizing limited segments of proteins with site-directed mutation and then enzymatically restitching these with native fragments obtained proteolytically. The approach also has proven useful for making the 4-fluoro-His-12 mutant of RNase A (Fig. 1). The 4-F-His mutation of SPeptide leads to a noncovalent semisynthetic complex with drastically lowered activity (Dunn and colleagues, 1974); but the mutant seems poised for activity, given its close-to-normal active site recognition of nucleotides and inhibitors (Taylor and Chaiken, 1977) and native-like three dimensional structure (Taylor and colleagues, 1981). The question has arisen of whether covalent reconstruction of this complex to an RNase-A derivative would stabilize the protein sufficiently to allow analysis of intrinsic enzymatic activity at low pH (closer to the pK_a of the 4-F-His imidazole),at which RNase A is conformationally stable but the S complex is not. [4-F-His 12]SRNase A has been made (Fig. 1) and its catalytic properties examined. Interestingly, the enzyme activity of the mutant does increase at low pH relative to native RNase A but does not approach that of the native protein; nonetheless, the data have helped to confirm the presence of intrinsic enzymatic function in the mutant.

A GENERALIZED APPROACH - USE OF MOLECULAR TRAPS

In spite of successes of enzymatic condensation of noncovalent fragment complexes and the expectation of improved understanding and thus predictability of limited proteolysis reactions which can make other protein systems amenable to this method, most protein fragments do not interact with one another noncovalently. Ultimately, the generalized usefulness of enzymatic fragment condensation will depend on the extent to which it can be used to join non-interacting polypeptides. One solution may be the use of non-substrate molecular interactions as kinetic traps.

We have examined the use of molecular traps in the formation of ribonuclease SPeptide from the non-interacting peptides (1-10) and (11-15), with the results shown in Fig. 4. Here, the Arg 10-Gln 11 bond was reformed by clostripain; the product, (1-15), was trapped by noncovalent binding to SProtein, which binds neither of the peptide substrates. The enzymatic reaction is dependent on

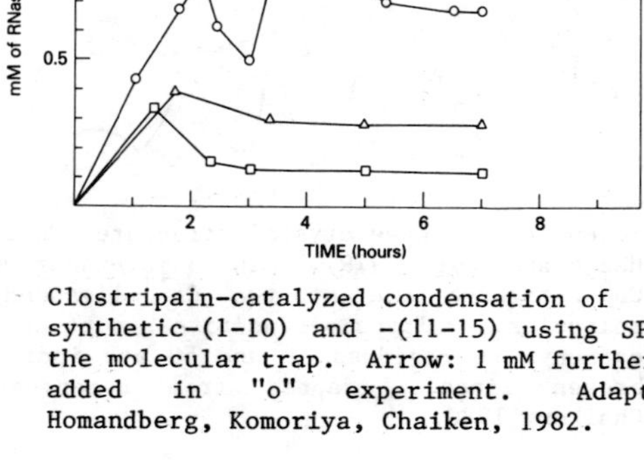

Fig. 4. Clostripain-catalyzed condensation of synthetic-(1-10) and -(11-15) using SProtein as the molecular trap. Arrow: 1 mM further SProtein added in "o" experiment. Adapted from Homandberg, Komoriya, Chaiken, 1982.

SProtein trapping of the product by complexation; this is reflected in the reduced condensation yield of ε-trifluoroacetyl-blocked (1-10), which forms a blocked (1-15) derivative having a lower affinity for SProtein than does unblocked (1-15).

The use of molecular traps for condensing non-interacting peptides may well have general applicability. Synthesis such as that of SPeptide is likely to be doable when the peptide binds to a complementary protein fragment (see Table I). A more general tactic may be the use of antibodies as traps when the product but not the substrates is antigenic at least for a selected antibody subpopulation. The use of non-protein traps, such as metals or other small molecular weight ligands, also could be helpful.

PROCESSING ENZYMES AS IN VITRO CATALYSTS AND THE CASE OF NEUROHYPOPHYSIAL HORMONE BIOSYNTHETIC PRECURSORS

While enzymatic fragment condensation results generally have emphasized the influence of substrate conformation in promoting selectivity in resynthesis while restricting proteolysis, the results also infer that substrate conformation may regulate and thus limit the usefulness of a particular protease to effect synthesis even when the sequence predicts its applicability. One such case which we have encountered recently is that of neurohypophysial hormone biosynthetic precursors, which we have attempted to reconstruct in order to study precursor assembly and regulation of enzymatic processing. As depicted in Fig. 5, the precursors for both oxytocin and vasopressin contain an amino terminal hormone domain and a neurophysin domain linked by a Gly-Lys-Arg spacer sequence (Ivell, Schmale and Richter, 1983). While we have been able (Kanmera and Chaiken, 1985a; Ando, Fassina and Chaiken, 1986) to reconstruct precursor analogs chemically by coupling both oxytocinyl-Gly-Lys-Arg and vasopressinyl-Gly-Lys-Arg to neurophysins I and II, respectively, enzymatic condensation has failed (yields are less than 1%) with either clostripain, Arg-specific submaxillary protease or several other general proteases, including trypsin and thrombin. This failure occurred even though the hormone-GKR and NP fragments were expected (Fig. 5) to form noncovalent complexes with the carboxyl terminal Arg of the spacer and the N-terminal Ala of NP juxtaposed. It seems quite plausible that the failure to produce the semisynthetic precursor enzymatically in significant yields is due to the difficulty of the enzymes we have tested to adapt to the conformational constraints of the precursor-derived noncovalent complexes; analogously with enzymatic fragment condensation of other products of limited proteolysis (see above), it is predicted that such a reaction will be more facile with the enzyme responsible for limited precursor endoproteolysis. For the NP/hormone precursor, the dibasic endoprotease (Fig. 5) has been identified (Loh, Parish and Tuteja, 1985; Clamagirand and colleagues, 1986) and eventually should be available at sufficient purity to be useful for precursor resynthesis.

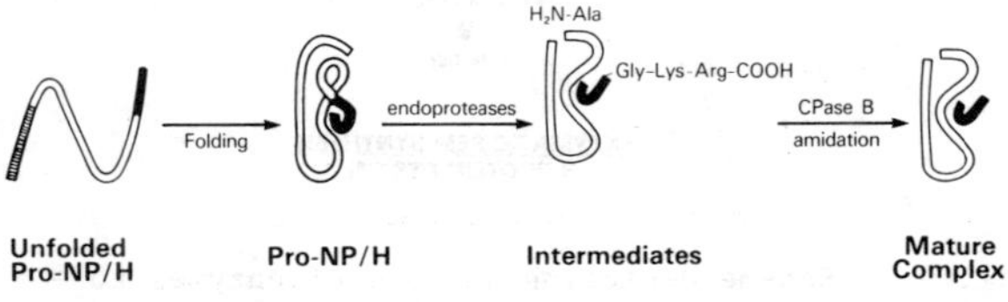

Fig. 5. Scheme of biosynthetic precursor structure and enzymatic processing events occurring in neurohypophysial hormone (H, filled lines)/neurophysin (NP, open lines) biosynthesis. Hatched line: C-terminal tail in precursors. Adapted from Chaiken, Kanmera, Sequeira, 1986.

A corollary of the above hypothesis of conformational constraints for enzymatic resynthesis of hormone precursor is that the action of enzymes in processing (including limited proteolysis) of precursor and precursor intermediates is regulated by the conformation of the latter substrates. This argument is experimentally accessible and is being examined at the present time by comparing enzymatic processing of the hormonogen oxytocinyl-GKR alone and with neurophysin noncovalently attached (see Fig. 5). For the carboxypeptidase B-like sequential removal of Arg and Lys, the rates of enzymatic hydrolysis are slowed significantly when neurophysin is noncovalently associated (Kanmera and Chaiken, 1985b). The concentration dependence of this neurophysin effect suggests that it is due to formation of peptide-bound neurophysin dimers. A similar rate reduction has been found for the enzymatic amidation which converts oxytocinyl-Gly to mature, C-terminal amidated oxytocin (S. Ando, A. Murthy, B. Eipper, and I. Chaiken, unpublished data). The high concentration of peptides and NP's found in the secretory granules in which enzymatic processing occurs suggests that the forms of precursor and precursor intermediates available for processing are likely to be dimeric. Thus, at least in the hormone/neurophysin case, the enzymatic processing enzymes are likely to be sensitive to substrate conformation and, as with general proteases, their use in vitro for enzymatic synthesis is likely to be regulated by conformation.

The neurophysin/hormone precursor processing pathway is a rather simple representative of an expanding collection of such pathways, each encompassing a set of processing enzymes (Chaiken, Kanmera and Sequeira, 1986; Mains and colleagues, 1983; Loh, Brownstein and Gainer, 1984; Douglass, Civelli and Herbert, 1984). Even assuming that some of the enzymes involved may be common for more than one pathway, the expectation is that the multiplicity of precursor processing pathways can yield an interesting collection of proteolytic and peptide-modifying enzymes of use in making peptides and proteins in vitro. The selectivity of particular enzymes in this potential library, control of this selectivity by local sequence and folded conformation, and ultimate use of such enzymes in producing active peptides by selective bond formation and cleavage all remain challenges for future experimental investigation.

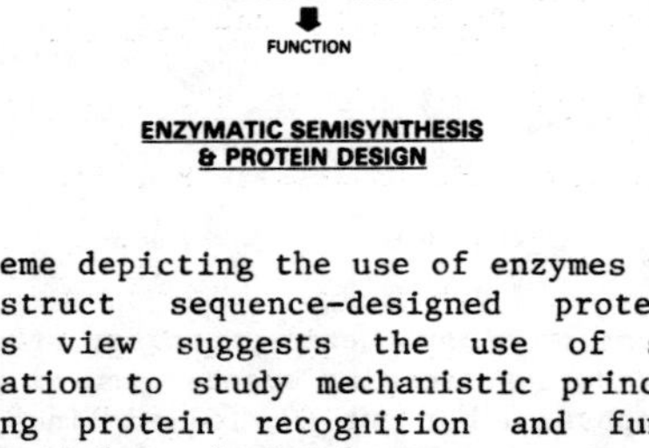

Fig. 6. Scheme depicting the use of enzymes to construct sequence-designed protein mutants. This view suggests the use of such protein mutation to study mechanistic principles underlying protein recognition and function. The potential to devise hormone precursor constructs (see Fig. 5) with altered enzymatic processing characteristics also seems feasible.

FINAL COMMENTS-PROSPECTS FOR ENZYMATIC SYNTHESIS AND PROCESSING TO MAKE PEPTIDES AND PROTEINS

Experience has shown that enzymes can be used in research and biotechnology to synthesize active polypeptides. Yet, the advent and continued development of genetic engineering has obviated the need for total synthesis *in vitro* by either chemical or enzymatic means alone, at least for large polypeptides and proteins. In this context, enzymatic fragment condensation can be seen (Fig. 6) as a means to form proteins and their mutants not only by semisynthetic recombination of native and *in vitro* synthesized fragments but also to incorporate polypeptides made by recombinant DNA cloning and expression. In addition, the increasing library of enzymes for limited proteolysis and modification can be used to convert cloned protein products to smaller, active pieces (both protein and peptide). Such enzymatic processing could be of use for releasing and purifying desired polypeptide products from cloned proteins in the laboratory or even to effect therapeutically useful timed release of active peptides from *in vivo* implanted precursors or precursor intermediates. The exciting prospect thus remains to use enzymes as both synthesis and processing vehicles, to produce active peptides and proteins from the combined resources of isolated native proteins, cloned products, and *de novo* polypeptide synthesis *in vitro*.

REFERENCES

Ando, S., Fassina, G., and Chaiken, I.M. (1986). *Fed. Proc.*, *45*, 1848.

Bennett, W.S. and Huber, R. (1984). *CRC Crit. Rev. Biochem.*, *15*, 291-384.

Chaiken, I.M. and Anfinsen, C.B. (1971). *J. Biol. Chem.*, *246*, 2285-2290.

Chaiken, I.M. and Sanchez, G.R. (1972). *J. Biol. Chem.*, *247*, 6743-6747.

Chaiken, I.M., Komoriya, A., Ohno, M., and Widmer, F. (1982). *Appl. Biochem. Biotechnol.*, *7*, 385-399.

Chaiken, I.M., Kanmera, T., and Sequeira, R.P. (1986). In R.S. Rapaka and R.L. Hawks (Eds.) *Medicinal Chemistry and Molecular Pharmacology of the Opioid Peptides*, NIDA press, Washington, pp. 3-20.

Clamagirand, C., Camier, M., Boussetta, H., Fahy, C., Morel, A., Nicolas, P., and Cohen, P. (1986) *Biochem. Biophys. Res. Commun.*, *134*, 1190-1196.

Cotton, F.A., Hazen, E.E., and Legg, M.J. (1979). *Proc. Natl. Acad. Sci. USA*, *76*, 2551-2555.

Douglass, J., Civelli, O., and Herbert, E. (1984). *Annu. Rev. Biochem.*, *53*, 665-715.

Dunn, B.M., DiBello, C., Kirk, L., Cohen, L.A., and Chaiken, I.M. (1974). *J. Biol. Chem.*, *249*, 6295-6301.

Fontana, A., Fassina, G., Vita, C., Dalzoppo, D., Zamai, M., and Zambonin, M. (1986). *Biochemistry*, *25*, 1847-1851.

Fruton, J. (1982). In A. Meister (Ed.), *Advances In Enzymology*, Vol. 53, Wiley, New York, 239-306.

Gattner, H.G., Danho, W., Knorr, R., Naithan, V.K., and Zahn, H. (1981). In K. Brunfeld (Ed.) *Peptides 1980*, Scriptor, Copenhagen, pp. 372-377.

Graf, L. and Li, C.H. (1981). *Proc. Nat. Acad. Sci. U.S.A.*, *78*, 6135-6138.

Hibler, D.W., Stolowich, N.J., Mehdi, S., and Gerlt, J.A. (1986). In P.A. Frey (Ed.), *Mechanisms of Enzymatic Reactions: Sterochemistry*, Elsevier, New York, pp. 101-114.

Homandberg, G.A., Mattis, J.A., and Laskowski, M., Jr. (1978). *Biochemistry*, *17*, 5220-5227.

Homandberg, G.A. and Laskowski, M., Jr. (1979). *Biochemistry*, *18*, 586-592.

Homandberg, G.A., Komoriya, A., Juillerat, M. and Chaiken, I.M. (1979). In E. Gross and J. Meienhofer (Eds.), *Peptides-Structure and Biological Function*, Pierce Chem. Co., Rockford, IL. pp. 597-600.

Homandberg, G.A. and Chaiken, I.M. (1980). *J. Biol. Chem.*, *255*, 4903-4909.

Homandberg, G.A., Komoriya, A., and Chaiken, I.M. (1982). *Biochemistry*, *21*, 3385-3389.

Inouye, K., Watanabe, K., Morihara, K., Tochino, Y., Kanaya, T., Emma, J., and Sakakibara, S. (1979). J. Amer. Chem. Soc., 101, 751-752.
Isowa, Y., Ohmori, M., Ichikowa, T., Mori, K., Nonaka, Y., Kihara, K.-I., Oyama, K., Satoh, H., and Nishimura, S. (1979). Tetrahedron Lett., 2611-2612.
Ivell, R., Schmale, H., and Richter, D. (1983). Neuroendocrinology, 37, 235-239.
Jakubke, H.-D., Kuhl, P., and Konnecke, A. (1985). Angew., Chem, Int. Ed. Engl., 24, 85-93.
Juillerat, M. and Homandberg, G.A. (1981). Internat. J. Pept. Prot. Res., 18, 335-342.
Kanmera, T. and Chaiken, I.M. (1985a). J. Biol. Chem., 260, 8474-8482.
Kanmera, T. and Chaiken, I.M. (1985b). J. Biol. Chem., 260, 10118-10124.
Komoriya, A., Homandberg, G.A., and Chaiken, I.M. (1980). Internat. J. Pept. Prot. Res., 16, 433-439.
Kullmann, W. (1984). Biochem. J., 220, 405-416.
Kullmann, W. (1985). J. Prot. Chem., 4, 1-22.
Loh, Y.-P., Brownstein, M.S., and Gainer, H. (1984). Annu. Rev. Neurosci., 7, 189-222.
Loh, Y.-P., Parish, D.C., and Tuteja, R. (1985). J. Biol. Chem. 260, 7194-7205.
Mains, R.E., Eipper, B.A., Glembotski, C.C., and Dores, R.M. (1983). Trends in Neurosciences, 6, 229-235.
Obermeier, R. and Seipke, G. (1984). In W. Voelter, E. Bayer, Y.A. Ovchinikov, and E. Wunsch (Eds.) Chemistry of Peptides and Proteins, Vol. 2, de Gruyter, Berlin, pp. 3-10.
Proudfoot, A.E.I. and Wallace, C.J.A. (1983). In K. Blaha and P. Malon (Eds.) Peptides 1982, de Gruyter, Berlin, pp. 353-356.
Rose, G.D., Gierasch, L.M., and Smith, J.A. (1985). Adv. Prot. Chem., 37, 1-109.
Rose, K., Gladstone, J., and Offord, R.E. (1984). Biochem. J., 220, 189-196.
Sealock, R.W. and Laskowski, M., Jr. (1969). Biochemistry, 8, 3703-3710.
Tager, H., Thomas, N., Assoian, R., Rubenstein, A., Salkow, M., Olefsky, J., and Kaiser, E.T. (1980). Proc. Nat. Acad. Sci. U.S.A., 77, 3181-3185.
Taylor, H.D. and Chaiken, I.M. (1977). J. Biol. Chem., 252, 6991-6994.
Taylor, H.C., Richardson, D.C., Richardson, J.S., Wlodawer, A., Komoriya, A., and Chaiken, I.M. (1981). J. Mol. Biol., 149, 313-317.
Westerhuis, L.W., Tesser, G.I., and Nivard, R.J.F. (1980). Recl. Trav. Chim. Pays-Bas, 99, 400-403.
Widmer, F., Breddam, K., and Johansen, J.T. (1981). In K. Brunfeld (Ed.) Peptides 1980, Scriptor, Copenhagen, pp. 46-55.
Widmer, F., Bayne, S., Houen, G., Moss, B.A., Rigby, R.D., Whittaker, R.G., and Johansen, J.T. (1984). In U. Ragnasson (Ed.) Peptides 1984, Almquist and Wiksell, Stockholm, pp. 193-200.
Zhang, Y-S. (1983). Trends Biochem. Sci., 8, 16-17.

Mechanism and Maximum Yield in Kinetically Controlled Peptide Synthesis

V. Kasche

AB Biotechnologie II, Technische Universität Hamburg–Harburg, PF 901403, D-2100 Hamburg 90, Federal Republic of Germany

ABSTRACT

The mechanism of peptide and penicillin synthesis using activated substrates - kinetically controlled synthesis - involves nucleophile binding before the deacylation of the acylenzyme. The yield depends on whether the substrate and nucleophile have overlapping binding sites or not.
In the latter case the yield can be increased by substrate recycling.

KEYWORDS

Enzyme catalyzed peptide synthesis; semisynthetic penicillins; α-chymotrypsins; trypsin; penicillin amidase; nucleophile binding.

INTRODUCTION

Proteinases and other hydrolases catalyze the hydrolysis of condensation products in biological systems. The reverse reaction, i.e. the formation of the condensation product from the hydrolysis products, can also be catalyzed by these enzymes. In this equilibrium controlled synthesis the yield of condensation product cannot exceed that given by the thermodynamic equilibrium constant for the synthesis at the existing experimental conditions. This cannot be influenced by the properties of the enzyme used to catalyze the condensation reaction.

In biological systems transferases are used to catalyze the formation of condensation products. They require that one of the substrates is activated (adenylated, esterified, phosphorylated etc.). The catalyzed reaction is a group transfer from the group used for activation to the other substrate. The yield in this kinetically controlled synthesis exceeds the yield obtained in the kinetically controlled reaction and depends on the properties of the enzyme (Kasche, 1986). Hydrolases are transferases that transfer groups to water. Some hydrolases can also be used as transferases to catalyze such kinetically controlled reactions. This is shown in Fig.1 for the equilibrium and kinetically controlled synthesis of benzylpenicillin.

The apparent ratio of the transferase to hydrolase activity $(\underline{k}_3'/\underline{k}_3)_{app}$ for a

AB VOL65-K

Advances in the Biosciences Vol. 65
© 1987 Pergamon Journals Ltd.
Printed in Great Britain

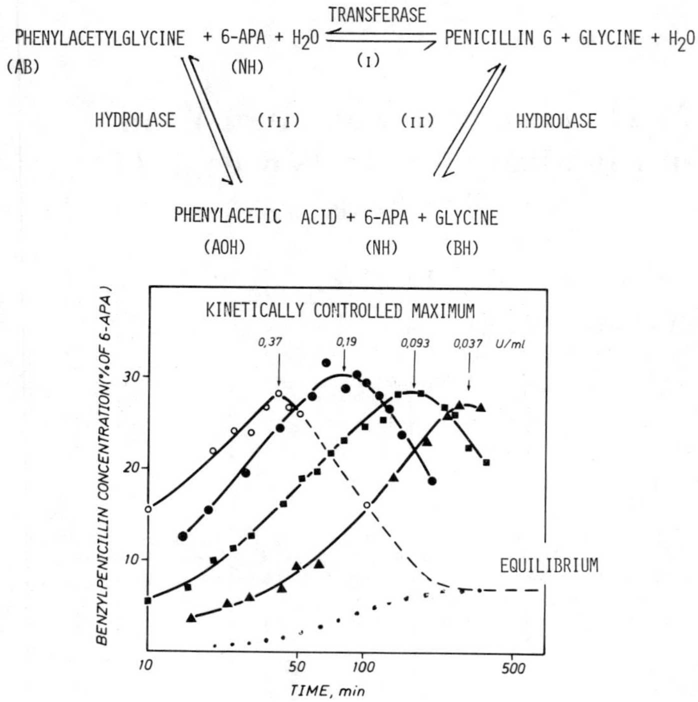

Fig.1. Kinetically (—— reaction (I), (II), and (III) and equilibrium (.... reaction (II)) controlled synthesis of benzylpenicillin at pH 6,0 and 25°C (Ionic strength I = 0.2). Initial concentrations of substrates 10 mM. The enzyme concentrations in U/ml are given above the kinetically controlled maxima of each curve.

hydrolase or a transferase is defined as

$$\left(\frac{k_3'}{k_3}\right)_{app} = \frac{v_{AN}}{v_{AOH}} \cdot \frac{[H_2O]}{[NH]} \qquad (1)$$

where v_{AN} and v_{AOH} are the initial rates of the formation of the condensation (transferase) and hydrolysis product, respectively; [NH] is the initial concentration of the group acceptor, generally a nucleophile. Experimental data for the ratio of the transferase to hydrolase activities for some transferases and hydrolases are given in Table I (Kasche, 1986; Cook and Peters, 1985). From this we conclude that transferases are poor hydrolases and hydrolases are poor transferases. This of course is essential for the biological function of these enzymes. The hydrolases may, however, be used as transferases in artificial and biotechnological systems when their ratio of the transferase to hydrolase rate exceeds 1. This can only be achived with nucleophile concentrations > 100 mM. These concentrations are far above those at which transferases catalyze the corresponding reactions in living systems.
For a rational analysis of the factors influencing the yield in the kinetically

TABLE 1 The ratio of apparent transferase to hydrolase activity $(k_3'/k_3)_{app}$ for different hydrolases with covalent intermediates and transferases.

Enzyme	Nucleophile	$(k_3'/k_3)_{app}$
TRANSFERASES		
DNA-polymerase	DNA	$\sim 10^7$
Hexokinase	Glucose	$\sim 10^6$
γ-Glutamyl-transferase	Gly-Gly	$\sim 10^5$
HYDROLASES		
Serine proteinases	Amino-acid amides	$10^3 - 10^4$
	D-Amino-acid amides	10
(α-chymotrypsin	Amino acid esters	10^2
trypsin	Amino acids	10
carboxypeptidase Y)	Alcohols	$10^2 - 10^3$
Thiol proteinases	Amino acid amides	$10^3 - 10^4 (10^6)$
(papain)	Alcohols	50
Amidases	6-Aminopenicillanic acid	$10^3 - 10^4$
(penicillin amidase)	Alcohols	1
β-Galactosidase	Lactose	$10^2 - 10^3$
	Alcohols	10^2

controlled synthesis of condensation products using hydrolases as catalysts the mechanism of this reaction must be studied in detail.

GENERAL MECHANISM FOR THE KINETICALLY CONTROLLED SYNTHESIS OF CONDENSATION PRODUCTS

The active site of transferases and hydrolases can be assumed to include specific binding sites for the groups adjacent to the chemical bond formed or hydrolyzed in the enzyme catalyzed reaction.
For the transferases in Table 1 the binding site of the activated substrate and the group accepting nucleophile have been shown to be nonoverlapping. For hydrolases this may not generally be the case. For COOH- and metalloproteinases that have no covalent acyl-enzyme intermediates the ratio of transferase to hydrolase activity has been shown to be neglizible (Fruton, 1982). This is probably due to that in this case the activated substrate and the acceptor nucleophile have overlapping binding sites. These hydrolases can thus not be used as transferases. Hydrolases with covalent acylenzyme intermediates can, however, always be used as transferases. The general mechanism for the kinetically controlled synthesis of condensation products catalyzed by transferases and hydrolases, derived from these considerations is given in Fig. 2.

MAXIMUM YIELD OF CONDENSATION PRODUCT

The highest concentration of the condensation product is obtained at the kinetically controlled maximum (Fig.1.). Then the rate of product formation equals the rate of hydrolysis of this product.

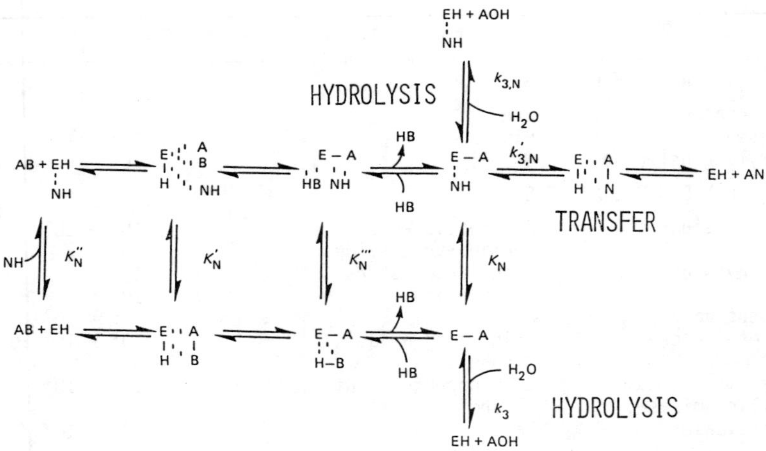

Fig.2. A general mechanism for the hydrolase/transferase catalyzed synthesis of condensations products - as peptides - in the group transfer reaction AB + NH → AN + BH. Dots indicate non-covalent binding. For overlapping binding sites of AB and NH the intermediates AB ··· EH ··· NH and HB ··· E- A ··· NH cannot be formed.

The following relation for the concentration of the condensation product at this maximum can be derived (Kasche, 1986).

$$[AN]_{max} = \frac{v_{AN}}{v_{AN} + v_{AOH}} \cdot \frac{\left(\frac{k_{cat}}{K_m}\right)_{AB,\ NH}}{\left(\frac{k_{cat}}{K_m}\right)_{AN,\ NH}} \cdot [AB] = \alpha \cdot \beta \qquad (2)$$

where the apparent (k_{cat}/K_m)-values apply for the disappearence of the substrate given in the subscript at the concentration of NH at this maximum. The quantity α depends on the deacylation rates and nucleophile binding constants given in Fig. 2. It increases with $(k_3' / k_3)_{app}$ in Eq.(1). Both α and β depend on whether the activated substrate and the nucleophile have overlapping or non-overlapping binding sites. With overlapping binding sites NH is a competitive or non-competitive inhibitor for AB, with non-overlapping binding sites a mixed inhibitor. Eq.(2) shows that the maximum yield in the kinetically controlled synthesis depends on the properties of the enzyme. This relation can be used to analyze the influence of other parameters that may influence the yield as substrate activation, pH, temperature, ionic strength, etc.

NUCLEOPHILE BINDING BEFORE DEACYLATION OF THE ACYL-ENZYME

Evidence for the nucleophile binding to the acyl-enzyme prior to deacylation can be derived from the determination of $(k_3'/k_3)_{app}$ for different nucleophiles. Such experiments must be performed with the same nucleophile content at otherwise identical experimental conditions. For peptide synthesis both the P_1' and P_2'

position can be varied in these experiments. Some experimental data that indicate nucleophile binding to the acyl-enzyme are given in Table 2.

TABLE 2 The apparent ratio of the transferase to hydrolase rate $(k_3'/k_3)_{app}$ for some serine proteinases using different nucleophiles

Enzyme	P_1'	P_2'	$pK_{NH,P_1'}$	$(k_3'/k_3)_{app}$	ref.
β-Trypsin	H-Val	- NH_2	7.75	650	a
	H-Val	- OMe	7.63	100	a
α-Chymotrypsin	H-Ala	- NH_2	8.02	2.400	a
	H-D-Ala	- NH_2	8.20	10	a
	H-Arg	- NH_2	7.5	58.000	b
	H-Arg	- OMe	-	2.500	b
	H-Ala	- NH_2	8.02	2.000	b
	H-Ala	- OMe	-	< 100	b

a) Riechmann and Kasche, 1985;
b) Schellenberger and Jakubke, 1985.

Without such a binding $(k_3'/k_3)_{app}$ should only depend on the pK-value for the nucleophile. The differences in the ratio given in the table cannot be explained by the small differences in the pK-values. The data also show that the P_2'-position also influences the binding of the nucleophile. Amides in P_2' give higher ratios and yields than ester groups. This has also been observed in the kinetically controlled synthesis of human insulin from porcine insulin (Obermeier and Seipke, 1984). The data also show that these investigations can be used analyze the P_1' and sequence specificity of the proteinases.
The results in Table 2 cannot alone be used to distinguish whether the nucleophile and activated substrate binding sites are overlapping or non-overlapping.

OVERLAPPING BINDING SITES; EQUILIBRIUM OR STEADY-STATE IN NUCLEOPHILE BINDING ?

For overlapping binding sites of the activated substrate and the nucleophile the steady-state concentration of E-A··NH (Fig.2.) is given by the relation

$$\frac{d\,[E\text{-}A\cdot\cdot NH]}{dt} = k_{1,N}\,[NH][E\text{-}A] - k_{-1,N}\,[E\text{-}A\cdot\cdot NH] - v_{AN} - v_{AOH} = 0 \quad (3)$$

where the v:s are the initial transferase and hydrolase product formation rates from E-A··NH and $K_N = (k_{-1,N}/k_{1,N})$. When E-A and E-A··NH are in equilibrium the sum of first two terms in Eq.(3) equals zero. Then $(v_{AN} + v_{AOH})$ must vanish in initial rate measurements. This is never observed and we must therefore generally assume that the nucleophile binding to the different enzyme forms in this case cannot be described as an equilibrium process.

In Fig. 3 $(k_3/k_3')_{app}$ and the initial rates, v_{AN} and v_{AOH} are given for the synthesis of ampicillin from D-phenylglycine-methyl ester at different concentrations of 6-aminopenicillanic acid. In this case the linear dependence of $(k_3/k_3')_{app}$ on 6-APA also indicates nucleophile binding to the acylenzyme (Kasche, Haufler and Zöllner, 1984). A maximum in v_{AN}, is only possible when nucleophile binding is a steady-state not an equilibrium process. This applies also when 6-APA and the activated substrate have non-overlapping binding sites (see below).

This finding has general implications for the analysis of two substrate systems by initial rate kinetics. Generally equilibrium in substrate binding is

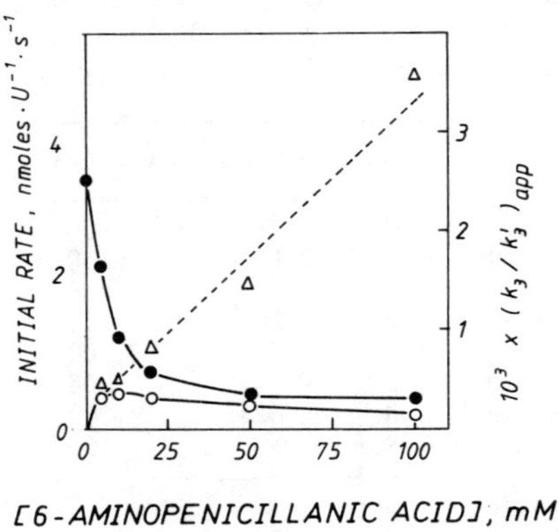

Fig. 3. Initial rates of product formation and the apparent ratio of the hydrolase to transferase rate for the penicillin amidase (E.Coli) catalyzed hydrolysis of 10 mM D-phenylglycin-methyl ester at different nucleophile concentrations

● $\underline{v}_{D\text{-phenylglycine}}$; O $\underline{v}_{Ampicillin}$; Δ $(\underline{k}_3/\underline{k}_3')_{app}$.

assumed (Laidler and Bunting, 1973; Fink and Bender, 1969). This can lead to errors in the derivations of mechanisms based on initial rate studies (Ingraham and Makower, 1954). The simultaneous measurements of all substrates and products using HPLC, as has been done here, yields more information for analysis of different mechanisms, than the measurement of only one substrate or product.

YIELD DETERMINING FACTORS

Substrate activation: In Table 3 the maximum yields in the kinetically controlled synthesis of ampicillin are given as function of substrate

TABLE 3 Influence on substrate activation on the maximal yield in the kinetically controlled synthesis of ampisillin catalyzed by penicillin amidase (E.Coli) at pH 7.5 25°C (initial substrate concentration 200 mM)

Activated substrate		Maximum yield (% of Nucleophile)
A	- B	
D-Phenyl-Glycine	- OMe	47
	- OMe + 20 % MeOH	56
	- OEt	36
	- NH_2	34
	- Gly	20

activation, esters generally give higher yields than amides. The factor α in Eq.(2) was, however, larger for the latter than for the esters. The higher yields must therefore be due to differences in the $(\underline{k}_{cat}/\underline{K}_m)$-values in the factor β in Eq.(2). For the small esters 6-APA was found to be a mixed inhibitor. For the amides it was found to be a competitive inhibitor. This gives a larger decrease in β than for the small esters at the large nucleophile concentrations used here (> 100 mM). These results also show that the determination of the initial transferase to hydrolase activity $(\underline{k}_3'/\underline{k}_3)_{app}$ alone cannot be used to predict the maximum yields in the kinetically controlled synthesis.

Substrate recycling: The increase in the yield in the presence of MeOH (Table 3) indicates the existence of different S_1' binding sites for -OMe and 6-APA on the penicillin amidase from E.Coli. This was verified by determination of the rate of transesterification with and without a second nucleophile (Table 4).

TABLE 4 The influence of a second nucleophile on the apparent ratio of the transferase to hydrolase rate $(\underline{k}_3'/\underline{k}_3)$ for the first nucleophile

Enzyme	Activated Substrate	First Nucleophile	(M)	Second Nucleophile	(mM)	$(\underline{k}_3'/\underline{k}_3)_{app}$, for Transesterification
Penicillin Amidase	D-Phenyl-	EtOH	4.4	6-APA	0	2
(E.Coli)	Glycine-OMe	EtOH	4.4	6-APA	100	10
α-Chymotrypsin	Ac-Tyr-OEt	MeOH	6.3	$Val-NH_2$	0	24
		MeOH	6.3	$Val-NH_2$	100	26

For the penicillin amidase the second nucleophile increased the rate of transesterification. Thus we may conclude that the increase in ampicillin yield in the presence of MeOH is due to substrate recycling. Substrate that would otherwise have been hydrolyzed are regenerated. Such possible substrate recycling was not observed for α-chymotrypsin. This indicates that the S_1' binding sites on this enzyme for -O-Me and the nucleophile are overlapping.

Ionic-strength: The kinetically controlled synthesis of condensation products involve ions that interact with charged active sites. A large ionic strength dependence in yields is thus expected and has been observed in different studies (Kasche, 1986; Kasche, Haufler and Zöllner 1984; Kasche and coworkers, 1983). More studies on the specific influence of different ions are, however, required.

Temperature: The yields have been found to decrease with temperature in the kinetically controlled synthesis of semisynthetic penicillins and peptides (Kasche, 1986). A rational explanation for this has only been given in qualitative terms. The same applies for the pH-dependence.

Organis solvents: Systematic studies on the influence of organic solvents on the yields in kinetically controlled synthesis of condensation products have not yet been published.

CONCLUSIONS

In this and other studies it has been shown that the mechanism of the kinetically controlled synthesis of peptides with hydrolases as transferases involves the specific binding of the nucleophile to the acyl-enzyme prior to the deacylation. Based on this mechanism a general equation the maximal product yield (Eq.(2)) has been derived.

This has been used for a rational analysis of different factors that influence the yield. It has been shown that the ratio of the initial transferase to hydrolase activity $(k_3'/k_3)_{app}$ alone cannot be used to predict the yields. The nature of the inhibition of the activated substrate by the nucleophile must also be considered.

The results also show the limitations in the assumption of equilibrium binding of substrates in the kinetic analysis of two substrate enzymes using initial rate measurements.

The general conclusion for kinetically controlled peptide and penicillin synthesis derived here that with the same amount of enzyme maximal yields are obtained much earlier than for equilibrium controlled reactions seem to contradict the results presented for the semisynthesis of insulin given by Morihara (1987). There very high ratios in [NH]/ [insulin] (10^2- 10^3) were used, compared to the corresponding rations of 1-10 used here.
For the general mechanism (Fig.2) with steady state in the nucleophile binding, maxima in the rate of transferase product formation (Fig.3) and in the maximal yields (Eq.(2)) are expected as a function of nucleophile concentration. Maxima in yields have only been observed in the semisynthesis of insulin at the large [NH]/[activated substrate] ratios used by Morihara (1987). At these ratios the rate of product formation is reduced to the same order of magnitude as in the equilibrium controlled synthesis due to the inhibition by the nucleophile. Thus the apparent contradiction is only caused by the large differences in the [NH]/ [activated substrate] ratios used in these studies.

ACKNOWLEDGEMENT

This work has been supported by the Deutsche Forschungsgemeinschaft (Ka 505/3-1 and 436 BUL-113/20/4) the Bundesminister für Forschung und Technologie (PTB 8478) and Fonds der chemischen Industrie.

REFERENCES

Cook, N.D., and T.J. Peters (1985) Biochim.Biophys.Acta 832, 142-147.

Fink, A.L., and M.L. Bender (1969), Biochemistry 8, 5109-5118.

Fruton, J.S., (1982) Adv.Enzymol.Relat.Areas Mol.Biol. 53, 239-306.

Ingraham, L.L., and B. Makower (1954), J.Phys.Chem. 58, 266-270.

Kasche, V., B. Galunsky, U. Haufler, and R. Zöllner, in Enzyme Technology (Lafferty, R.M. ed.), Springer Verlag, Heidelberg, 1983, p.297.

Kasche, V., U. Haufler, and R. Zöllner (1984), Hoppe Seyler's Z.Physiol.Chem. 365, 1435-1443.

Kasche, V., (1986) Enzyme Microb.Technol. 8, 4-16.

Laidler, K.J., and P.S. Bunting, The chemical kinetics of enzyme Action, Clarendon Press, Oxford, 1973.

Morihara, K., This volume p.

Obermeier, R., and G. Seipke (1984), Process Biochem. 19, 29-33.

Riechmann, L., and V. Kasche (1985), Biochim.Biophys.Acta 830, 164-172.

Schellenberger, V., and H.D. Jakubke (1986), Biochim.Biophys.Acta 832, 54-60.

Attempts for Quantifying the S′ Subsite Specificity of Serine Proteases

V. Schellenberger, U.-K. Hanisch, A. Könnecke and H.-D. Jakubke

Karl Marx University, Section of Biosciences, Department of Biochemistry, Leipzig, German Democratic Republic

ABSTRACT

The characterization of the S' subsite specificity of serine proteases has been studied on the basis of kinetically controlled synthesis reactions. Starting from suitable chromophoric acylamino acid donor esters the partitioning of the resulting acyl-enzyme intermediate between amino acid-derived nucleophiles and water was determined spectrophotometrically. Graphical analysis of the experimental data gave evidence of a extremely high specificity of chymotrypsin towards arginine in the P_1' position. A preference of trypsin for the same residue - less strongly but quite similar to chymotrypsin - could be established. In general, amides of amino acids show a greater nucleophile efficiency than the corresponding methyl esters. From trypsin-catalyzed reaction it can be deduced that dipeptide amides are fixed within the active site in a conformation which favours the formation of the peptide bond.

KEYWORDS

S' characterization of serine proteases; enzymatic peptide synthesis; partitioning of acyl-enzymes; chymotrypsin; trypsin.

INTRODUCTION

The usual approach for the elucidation of interactions between the binding subsites of the protease active site and the amino acid residue of peptide substrates involves both crystallographic studies and kinetic analysis of the enzyme-catalyzed reaction (Baumann, Bizzozero and Dutler, 1973; Bizzozero, Baumann and Dutler, 1975; Kurachi, Powers and Wilcox, 1973; Morihara and Oka, 1973; Segal, 1972). According to the notation of Schechter and Berger (1967) the scissile bond of the peptide substrate is located across to S_1 to S_1' subsites with its C-terminal side occupying the S_1'-S_n' area and the N-terminal site occupying S_1 to S_n. In contrast to the mostly well-known primary specificity of the serine proteases the interactions of the P' amino acid residues of suitable peptide substrates have been investigated less extensively (Bauer, 1976; Baumann, Bizzozero and Dutler, 1970;

© 1987 Pergamon Journals Ltd.
Printed in Great Britain

Bizzozero, Baumann and Dutler, 1982; Morihara and Oka, 1977). Since the same structural features which influence the rate of the hydrolytic cleavage of a peptide bond are involved in the formation of this bond, a systematic approach to the elucidation of the S' subsite specificity can therefore be conceived on the basis of the reverse reaction. The different reactivity of nucleophilic amino components in synthesis reactions should better reflect the S' subsite specificity of proteases than enzymatic hydrolysis of peptide substrates since the kinetic parameters are determined to a great extend by the P' region of the substrates. On practical importance is the fact that studies based on synthesis reactions do not require complete artificial peptide substrates. Furthermore, peptide-derived nucleophiles containing amino acid sequences according to the substrate specificity of the protease under investigation can be involved in the studies. In the present paper we describe a spectrophotometric assay for the characterization of the S' subsite specificity of chymotrypsin and trypsin.

RESULTS AND DISCUSSION

The Partition Constant as Efficiency Parameter

Protease-catalyzed peptide synthesis can be carried out either as an equilibrium controlled or a kinetically controlled process (Jakubke, Kuhl and Könnecke, 1985). The equilibrium controlled formation of peptide bonds represents the reverse of hydrolysis reaction for which the protease evolved. Such an equilibrium controlled process does not allow a spectrophotometric determination of the partitioning ratio since non-equilibrium concentrations of the peptide product cannot be obtained. In contrast to the equilibrium approach, the non-equilibrium or kinetic approach is based on the rate-limiting partitioning of rapidly formed acyl-enzyme intermediates between water and the amino acid- or peptide-derived nucleophile (Scheme 1):

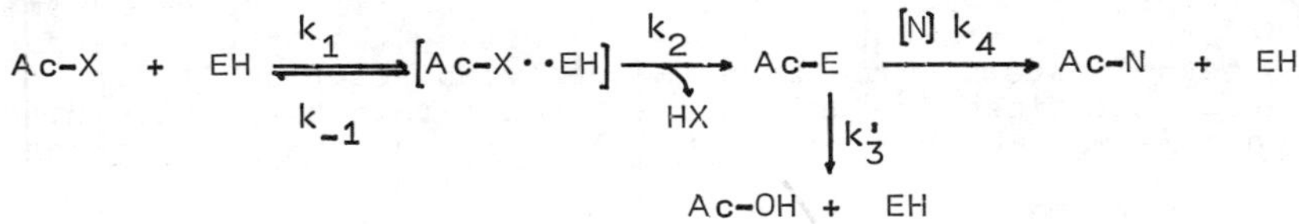

In this case the yield of peptide product (Ac-N) is influenced by the properties of the protease acting as a transferase in this reaction. This process requires the use of weakly activated acyl donor derivatives, but higher peptide product yields may be obtained much more rapidly in comparison to equilibrium controlled processes. From Scheme 1 it follows that the departure of the leaving group (HX) from the active site is the first step of the acyl transfer. The resulting acyl-enzyme is partitioned between aminolysis and hydrolysis reactions forming the target product and the commonly more stable acylamino acid (Ac-OH). After the complete consumption of the ester substrate (Ac-X) the product will be hydrolyzed by the protease. Therefore, the reaction should be stopped at the so-called kinetic maximum. The partitioning ratio between aminolysis and hydrolysis depends primarily on the ability of the protease to bind the amino acid-derived nucleophile and to exclude water, which competes with the nucleophile in the deacylation step. For this reason, the kinetic approach fulfils the requirements for investigations to quantify the S' subsite specificity of serine proteases.
As mentioned above, the success of kinetically controlled peptide

synthesis is decisively determined by the specificity of the serine protease for the nucleophile used as amino component in this process. The different nucleophile reactivity against acyl-enzyme intermediates can be both expressed by k_4 (Fastrez and Fersht, 1973) and by the rate constant ratio k_4/k_3 (Petkov and Stoineva, 1984). According to Scheme 1, the competitive partitioning of the acyl-enzyme intermediate via hydrolysis and aminolysis is expressed by the reaction velocities v_3 and v_4, respectively, as follows:

$$\frac{v_3}{v_4} = \frac{\frac{d[Ac\text{-}OH]}{dt}}{\frac{d[Ac\text{-}N]}{dt}} = \frac{k_3'[Ac\text{-}E]}{k_4[N][Ac\text{-}E]} \qquad (1)$$

This expression can be simplified to Eqn. 2

$$\frac{v_3}{v_4} = \frac{d[Ac\text{-}OH]}{d[Ac\text{-}N]} = \frac{k_3'}{k_4\,[N]} = \frac{p}{[N]} \qquad (2)$$

with $p = k_3'/k_4$. According to our definition (Könnecke and others, 1984) the partition constant p corresponds to the nucleophile concentration necessary for $d[Ac\text{-}OH] = d[Ac\text{-}N]$. The partition constant is a helpful parameter for synthesis planning and, furthermore, it allows the characterization of the S' subsite specificity of serine and cysteine proteases.

Eqn. 2 is strictly valid only for the case where p is independent of the nucleophile concentration. Deviations from linearity cannot be interpreted in terms of Scheme 1 (Brubacher and Bender, 1966; Fink and Bender, 1969; Riechmann and Kasche, 1984, 1985). Brubacher and Bender (1966) were the first who discussed the formation of an acyl-enzyme-nucleophile complex prior to aminolysis. The mechanism in Scheme 2 can be used to explain that water can hydrolyze both the acyl-enzyme with and without bond nucleophile (Riechmann and Kasche, 1984):

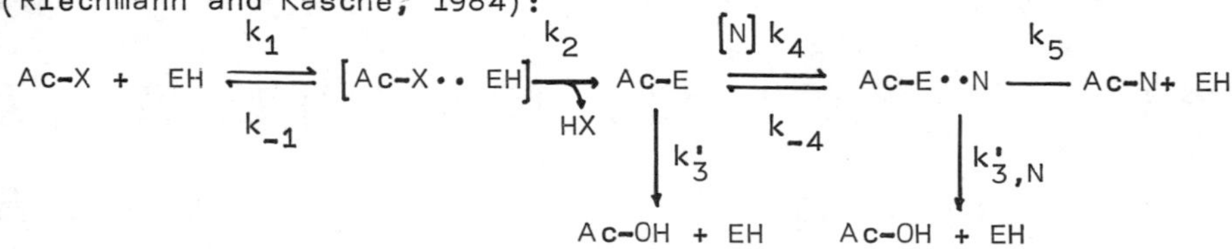

According to Riechmann and Kasche (1984) the relation between the velocities of hydrolysis and synthesis can be deduced from Scheme 2 as follows:

$$\frac{v_3}{v_4} = \frac{(k_3' k_{-4})/k_4 + [N]\,k_{3,N}'}{[N]\,k_5} \qquad (3)$$

Taking into account the definition of the partition constant as given in Eqn. 2, we get the following dependence of p on the different rate constants:

$$p = \frac{k_{3,N}'}{k_5}\,[N] + \frac{k_3'\,k_{-4}}{k_4\,k_5} \qquad (4)$$

The linear dependence of p on[N] allows the determination of the quotient $k'_{3,N}/k_5$ which corresponds to the ratio between the formation of hydrolysis product and peptide product when the condition of saturation of the acyl-enzyme with nucleophile is met. Therefore, it can be deduced from this value whether the nucleophile is fixed within the active site in a conformation favourable for peptide bond formation or not. The term $k'_3\, k_{-4}/k_4\, k_5$ reflects the affinity between the acyl-enzyme and the nucleophile in the sence of the Michaelis constant.
Since Eqn. 4 allows the characterization of the S' subsite specificity of serine and cysteine proteases, it is necessary to find a quick and easy to handle assay for the determination of p. The commonly used method for the analysis of enzymatic peptide synthesis is HPLC which is usable for this purpose. The partitioning of acyl-enzymes can also be determined spectrophotometrically as recommended by Petkov and Stoineva (1984).

Spectrophotometric Determination of the Partition Constant

The spectrophotometric analysis of the products from the partitioning of an acyl-enzyme is based on differences in the absorption coefficients. In contrast to the approach of Petkov and Stoineva (1984) based on Phe(NO_2)-containing peptide substrates, we have preferred amino acid esters as acyl donors bearing chromophoric residues in the α-amino group. Beside the maleyl-residue (Schellenberger and Jakubke, 1985) for the substitution of the α-amino group of phenylalanine methyl ester we have also used the o-nitrophenylsulfenyl residue as chromophor for arginine alkyl ester in trypsin-catalyzed reactions.
Using a constant concentration of the appropriate acyl donor and various concentrations of the nucleophile under investigation every experiment of a series is started by addition of the enzyme. The time course of absorbance is monitored until a constant level is reached corresponding to the time of complete consumption of the acyl donor ester E.
The difference of absorbance ΔA is given by

$$\Delta A = [H]\varepsilon_H + [P]\varepsilon_P - [E]\varepsilon_E \tag{5}$$

where H = hydrolysis product and P = peptide product. Since the absorption is measured after the consumption of the acyl donor ester it holds

$$[P] = [E] - [H] \tag{6}$$

From Eqns. 5 and 6 it follows that

$$\Delta A = [E](\varepsilon_P - \varepsilon_E) + [H](\varepsilon_H - \varepsilon_P) \tag{7}$$

Assuming that the term $[E](\varepsilon_P - \varepsilon_E)$ is constant within on series we get

$$\Delta A' = \Delta A - [E](\varepsilon_P - \varepsilon_E) = \Delta A - \Delta A_0 \frac{\varepsilon_P - \varepsilon_E}{\varepsilon_H - \varepsilon_E} \tag{8}$$

where ΔA_0 is the absorption difference in the absence of nucleophile.
From Eqns. 7 and 8 it follows that

$$\Delta A' = [H](\varepsilon_H - \varepsilon_P) \tag{9}$$

$\Delta A'$ is directly proportional to the concentration of hydrolysis product formed. The relations between the absorbance terms are illustrated schematically in Fig. 1.

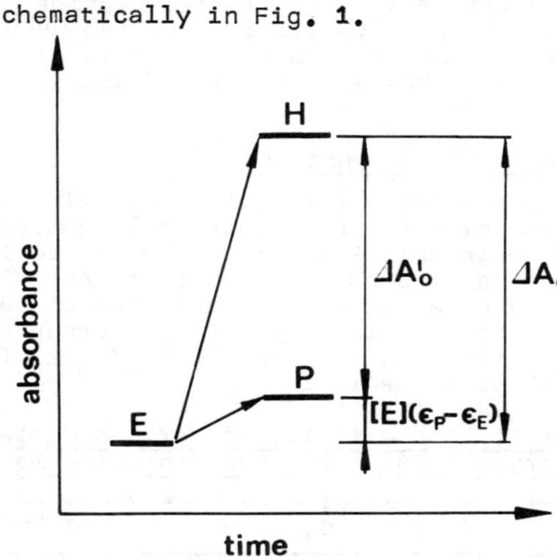

Fig. 1. Schematic representation of the relations between the absorbance terms.

Starting from Eqns. 2,6 and 9 one obtains after rearrangement and simplification:

$$\Delta A' = [E](\varepsilon_P-\varepsilon_E)/([N]/p + 1) \quad (10)$$

Replacing $[E](\varepsilon_P-\varepsilon_E)$ by $\Delta A'$ gives finally

$$\frac{1}{\Delta A'} = \frac{1}{p\,\Delta A'_0}[N] + \frac{1}{\Delta A'_0} \quad (11)$$

According to Eqn. 11, plots of $1/\Delta A'$ versus [N] yield straight lines with an intercept on the x-axis = − p when p is independent of [N Otherwise, the p values can be drawn from the plot separately for each nucleophile concentration. Typical plots of $1/\Delta A'$ versus [N] for p values which are both independent and dependent on [N] are shown in Fig. 2.

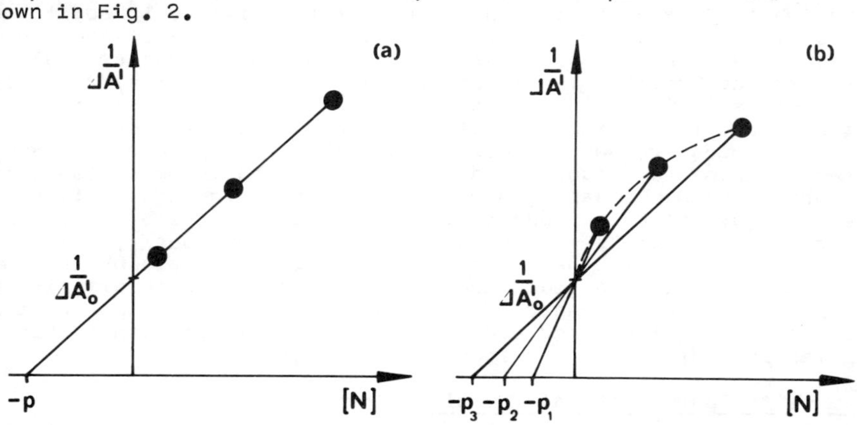

Fig. 2. Plots of 1/ A' versus the nucleophile concentration for p values independent on [N] (a) and dependent on [N] (b).

If the absorption coefficient of the formed peptide product is unknown, it is possible to calculate ε_p from the HPLC chromatogram of the product mixture. More information on this assay including a computer program in BASIC has been published elsewhere (Schellenberger and Jakubke, 1985).

S' Subsite Specificity of α-Chymotrypsin

Using maleyl-L-phenylalanine methyl ester as acyl donor we have studied the partition of the appropriate acyl-chymotrypsin between amino acid-derived nucleophiles and water by monitoring the changes in the absorbance at 260 nm. The substrate ester Mal-Phe-OMe is soluble in buffers of alkaline pH. Experimental series were run as described above. The calculations were performed using a computer program. The quotients $k'_{3,N}/k_5$ and $k'_3\, k_{-4}/k_5\, k_4$ for various nucleophiles are listed in Table 1.

TABLE 1 Reactivity and Affinity of Various Nucleophiles in Chymotrypsin-Catalyzed Reaction With Mal-Phe-OMe

Nucleophile	$k'_{3,N}/k_5$	$k'_3\, k_{-4}/k_4\, k_5$
H-Arg-NH_2	0	0.96
H-Leu-NH_2	0	20
H-Arg-OMe	0.67	22
H-Ala-NH_2	0.06	27
H-Gly-NH_2	0.33	132
H-Leu-OMe	-	>600
H-Ala-OMe	-	>600

From these data significant differences in nucleophile reactivity can be recognized. The extremely high specificity of chymotrypsin for arginine residues in the P'_1 position has never been observed before. Furthermore, within the series of amides of neutral amino acids, a preference for residues with hydrophobic side chains could be established. These results are in accordance with data of Petkov and Stoineva (1984) although the acyl-enzymes used in their experiments show a significantly smaller reactivity. In addition, chymotrypsin-catalyzed hydrolysis studies (Bauer, 1976; Baumann, Bizzozero and Dutler, 1970; Bizzozero, Baumann and Dutler, 1982; Morihara and Oka, 1977) demonstrated a preference for hydrophobic side chains in P'_1 position confirming the expected correlation between specificity in synthesis and hydrolysis reactions. Amides of amino acids are better nucleophiles than the appropriate methyl esters. Bizzozero, Baumann and Dutler (1982) could establish that a hydrogen bond between O(Phe-41) of chymotrypsin and NH(P'_2) of the substrate plays an important role for an effective binding of the peptide substrate within the active site. Obviously, this hydrogen bond is also necessary for an efficient nucleophile binding in synthesis reactions. Despite the observed high affinity of H-Arg-OMe to the acyl-enzyme, a non-efficient orientation based on the missing hydrogen bond seems to be the main reason that this complex is hydrolyzed to a significant extent.

S' Subsite Specificity of Trypsin

The S' subsite specificity of bovine trypsin was determined using the described spectrophotometric assay. o-Nitrophenylsulfenyl-L-arginine methyl ester was used as a chromophoric acyl donor. The compound is readily soluble in buffers and the change of absorbance at 440 nm was used for monitoring the ratio between hydrolysis and synthesis product. The results are compiled in Table 2. In contrast to the chymotrypsin-catalyzed reactions all nucleophiles investigated form complexes with Nps-Arg-trypsin which are sensitive to hydrolysis to a different extent. It can be deduced from the values in Table 2 that the amides of arginine, norvaline and the two dipeptides are fixed within the active site of trypsin in a conformation which favours peptide bond formation. In these cases the degree of hydrolysis is rather low.

TABLE 2 Reactivity and Affinity of Various Nucleophiles in Trypsin-Catalyzed Reaction With Nps-Arg-OMe

Nucleophile	$k'_{3,N}/k_5$	$k'_3\ k_{-4}/\ k_5\ k_4$
H-Arg-NH_2	0.05	66
H-Lys(Ac)-Leu-NH_2	0.06	7
H-Lys-Leu-NH_2	0.15	22
H-Nva-NH_2	0.09	30
H-Leu-NH_2	0.27	72
H-Phe-NH_2	0.33	50
H-Met-NH_2	0.34	12
H-Thr-NH_2	0.42	90
H-Lys-NH_2	0.44	104
H-Val-NH_2	0.46	129
H-Met-OMe	1.60	180
H-DLeu-NH_2	1.90	290
H-Glu-NH_2	2.20	175
H-Asp-NH_2	2.30	274
H-Ala-OMe	3.20	450
H-Leu-OMe	-	$>$ 700

It seems likely that the correct nucleophile binding within the active site is sterically hindered for most of the nucleophiles. Especially, the significant difference in the reactivity between the amides of valine and norvaline strenghtens this thesis. The sterical hindrance can be overcome only by the formation of additional binding sites of the nucleophile to the S' area of the active site such as P'_2-S'_2 interactions for the dipeptide amides or special ionic interactions with the guanido group of arginine. The low binding tendency of amino acid methyl ester should be interpreted by similar reasons as described for chymotrypsin. The interpretation of the differences in reactivity of the various nucleophiles in accordance with the three-dimensional structure of the active site will undoubtely lead to a better understanding of the enzyme-substrate interactions in trypsin-catalyzed reactions.

CONCLUSIONS

Kinetically controlled peptide synthesis is a useful tool for the characterization of the S' subsite specificity of serine proteases. In comparision to investigations based on hydrolysis of peptide substrates the approach described here affords less time consuming synthesis of artificial peptide substrates. Furthermore, the advantage of this method lies in the fact that, in principle, peptide-derived nucleophiles can be studied which contain amino acid sequences according to the primary specificity of the protease. Since hydrolysis and aminolysis of the acyl-enzyme intermediate are influenced in the same manner, inhibition of the acylation process of the enzyme does not influence the results of this assay. Using suitable chromophoric acyl donor esters of amino acids which met the primary specificity of the protease it is possible to measure the relation between hydrolysis and peptide bond formation very quickly. The quantitative treatment provides useful data on the interactions of amino acid and peptide derivatives with the S' subsites of serine and also cysteine proteases. Beside the importance of the results for the prediction of favourable experimental conditions in protease-catalyzed peptide synthesis they allow simple estimation of nucleophile binding parameters. In addition, the approach should be a helpful tool for efficiency control in the search for better proteases for enzymatic peptide synthesis either by screening or by protein design.

REFERENCES

Bauer, C. A. (1976). Biochim. Biophys. Acta, 438, 495-501.

Baumann, W. K., S. A. Bizzozero, and H. Dutler (1970). FEBS Lett. 8, 257-260.

Baumann, W. K., S. A. Bizzozero, and H. Dutler (1973). Eur. J. Biochem., 39, 381-391.

Bizzozero, S. A., W. K. Baumann, and H. Dutler (1975). Eur. J. Biochem., 58, 167-176.

Bizzozero, S. A., W. K. Baumann, and H. Dutler (1982). Eur. J. Biochem., 122, 251-258.

Brubacher, L. J., and M. L. Bender (1966). J. Am. Chem. Soc., 88, 5871-5899.

Fastrez, J., and A. R. Fersht (1973). Biochemistry, 12, 2025-2034.

Fink, A. L., and M. L. Bender (1969). Biochemistry, 8, 5109-5118.

Jakubke, H.-D., P. Kuhl, and A. Könnecke (1985). Angew. Chem. Int. Edn. Engl., 24, 85-93.

Könnecke, A., V. Schellenberger, H.-J. Hofmann, and H.-D. Jakubke (1984). Pharmazie, 39, 785-786.

Kurachi, K., J. C. Powers, and P. E. Wilcox (1973). Biochemistry, 12, 771-777.

Morihara, K., and T. Oka (1973). FEBS Lett., 33, 54-56.

Morihara, K., and T. Oka (1977). Arch. Biochem. Biophys., 178, 188-194.

Petkov, D. D., and I. B. Stoineva (1984). Biochem. Biophys. Res. Commun., 118, 317-323.

Riechmann, L., and V. Kasche (1984). Biochem. Biophys. Res. Commun., 120, 686-691.

Riechmann, L., and V. Kasche (1985). Biochim. Biophys. Acta, 830, 164-172.

Schechter, I., and A. Berger (1967). Biochem. Biophys. Res. Commun., 27, 157-167.

Schellenberger, V., and H.-D. Jakubke (1985). Biochim. Biophys. Acta, 869, 54-60.

Segal, D. M. (1972). Biochemistry, 11, 349-356.

The Use of Enzymes in the Semi-synthesis of Insulin Analogues

H. Zahn, H.-G. Gattner, W. Fischer,
A. Jonczyk and A. Wiese

Deutsches Wollforschungsinstitut, Veltmanplatz 8,
D-5100 Aachen, Federal Republic of Germany

ABSTRACT

Trypsin catalysed couplings with the carboxyl components Boc_2-des-octapeptide (B23-30) insulin and des-Ala(B30) insulin under thermodynamic control with synthetic peptides, modified in the range B24-26 and amino acid derivatives were performed.
Exchange of the C-terminal Alanine (B30) by transpeptidation, with Threonine-tertbutylester and protected hydrazides for preparation of human insulin and elongated insulin is described.
The N-terminal elongation of insulin was achieved by kinetically controlled peptide synthesis with amino acid ester derivatives.
As carboxyl component for unspecific enzymes we used despentapeptide (B26-30) insulin under conditions where no secondary hydrolysis of the insulin molecule occures.

KEYWORDS

Insulin analogues, enzyme-assisted peptide synthesis, semisynthesis, selective enzymatic hydrolysis.

INTRODUCTION

Protein semisynthesis uses fragments of natural proteins as ready-made intermediates in the construction of proteins of novel structure with synthetic peptides or amino acid derivatives. First step of this procedure is the selective enzymatic hydrolysis of the natural protein fragments for combination with the synthetic part. The requirement of protecting groups is diminished widely compared to total synthesis, especially in the enzyme-catalyzed peptide bond formation no side chain protection is necessary.
Homogeneous analogues and derivatives are essential for studies which try to relate the primary and tertiary structure of insulin and specific molecular interactions. From the complex primary structure of the two-chain protein it is obvious that synthesis is a very demanding task. Hence, the development of procedures for systematic alterations

AB VOL65-L

© 1987 Pergamon Journals Ltd.
Printed in Great Britain

of the native molecule appeared to be a necessary and attractive alternative.
Biological activity and aggregation of insulin are known to be closely related to the C-terminal region of insulin. The B-chain sequence B24-26, Phe-Phe-Tyr, has been remained invariant during evolution and therefore the residues have been assumed to be part of the receptor binding region of the hormone (Pullen, 1976). Chemical modification of specific amino acid residues in this region should give more information about the role of these amino acids play regarding the structure-activity,receptor binding and aggregation behaviour of the insulin molecule.
Insulin has two peptide bonds susceptible to tryptic digestion, Arg-Gly (B22-23) and Lys-Ala (B29-30). Recently Inouye (1979) described the trypsin assisted semisynthesis of human insulin from des-octapeptide (B23-30) insulin (DOI) and a synthetic octapeptide. Using the procedure of Inouye we prepared a series of modified insulins.

RESULTS

SYNTHESIS WITH DES-OCTAPEPTIDE (B23-30) INSULIN (DOI)

The shortened insulin derivative is obtained by tryptic digestion. To prevent inter-and intramolecular selfcondensation during trypsin catalyzed peptide synthesis the α-amino groups should be protected, mainly with tert-butyloxycarbonyl-group (Boc-) (Kubiak, 1986a, 1986b). TABLE 1 shows a series of peptides coupled with Boc_2-DOI by trypsin catalysis.

TABLE 1 Coupling of Synthetic Peptides to Boc_2-des-octapeptide (B23-30) insulin

H-Gly·Phe-Phe-Tyr-Thr·Pro-Lys-Ala-Thr (24 25 26)	%	Activity %	Ref.
—— Leu ——————	25	12	(Tager, 1980)
——— Leu —————	10	3	(Gattner, 1980)
—— Leu-Leu ————	25	0,6	(Jonczyk, 1981)
—— Tyr ——————	10	7	
——— Tyr —————	50	5	
———— Phe ————	10	30	(Glatter, 1983)
—— D-Phe —————		140	(Kobayashi, 1982)
——— D-Phe ————	40	4	

The first example of human diabetes mellitus due to a defective insulin molecule was reported by Olefsky (1980) and Tager (1979). With small amounts of insulin extrated from a specimen of the patient's pancreas a leucine-for-phenylalanine substitution could be localized to either position B24 or B25. Reduced receptor binding and biological activity and, even more, inhibition of insulin action make this insulin extremely important for molecular mechanism of insulin action.

SHORTENED INSULINS

One of the aims of structure-function studies with biolocigally active substances is to find the minimal molecular size essential for full potency. Des-pentapeptide (B26-30) insulin (DPI) prepared by a limited peptic hydrolysis of insulin has a potency of about 20% (Gattner, 1975). Although lacking the last five amino acids of the B-chain DPI still retains the principal structural features of the intact insulin molecule (Bi. 1984; Stuart, 1984). The absence of the five C-terminal amino acids of the B-chain leaves the molecule with the full structural properties required for recognition and binding of the receptor. These are, however, impaired by the presence of the charged carboxylate group in an otherwise hydrophobic surface region. The negative effect is neutralized in the amidated molecule (Table 2).

TABLE 2 Synthesis and Biological Activities of shortened Insulin Analogues

C-component	Coupled Peptide	Yield %	Biol. Activity %
Boc_2 DOI	Gly-Phe-Phe-OH	20	15 ± 3
MSC^2 DOI	———————— OBu^t	31	70 ± 10
Boc_2 DOI	———————— NH_2	33	105 ± 14
Boc_2 DOI	———————— $Tyr-NH_2$	48	110 ± 16
Boc_2 DOI	———— Leu-OH	30	3 ± 1
MSC_2 DOI	———— $Leu-OBu^t$	45	12 ± 2
Boc_2 DOI	———— $Leu-NH_2$	46	20 ± 3

Reaction conditions: 60 mg Boc_2 DOI + 60 mg peptide in 80% DMF/ glycerol (1:1), 20 mM $CaCl_2$, 6 mg trypsin, pH 6,5, 16 h, RT.

EFFECT OF AMINE COMPONENT

Coupling of synthetic peptides by trypsin to DOI gave variing yields of insulin analogues. To investigate the influence of different amine components on the tryptic mediated synthesis we compared the yields with a varity of amino acid derivatives (TABLE 3).

For an equilibrium-dependent semisynthesis of insulin-derivatives one has to work with high concentration of the components. Addition of organic solvents increases the solubility of the reactants which governs the extend of peptide bond formation by mass action. Further addition of organic solvents shifts the equilibrium more to synthesis than to hydrolysis (Homandberg, 1978).

Among the tested organic cosolvents mixtures of dimethylsulfoxide or dimethylformamide with 1,4-butanediol up to conantrations of 80% (V/V) proved to be most suitable for insulin semisynthesis.

REPLACEMENT OF ALA (B30)

In the treatment of diabetes mellitus insulin preparations derived from porcine or bovine insulin have generally been used. Bovine, procine and human insulin exhibit minor differences with respect to their amino acid composition, the difference between human and porcine insulin being confined to a single amino acid in that the B30 amino acid of human insulin is threonine whereas that of porcine is alanine. However it could be argued that the ideal insulin preparation for treatment of human beings would be an insulin having exactly the same chemical structure as that of human insulin.

First for replacemants of Ala(B30) we have used des-Ala(B30) insulin prepared by limited hydrolysis of insulin with Carboxypeptidase A

TABLE 3 Coupling of Amino acid Derivatives to Boc_2-des-octapeptide (B23-30) Insulin: 60 : 1 (molar ratio) 60% dimethylformamide, pH 6,5, 20 h, RT

Carboxyl Component	Amine Component	Yield %
Boc_2-DOI	Tyr-OMe	10
"	Leu-OMe	17
"	Trp-OMe	19
"	Ala-OBu^t	25
"	Phe-OMe	26
"	Leu-NH_2	43
"	Lys(Boc)-OMe	52
"	Glu(OMe)-OMe	63
"	H_2N-NH-Boc$^+$	50

$^+$ Boc_2-DOI : H_2N-NH-Boc = 1 : 120

(Schmitt, 1978). In a condensation reaction we have prepared a series of insulin analogues modified in position B30 (TABLE 4). For coupling des-Ala(B30)insulin (procine) with a 100-fold molar excess of threonine methyl ester was incubated with trypsin in 60% dimethylformamide at pH 6,5. Maximum yield was achieved after 30 min.

TABLE 4 Trypsin-catalyzed Coupling of Amino Acid Esters to Des-Ala (B30) Insulin: 1 : 70, 50 % Dimethylformamide, pH 6,5, 16 h

Amine Component	Yield %	Amine Component	Yield %
Thr-OMe	85	Ala-OMe	65
Thr-NH_2	80	Ala-OBu^t	40
Tyr-OMe	70	Ala-OBzl	45
Trp-OMe	70	Ala-NH_2	50
p-NO_2-Phe-OMe	40	His-OMe	40
3-NH_2-Tyr-OMe	35	Cys-OMe	25
Glu(OMe)-OMe	90	O_3^-	
		H_2N-NH-Boc	80

The study of transformation of porcine to human insulin gave soon the indication that the condensation could be replaced by a transpeptidation using the components insulin, threonine ester and trypsin. In this case the preparation of the intermediate des-Ala(B30) insulin could be omitted. The transpeptidation runs more slowly than the condensation and therefore side reactions at Arg(B22) should be taken in account. Beside the formation of human insulin mono ester the amount of desoctapeptide (B23-30) insulin can be considered as a criterion of the reaction conditions (Jonczyk, 1981; Markussen, 1985) Example: 200 mg porcine insulin and 200 mg threonine-tertbutylester were dissolved in 2 ml 60 % dimethylformamide - 0,25 M Tris/acetate buffer. The pH was adjusted to 6,6 and 20 mg trypsin were added. After 4 h the reaction was stopped by addition of 50 % acetic acid. Yield of human insulin: 55 - 60 %. Recently various groups have independently found an essential improvement of the one step-transformation of porcine to human insulin (Obermeier, 1981; Markussen, 1981; Rose, 1983).

ENZYMATIC-CHEMICAL SYNTHESES

The optimal and most advantageous ration of chemical and enzymatic steps is determined by the amino acid composition and the sequence of the peptide to be produced. Peptide α-hydrazides wether substituted or not in the hydrazo function are useful intermediates in the semissnthesis. The α-hydrazide is the required starting material for the azide method for the formation of peptide bonds. The amino group of the peptide hydrazide should be reversible protected. Recently some hydrazides of peptides were prepared by the use of enzymes (Canoca-Davis, 1980; Jones, 1982).
In TABLE 5 some hydrazides of insulin are presented as suitable starting compounds for further chemical syntheses.

TABLE 5 Proteinase-catalysed Preparation of Insulin-hydrazide Derivatives

Hydrazides	Preparation
DAI-NH-NH Boc	insulin + H_2N-NH-Boc trypsin, pH 4,5
DOI-NH-NH-Boc	insulin + H_2N-NH-Boc trypsin, pH 6,5 - 7,0
DPI-NH-NH-Boc	DPI + H_2N-NH-Boc chymotrypsin, pH 5,5
DPI-NH-NH-Ph	DPI + H_2N-NH-Ph papain, pH 4,5
Human insulin (B30)Thr(Bu^t)-NH-NHBoc	insulin + Thr(Bu^t)-NH-NH-Boc + trypsin, pH 6,2

DAI : des-Ala(B30) insulin
DPI : des-pentapeptide (B26-30) insulin
DOI : des-octapeptide (B23-30) insulin

Transpeptidations with Boc-hydrazine were performed with 200-fold excesses of reagent. By variation of pH the position of reaction is determined.
To demontrate the enzymatic-chemical approach des Ala(B30) insulin (DAI) was elongated with disulfide-A-chain ($A(SS)_2$). The amino groups of DAI-NH-NH-Boc were protected with methylsulfonylethyl-oxycarbonyl-groups (Msc-). Then the hydrazide was deprotected by treatment with trifluoroacetic acid and an azide coupling was performed with $A(SS)_2$. The yield of DAI-$A(SS)_2$ was 12 %. The product has a biological activity of 40 % (corrected value according to molecular weight). For syntheses of C-terminal elongation human insulin transpeptidation with Thr(Bu^t)-NH-NHBoc as performed in high yield (about 90 %).

ENZYME-CATALYZED SYNTHESES WITH DES-(B26-30) PENTAPEPTIDE INSULIN (DPI)

So far enzyme-catalyzed peptide syntheses with insulin were performed with the specific trypsin at Arg (B22) and Ly(B29). We have investigated the use of unspecific enzymes in the semisynthesis of insulin analogues (TABLE 6). These enzymes split under conditions of hydrolysis insulin in small peptides. By selected conditions - pH and organic cosolvent - secondary hydrolysis should be prevented. As carboxyl component des-pentapeptide insulin was used which is prepared by limited hydrolysis of insulin by pepsin (Gattner, 1975).

TABLE 6 Enzyme-catalysed Peptide Syntheses with Des-pentapeptide (B26-30) Insulin (DPI)

Components	Enzyme	Product	Yield %
DPI + H_2N-NH-Boc	Papain	DPI-NH-NH-Boc	40
DPI + H_2N-NH-Ph	Papain	DPI-NH-NH-Ph	80
DPI + MeOH	Papain	DPI-OMe	50
DPI + H_2N-NH-Boc	Chymotrypsin	DPI-NH-NH-Boc	75
DPI + Ala-NH_2	Chymotrypsin	DPI-Ala-NH_2	70
DPI + Leu-NH_2	Thermolysin	DPI-Leu-NH_2	80

Papain: 0,1 M acetate buffer, pH 4,5, MeOH, pH 4,5 - buffer (1:1) for esterification

Chymotrypsin: 85% 1,4-butanediol 15% water, pH 5,5

Thermolysin: 25% DMF, 75% Tris/acetate buffer, pH 6,5

N-TERMINAL ELONGATION

The described enzyme-catalyzed reactions with insulin were performed with insulin or insulin derivatives as carboxyl components. For N-terminal elongation we used the kinetic controlled approach: Carboxyl component is esterified, pH of the reaction medium is high (9-10), enzyme concentration is low and the reaction times are short. To elongate insulin at Gly(A1) we protected Phe(B1) by reaction of insulin with tert-butyl-amino-carbonate (Harris, 1984) at pH 6,5 in 5o% dioxane. Conditions for trypsin catalysed synthesis were: 10 µMol B1-Boc-insulin, 1mMol Boc-ArgOMe dissolved in 20 ml 20% dimethylformamide pH 9,5, + 0,02 µMol trypsin in 10 µl 10^{-3} N HCL. After 3 min the reaction is stopped by lowering the pH to 2-3. Yield of Arg(A0)-insulin was about 30%.
Unprotected insulin was substituted to about 75% at Gly(A1) and to 25% at Phe(B1) with a total yield of 40%.
Unfortunately the common N-protected amino acid esters have a low solubility in water. This disadvantage can be overcome by addition of organic solvents but at high concentrations proteases are often inactivated. An alternative is the use of hydrophylic N-protected amino acid esters (TABLE 7).
We have used two types of esters: 1. 4-sulfobenzylester (Hubbuch, 1980). 2. α-glycerylester (Wiese, 1985).

TABLE 7 Chymotrypsin-catalyzed Coupling of Water-soluble Esters to Insulin

Ester	Yield of elongated Insulin %
Boc-Gly-Phe-OBzl-SO_3^- Na^+	30
Boc-Leu-OBzl-SO_3^- Na^+	25
Z-Phe-OGlyc	35

1 m-mol ester, 10 µ-mol, insulin, 0,02 µ-mol chymotrypsin, pH 9,5, 10 min

Also in chymotrypsin-catalyzed couplings Gly(A1) is the prefered position of reaction.

REPLACEMENT WITHIN THE CHAINS

As mentioned above the first step in the use of enzymes in semisynthesis is the definite hydrolysis of a peptide bond of a protein to get a pure ready-made fragment for further chemical or enzymatic reactions.

We want to report our results with the B-chain to replace Tyrosine (B16). This amino acid is an invariant residue, it is involved in dimer formation and constitutes part of the receptor binding region. Di-S-sulfonated B-chain was prepared by oxidative sulfitolysis of insulin. Reduction of the S-sulfonate groups followed by iodine oxidation in 50% acetic acid gave the disulfide-B-chain. For protection the amino groups were blocked by reaction with methylsulfonylethyloxycarbonyl-succinimid ester (Msc-ONSu) and the carboxyl groups were esterified with BF_3/methanol. The protected B-chain was then incubated with proteinase K at pH 4,5 in 4M urea which resulted in a nearly quantitative hydrolysis of the Leu(B15)-Tyr(B16) peptide bond. Purification at this and all following reaction steps was achieved by ion exchange chromatography on SP-Sephadex at pH 3. Tyrosine (B16) was then removed by Edman degradation. Phenylalanine was coupled as Boc-Phe-ONSu to Leu(B17). After treatment with trifluoroacetic acid the free amino group of Phe(B16) could be coupled with the only free carboxyl group of Leu(B15) by the DCC/HOBt-method. Amino and carboxyl protecting groups were removed in one step by treatment with O,1 N NaOH at 2°C for 30 min. During the whole synthesis amino- and carboxylcomponent were connected by a disulfide bridge which facilitates purification and coupling steps.

REFERENCES

Bi, R.C., Z. Dauter, E. Dodson, G. Dodson, F. Giordano and C. Reynolds (1984). Biopolymers, 23, 391-395

Canova-Davis, E. and F. Carpenter (1980). In D. Brandenburg and A. Vollmer (Eds.). Insulin-Chemistry, Structure and Function of Insulin and related Homones. Walter de Gruyter, Berlin, pp 107-115

Gattner, H.-G. (1975). Hoppe-Seyler's Z. Physiol. Chem., 356, 1397-1404

Gattner, H.-G., W. Danho, C. Behn and H. Zahn (1980). Hoppe-Seyler's Z. Physiol. Chem., 361, 1135-1138

Glatter, U. (1983). Dissertation RWTH Aachen.

Harris, R.B. and J.B. Wilson (1984). Int. J. Peptide Protein Res., 23, 55-60

Homandberg, G.A., J.A. Mattis and M. Laskowski, Jr., (1978). Biochemistry, 17, 5220-5227

Hubbuch, A., R. Bindewald, J. Föhles, V.K. Naithani and H. Zahn (1980). Angew. Chem., 92, 394

Inouye, K., K. Watanabe, K. Morihara, Y. Tochino, T. Kanaya, J. Emura and S. Sakakibara (1979). J. Am. Chem. Soc, 101, 751-752

Inouye, K., K. Watanabe, Y. Tochino, T. Kanaya, M. Kobayashi and Y. Shigeta (1981). Experientia (Basel), 37, 811-813

Jonczyk, A., L.M. Keefer, V.K. Naithani, H.-G. Gattner, P. DeMeyts and H. Zahn (1981). Hoppe-Seyler's Z. Physiol. Chem., 362,557-561

Jonczyk, A. and H.-G. Gattner (1981). Hoppe-Seyler's Z. Physiol. Chem., 362, 1591-1598

Jones, R.M.L. and R.E. Offord (1982). Biochem. J., 203, 125-129

Kobayashi, M., S. Ohgaku, M. Iwasaki, M. Maegawa, Y. Shigeta and K. Inouye (1982). Biochem. Biophys. Res. Commun., 1o7, 329-336.

Kubiak, T. and D. Cowburn (1986a). Int. J. Peptide Protein Res., 27, 514-521

Kubiak, T. and D. Cowburn (1986b). Biochem. J., 234, 665-670

Markussen, J. (1981) In L. Kerp, K.J. Schlüter and K.G. Petersen (Eds.). Neue Insuline, Freiburger Graphische Betriebe. pp 38-44

Markussen, J., K.H. Jørgensen, A.R. Sørensen and L. Thim (1985). Int. J. Peptide Protein Res., 26, 70-77

Obermeier, R., G. Seipke and U. Grau (1981). In L. Kerp, K.J. Schlüter and K.G. Petersen (Eds.). Neue Insuline, Freiburger Graphische Betriebe. pp 33-37

Olefsky, J.M., M. Seakow., M. Kobayashi, O.G. Kalterman, H. Tager, B. Given, D. Baldwin, M. Mako, J. Markese, A.H. Rubenstein and R. Poucher (1980). In D. Brandenburg and A. Wollmer (Eds.), Insulin-Chemistry, Structure and Function of Insulin and related Homones, Walter de Gruyter, Berlin, pp 393-402

Pullen, R.A., D.G. Lindsay, S.P. Wood, I.J. Tickle, T.L. Blundell, A. Wollmer, G. Krail, D.Brandenburg, H. Zahn, J. Gliemann and G. Gammeltoft (1976). Nature (London), 259, 369-373

Schmitt, E.W. and H.-G. Gattner (1978). Hoppe-Seyler's Z. Physiol. Chem., 359, 799-802

Stuart, D., D. Jinbi, M. Lou, L. You, R. Todd, L. Li, J. Wang and D. Liang (1984). Sci. Sin. B., 27, 572-582

Tager, H., B. Given, D. Baldwin, M. Mako, J. Markese, A. Rubenstein, J.M. Olefsky, M. Kobayashi, O. Kaltermann and R. Poucher (1979). Nature (London), 281, 122-125

Tager, H., N. Thomas, R. Assoian, A. Rubenstein, M. Saekow, J. Olefsky and E.T. Kaiser (1980). Proc. Natl. Acad. Sci. USA, 77, 3181-3185

Wiese, J. (1985). Dissertation RWTH Aachen

INHIBITION OF PROTEASES

Enthalpy–Entropy Compensation Phenomena in Serine (Pro) Enzyme: Inhibitor Complex Formation

G. Amiconi*, P. Ascenzi*, M. Bolognese†, M. Guarneri‡ and E. Menegatti‡

*C.N.R., Center for Molecular Biology, Department Biochemical Sciences, I Rome University, 00185 Rome;

†Department Genetics and Microbiology, Pavia University, 27100 Pavia;

‡Department Pharmaceutical Sciences, Ferrara University, 44100 Ferrara, Italy

ABSTRACT

Thermodynamics concerning the interaction of the bovine basic pancreatic trypsin inhibitor as well as bovine and porcine pancreatic secretory trypsin inhibitor with some serine (pro)enzymes has been obtained at pH 8.0 (I = 0.1 M) between 5°C and 45°C. Results have been analyzed by the enthalpy-entropy compensation plot (ΔH^o versus ΔS^o) and linear relationships obtained, suggesting the involvment of protein conformation and possibly hydration in the process(es) of protein-protein interaction(s).

KEYWORDS

Serine (pro)enzymes; macromolecular inhibitors; (pro)enzyme:inhibitor complexes; thermodynamics; enthalpy-entropy compensation phenomena.

INTRODUCTION

It is becoming increasingly apparent that non-active-site regions of enzymes behave as chemodynamical machines capable of coupling the active-site reactivity with the molecular chaos of the environment. In other words, both chemical and thermal properties of the medium are conveyed to the enzyme by various ways, such as direct collision of environmental solvent and solute particles, binding relaxation of ions and so on (Samogyi, Welch and Damjanovich, 1984). The idea of enzymes as energy transducers is not new, and at the root of the coupling of chemical processes to the properties of the environment has been proposed (Lumry and Rajender, 1970) the compensation of enthalpy changes by entropy changes in the Gibbs energy changes associated to the chemical event. Therefore, in order to investigate effects of enthalpy-entropy compensation phenomena in protein-protein interactions, thermodynamics concerning the formation of adducts of serine (pro)-enzymes with the bovine basic pancreatic trypsin inhibitor (BPTI) as well as bovine (b) and porcine (p) pancreatic secretory trypsin inhibitor (PSTI) has been determined at pH 8.0 (I = 0.1 M) between 5°C and 45°C. Results are in line with the idea that no unique source may be established to account for the enthalpy-entropy compensation phenomenon observed in complex formation between serine (pro)enzymes and macromolecular inhibitors; both conformational rearrangement and partecipation of water molecules (i.e., hydration) appears to play a relevant role.

© 1987 Pergamon Journals Ltd.
Printed in Great Britain

MATERIALS AND METHODS

Crystalline bovine trypsinogen (trypsinogen) was purchased from Merck AG (Darmstadt). Bovine α-chymotrypsin (α-chymotrypsin) was obtained from Sigma Chemical Co. (St. Louis, U.S.A.). Bovine β-trypsin (β-trypsin) was purified from commercial enzyme preparations (from Sigma Chemical Co., St. Louis, U.S.A.) according to Luthy and co-workers (1973). Human Lys^{77}-plasmin (Lys^{77}-plasmin) was prepared from the Lys^{77}-zymogen as previously detailed (Ascenzi and co-workers, 1985). Human urinary kallikrein (h u kallikrein) was prepared according to Geiger, Stuckstedte and Fritz (1980). Neuraminidase-treated β-kallikreins A and B from porcine pancreas (β-kallikreins A and B) were isolated as previously reported (Fiedler and co-workers, 1981). BPTI was obtained from Lepetit S.p.A. (Milano, Italy). Both b and p PSTI were prepared as previously detailed (Burck, 1971; Menegatti and co-workers, 1982). The characterization of (pro)enzymes and inhibitors has been previously reported (Antonini and co-workers, 1983; Ascenzi and co-workers, 1985; Burck, 1971; Fiedler and co-workers, 1981; Geiger, Stuckstedte and Fritz, 1980; Luthy and co-workers, 1973; Menegatti and co-workers, 1982, 1984).

Thermodynamics concerning the interaction of serine (pro)enzymes with BPTI as well as b and p PSTI has been determined at pH 8.0 (I = 0.1 M) between 5°C and 45°C (Antonini and co-workers, 1983; Menegatti and co-workers, 1984, 1986).

RESULTS AND DISCUSSION

Under all the experimental conditions, BPTI as well as b and p PSTI binding to serine (pro)enzymes is a simple equilibrium; moreover, association equilibrium constants (K_a) are independent of (pro)enzyme concentration.

Values of $\log K_a$ for inhibitor association to serine (pro)enzymes, obtained between 5°C and 45°C at pH 8.0 (I = 0.1 M), show a linear dependence with respect 1/T (indicating ΔCp^o value, between the complex and its components, equal to 0) and allow the determination of enthalpy changes associated with the formation of protein-protein adducts.

When ΔH^o values are plotted against ΔS^o values, a linear relationship for BPTI as well as b and p PSTI binding to the various serine (pro)enzymes is observed, with a correlation coefficient of 0.968 and 0.993, respectively (see Fig. 1); this leads to guess the existance of an enthalpy-entropy compensation effect in one or more part processes of the protein-protein interaction. Even though the observed correlation for BPTI is less rigorous than that for both b and p PSTI, in both cases the compensation behaviour appears to be a real and not a spurious one. In fact, the slope of the line (the so-called compensation temperature, T_c (Lumry and Rajender, 1970)) is found for both BPTI and PSTI systems to be about 170°K, a value different from the mean experimental temperature. It has been stated (Krug, Hunter and Grieger, 1976) that in the spurious compensation phenomena the propagation of experimental errors will tend to distribute enthalpy and entropy estimates along a line characterized by a slope equal to the harmonic mean of the experimental temperatures.

The observed compensation temperature in the formation of serine (pro)enzyme:inhibitor complexes is different from previous results (Lumry and Rajender, 1970), not only because of the low value (in general between 260°K and 320°K or between 430°K and 525°K) but also of negative relationship; it should however be reminded that literature does not include association of macromolecules. In any case, the T_c value here reported ($T_c \simeq 170$°K; see Fig. 1) supports the Exner's view (Exner, 1964), according to which there is no tendency toward a common compensation temperature.

As a whole, data shown in Fig. 1 are in line with the idea that both water and protein cooperate to produce compensation changes in protein-protein interactions, and suggest that protein conformation and solvent act as a unity generating high free-energy events at the (pro)enzyme active center.

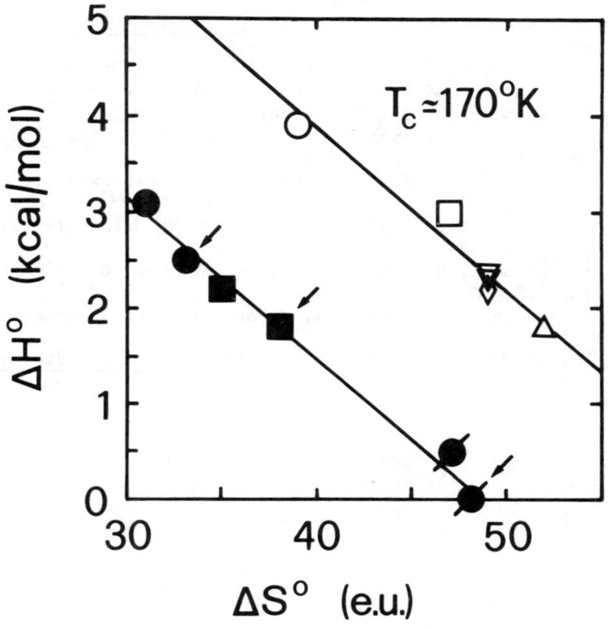

Fig. 1. Compensation plot for the binding of trypsinogen (○, ●), α-chymotrypsin (□, ■), p p β-kallikreins A and B (▽, ▽), Lys^{77}-plasmin (◇), h u kallikrein (△) and β-trypsin (●) to BPTI (open symbols), b PSTI (filled symbols with arrows) and p PSTI (filled symbols without arrows). Straight lines show a slope (i.e., T_c value) of about 170°K. Data were obtained at pH 8.0 (I = 0.1 M). For details, see the text.

Such view agrees with results of X-ray studies for some of the systems examined (Huber and Bennett, 1983) which indicate that: (i) different water shells are in free (pro)enzymes and in their adducts with macromolecular inhibitors; and (ii) structural differences are in (pro)enzyme:inhibitor components of the adducts, compared to free (pro)enzymes and inhibitors, small in the case of active enzymes with respect to trypsinogen.

However, the unusual compensation temperature and the observed negative slope suggest that the behaviour known as enthalpy-entropy compensation should be considered an empirical relationship (Lumry and Gregory, 1986) with a number of distinct, unrelated sources, rather than a general phenomenon.

REFERENCES

Antonini, E., P. Ascenzi, M. Bolognesi, G. Gatti, M. Guarneri, and E. Menegatti (1983). J. Mol. Biol., 165, 543-558.

Ascenzi, P., A. Torroni, E. Menegatti, M. Guarneri, and G. Amiconi (1985). Biochim. Biophys. Acta, 832, 215-219.

Burck, P. J. (1971). Methods Enzymol., 19, 906-914.
Exner, O. (1964). Nature, 201, 488-489.
Fiedler, F., E. Fink, H. Tschesche, and H. Fritz (1981). Methods Enzymol., 80, 493-532.
Geiger, R., U. Stuckstedte, and H. Fritz (1980). Hoppe-Seyler's Z. Physiol. Chem., 361, 1003-1016.
Huber, R., and W. S. Bennett Jr (1983). Biopolymers, 22, 261-279.
Krug, R. R., W. G. Hunter, and R. A. Grieger (1976). J. Phys. Chem, 80, 2335-2341.
Lumry, R., and S. Rajender (1970). Biopolymers, 9, 1125-1227.
Lumry, R., and R. B. Gregory (1986). In G. R. Welch (Ed.), The Fluctuating Enzyme, J. Wyley and Sons, New York., pp. 1-190.
Luthy, J. A., M. Praissman, W. R. Finkenstadt, and M. Laskowski Jr (1973). J. Biol. Chem., 248, 1760-1771.
Menegatti, E., F. Bortolotti, L. Minchiotti, and A. De Marco (1982). Biochim. Biophys. Acta, 707, 50-58.
Menegatti, E., M. Guarneri, M. Bolognesi, P. Ascenzi, and G. Amiconi (1984). J. Mol. Biol., 176, 425-430.
Menegatti, E., M. Guarneri, M. Bolognesi, P. Ascenzi, and G. Amiconi (1986). J. Mol. Biol., 191, 295-297.
Samogyi, B., G. R. Welch, and S. Damjanovich (1984). Biochim. Biophys. Acta, 768, 81-112.

Dipeptidyl Peptidase IV Catalyzed Hydrolysis of Azapeptide Substrate Analogs

H. U. Demuth, U. Neumann and A. Barth
Department of Biochemistry, Division of Biosciences,
Martin Luther University of Halle, Halle (Saale),
German Democratic Republic

ABSTRACT

During hydrolysis of Xaa-Pro-arylamides by dipeptidyl peptidase IV (DP IV) the deacylation of the intermediate formed acylenzyme is rate limiting. On the other hand, some Xaa-Pro derivatives act as irreversible inhibitors, the inactivation event obviously occuring before formation of the Xaa-Pro-enzyme. The design of effective inactivators of this class will require substrat-analog structures which induce rate-limiting acylation during the catalytic cycle.
Azapeptide esters have been found to form stable acyl enzymes with several serine proteases (Powers, 1984). In contrast, azapeptide amides seemed to force a rate-limiting acylation step during their hydrolysis.
Therfore we synthesized several azapeptide derivatives as analogs of DP IV substrates. Stability in aqueous solutions and enzyme-catalyzed hydrolysis were investigated.

KEYWORDS

Dipeptidylpeptidase IV; azapeptide substrates; stability; enzyme catalyzed hydrolysis; inhibitor design.

INTRODUCTION

Dipeptidylpeptidase IV (DP IV, EC 3.4.14.5) is a membrane-bound enzyme anchored by a hydrophobic peptide chain into the bilayer (for a review see Walter, 1980). The enzyme, a serine aminopeptidase with a molecular weight of 230000 and two identical but catalytically inde-

© 1987 Pergamon Journals Ltd.
Printed in Great Britain

pendent subunits exhibits a post- proline- cleaving specificity. Thus, it is believed the enzyme is involved in processing of peptide hormones. The hydrolysis of β-casomorphine and other neuroactive peptides has been reported (Hartrodt, 1982, for review see Barth, 1987). Recently , DP IV participation in T-lymphocyte proliferation was reported (Schoen, 1985).

Specific inhibitors of this enzyme should be an important tool for further investigations of its biological functions.
In a previous paper we introduced N-peptidyl-O-(4-nitrobenzoyl) hydroxylamines as inactvators for serine proteases including DP IV (Fischer, 1983).
Studies on chemistry and biochemistry of these inhibitors demonstrated that DP IV inactivation occufred in a "suicide-type" reaction (see Scheme 1, Demuth,unpublished results).
The acylenzyme generated by these compounds is a simple dipeptidyl enzyme. Therefore the inactivation event must occur before acyl enzyme is formed.

Scheme 1. Proposed mechanism of reaction between DP IV and diacylhydroxylamine substranalogs

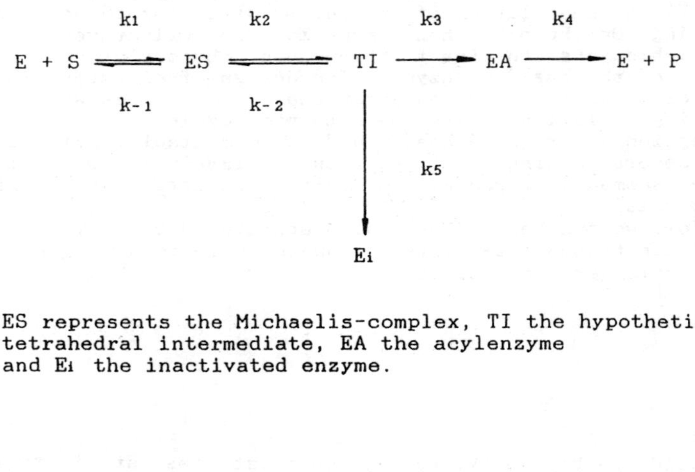

ES represents the Michaelis-complex, TI the hypothetical tetrahedral intermediate, EA the acylenzyme and E_i the inactivated enzyme.

Kinetic analysis of the reaction between enzyme and inhibitor shows enzyme catalyzed hydrolysis as well as inactivation of the enzyme according to scheme 1. The reaction is characterized by a ratio k_4 over k_5 of 30000. Like with other proline-containing substrates of DP IV, the deacylation is rate determining (Barth, 1987).
If the amount of intermediate TI (Scheme 1) which accumulates during the catalytic cycle is responsible for the inactivation path, substrate analogs which have rate-limiting acylation should be better inhibitors. Azapeptide derivatives are known to acylate serine proteases slowly, forming stable carbazyl-enzymes (Elmore, 1968). In this paper we describe, first studies on the reaction of DP IV reaction with azapeptides substrate analogs.

RESULTS AND DISCUSSION

The compounds were synthesized using pyrazolidine as starting material (Scheme 2, analytic data are given in the appendix).

Scheme 2. Synthesis of Azapeptide derivatives

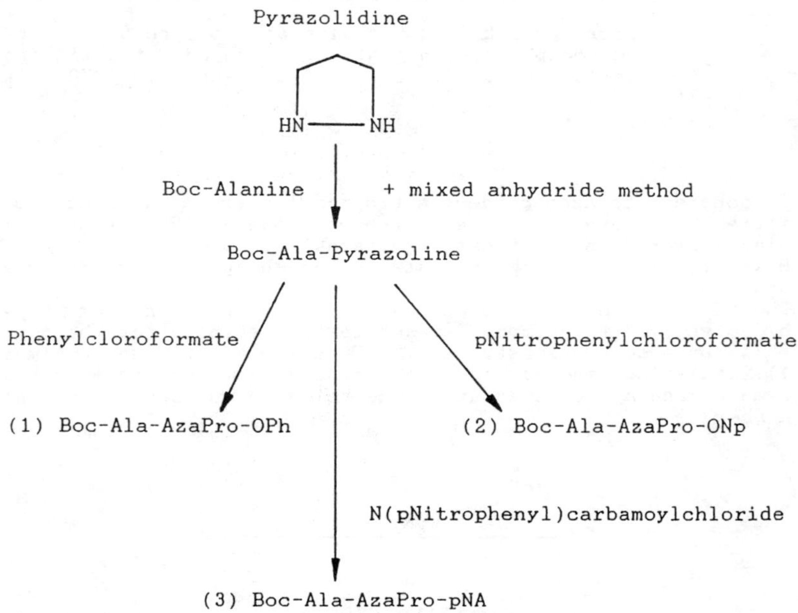

Deprotection of (1), (2) and (3) with HCl in glacial acetic acid gave the appropriate azapeptide hydrochlorides (see appendix).

The degradation of the compounds was studied in aqueous solution at different pH-values by following the change in their UV-vis-spectra. The pseudo-first-order rate constants are given in Table 1.

Table 1. Spontaneous degradation of Azapeptide derivatives in aqueous solution

pH	5.2	6.0	6.5	7.0	7.6
compound		10^3 k_{obs} (sec^{-1})			
Ala-AzaPro-pNA	-	-	-	-	0.18 *
Ala-AzaPro-OPh	-	0.93	2.9	7.6	28
Ala-AzaPro-ONp	29.0+	240+	-	-	-

buffer: pH 5.2 0.04 M sodium acetate, pH 6.0 - 7.6 0.04 M sodium phosphate, ionic strength was adjusted with potassium chloride to 0.125, temp.: 30°C, exept for * 70°C and + 25°C

In contrast to Azapeptide-p-Nitrophenylesters used as active site titrants of several serine proteases (Powers, 1984) Ala-AzaPro-ONp and Ala-AzaPro-OPh exhibit extreme instability under conditions tested. However, Ala-AzaPro-pNA was stable enough to perform enzyme kinetic studies.

The pH-dependence of the second-order rate constant for DP IV catalyzed hydrolysis of the compound resembles those known for the acylation of DP IV by α-carbon substrates. A pK_a-value of 7.13 was estimated (Fig. 1).This value, indicating base assisted nucleophilic attack of the enzyme serine hydroxyl group on the scissile carbonyl atom, is slightly higher

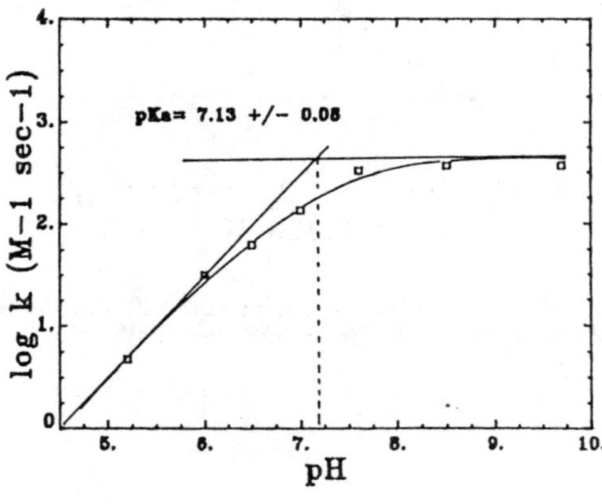

Fig. 1. pH-dependence of the second order rate constant of DP IV catalyzed hydrolysis of Ala-AzaPro-pNA, Eo: 0.31 μM, substrate concentration: 1.2 mM, temp.: 30°C, buffer: pH 9.1 sodium borate, pH 8.5-6.0 sodium phosphate, pH 5.2 sodium acetate 0.04 M, μ: 0.125 maintained with potassium chlorid

then the value of 6.7 found in DP IV catalyzed hydrolysis of several substrates (Barth, 1983). The nature of this enzymic base is unknown, since alkylation of all four histidyl residues per subunit leads to modified enzyme with 80 % remaining activity (Harada, 1984). Therefore the origin of the pH dependence and of the shift still remains unclear
Because DP IV catalyzed substrate hydrolysis follows Michaelis-Menten-kinetics, the lack of saturation in the enzyme-catalyzed hydrolysis of Ala-AzaPro-pNA was unexpected (Fig.2). However studies on azapeptide conformations show alterations in structure due to the semicarbazide system in the peptide backbone (Greenlee, 1984). Those changes seems likely to reduce the affinity of Ala-AzaPro-pNA fo: DP IV. The second order rate constant of 334 $M^{-1} sec^{-1}$ is in fact abou 20000 times smaller then the second order rate constant obtained with Ala-Pro-pNA.
For elastase, differences in reactivity of "normal" and azapeptide substrates up to 40000 are reported. This poor acylating activity of azapeptide substrates is commonly discussed in electronic and steric terms (Powers, 1984).
In carbazates, the carbonyl carbon is less nucleophilic due to of possible resonance delocalisation of electrons on the attached nitrogene. Second, is the α-nitrogene in azapeptides closer to trigonal than to tetrahedral as the α-carbon would be. Therefore the steric arrangement of parts of the substrate molecule at the cleavage site is different from that in α-carbon substrates. The resulting conformation should be different from the

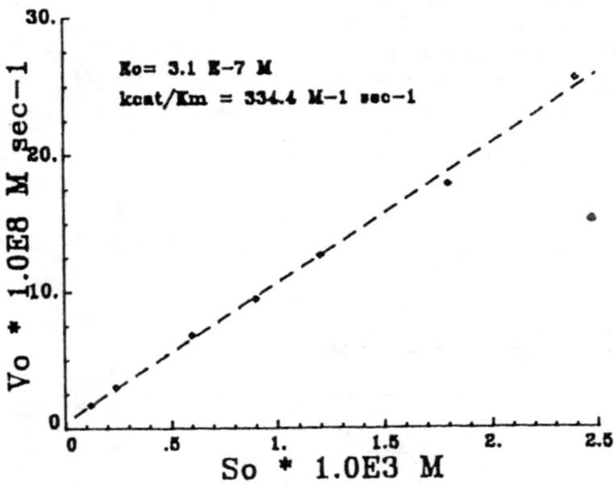

Fig. 2. DP IV catalyzed hydrolysis of Ala-AzaPro-pNA dependence of the reaction rate from substrate concentration
pH 7.6, 0.04 M sodium phosphate buffer, μ: 0.125
temp.: 30°C, Eo: 0.31 μM

optimal conformation nessesary for effective acylation and deacylation of the enzyme.
In order to evaluate the influence of the ester derivatives which were

synthesized the following experiment was designed: DP IV was incubated with the anilide Ala-AzaPro-pNA at pH 6.0 and with the esters Ala-AzaPro-OPh at pH 6.0 and Ala-AzaPro-ONP at pH 5.2. After one minute incubation time, aliquots of the solution were withdrawn and added to thermally equilibrated UV-cells containing buffer and the substrate Gly-Pro-pNA. The increase in absorbance with time due to DP IV catalyzed liberation of p-nitroaniline was monitored at 390 nm.
In those experiments where DP IV was preincubated with the azapeptide ester derivatives, the absorbance change was nonlinear in time, exhibiting a lag-phase after addition of the preincubated enzyme to the substrate solution. Obviously, the enzyme is becoming reactivated. The data fit a model of the type Absorbance = A + Bt + C * e^{-kt} where the exponential term represents the increase of activity during the lag-phase, and B the zero-order rate after complete reactivation of the enzyme.
The reactivation rate constant was not affected by different concentrations of enzyme, azapeptide, or substrate (Table 3).
Furthermore it was identical for the two different esters. But when the enzyme was incubated with the anilide substrate Ala-AzaPro-pNA, full activity was observed from the beginning of the experiment.

Table 3. Reactivation of DP IV after preicubation with azapeptide substrates

Testsubstrate: Gly-Pro-pNA, incubation see text, conditions see Fig. 2

Enzyme (μM)	Azapeptide (mM)	Testsubstrate (mM)	10^3 kobs (sec^{-1})
	Ala-AzaPro-ONp		
0.93	1.5	0.4	5.1
0.47	2.2	0.4	4.8
0.47	1.5	1.2	4.6
	Ala-AzaPro-OPh		
0.47	1.9	1.2	4.75
0.47	1.9	0.4	4.6
0.93	1.9	0.4	4.3
0.47	3.8	0.4	4.73
	Ala-AzaPro-pNA		
0.47	2.2	0.4	-
0.47	4.4	0.4	-

From these results, we believe that the reactivation process is water or hydroxid-ion catalyzed hydrolysis of carbazyl-DP IV (see Scheme 3). For the deacylation of several carbazyl-subtilisins, halflives between 1 and 60 minutes are reported (Gupton, 1984),thus in the same range as our reactivation rate constants.

Scheme 3. Proposed reaction mechanism between DP IV and azapeptide ester substrate analogs

$$E + S \rightleftharpoons ES \rightleftharpoons TI \longrightarrow EA \xrightarrow[\text{slow}]{} E + P$$

EA represents the carbazyl enzyme

If the formation of carbazyl-DP IV were rate determining in the incubation, most of the enzyme would be free and no lag-phase would be seen. This is in fact the case after preincubation of DP IV with Ala-AzaPro-pNA.
From the outcome of these experiments, we conclude that DP IV catalyzed hydrolysis of Ala-AzaPro-pNA proceeds with rate determining acylation, while hydrolysis of the esters has deacylation as rate-determining step.
Therefore currently efforts are being undertaken to synthesize N-Azapeptidyl-O-Acylhydroxylamines as possible potent DP IV inhibitors.

APPENDIX

The substrate analo g compounds used for kinetic studies were obtained according to Scheme 1 after removal the Boc-protecting group in glacial acetic acid / HCl.
The course of the synthesis was monitored using Proton-NMR and elemental analysis.

<u>Alanyl-AzaProlyl-Phenylester</u> * HCl (Ala-AzaPro-OPh) Fp. 121-122 °C
MW. 299.76 Anal. Calcd.: C:52.09 H:6.05 N:14.02
Found : C:51.34 H:6.09 N:13.91
<u>Alanyl-AzaProlyl-pNirophenylester</u> * HCl (Ala-AzaPro-ONp) Fp. 133-134 °C
MW. 344.73 Anal. Calcd.: C:45.29 H:4.97 N:16.25
Found : C:45.05 H:5.04 N:16.71
<u>Alanyl-AzaProlyl-pNitroanilide</u> * HCl (Ala-AzaPro-pNA) Fp. 168-169 °C
MW. 343.77 Anal. Calcd.: C:45.42 H:5.23 N:20.37
Found : C:44.31 H:5.54 N:19.86

The dipeptidyl peptidase IV (DP IV) used in this study was prepared according to (Wolf, 1978) and had a specific activity of 33 U/mg against Glycyl-Prolyl-pNitroanilide (Gly-Pro-pNA).

REFERENCES

Powers, J. C., R. Boone, D. L. Caroll, B. F. Gupton, C.-M. Kam, N. Nishino, M. Sakamoto, and P. M. Tuhy (1984). J. Biol. Chem., 259, 4288-4294

Walter, R., W. H. Simmons, and T. Yoshimoto (1980). Mol. Cell. Biochem., 30, 111-127

Hartrodt, B., K. Neubert, G. Fischer, H. U. Demuth, T. Yoshimoto, and A. Barth (1982). Pharmazie, 37, 72-73

Barth, A. and R. L. Schowen (Eds.), "Dipeptidyl Peptidase IV", VEB Deutscher Verlag der Wissenschaften, Berlin, 1987

Schoen, E., H. U. Demuth, A. Barth, and S. Ansorge (1984). Biochem. J., 223, 255-258

Fischer, G., H. U. Demuth, and A. Barth (1983). Pharmazie, 38, 249-250

Elmore, D. T., and J. J. Smyth (1968). Biochem. J., 107, 103-107

Barth, A., and G. Kempter (1983). Wiss. Z. Martin-Luther-Univ. Halle-Wittenberg, 4, 3-56

Harada, M., B. Y. Hiraoka, K. M. Fukasawa, and K. Fukaswa (1984). Arch. Biochem. Biophys., 234, 622-628

Greenlee, W. J., E. D. Thorsett, J. P. Springer, and A. A. Patchett (1984). Biochem. Biophys. Res. Commun., 122, 791-797

Gupton, F. B., D. L. Caroll, P. M. Tuhy, C.-M. Kam, and J.C. Powers (1984). J. Biol. Chem., 259, 4279-4287

Wolf, B., G. Fischer , and A. Barth (1978). Acta biol. med. german., 37, 409-413

Chemical Gene Synthesis and cDNA Cloning in Competition: Cloning of Genes Encoding Proteinase Inhibitors

J. Dodt, C. Bergmann, H. Appelhans, R. Heinzel and H. G. Gassen
Department of Biochemistry, Technical University Darmstadt, Federal Republic of Germany

ABSTRACT

The genes of two proteinase inhibitors of medical interest were cloned and expressed in E. coli. The gene for the thrombin inhibitor hirudin was synthesized chemically via oligonucleotide synthesis according to known amino acid sequence data. Fusing the hirudin gene to the alkaline phosphatase signal peptide of E. coli a high level of expression of biologically active hirudin could be demonstrated. The gene for human antileukoprotease, an inhibitor with antitryptic, antielastolytic and antichymotryptic activity, was cloned via cDNA synthesis knowing only partial protein sequence data. The cDNA sequence revealed the entire amino acid sequence and an additional 25 amino acid leader sequence. Expression of antileukoprotease in E. coli produced little but actively inhibiting protein

KEYWORDS

Gene synthesis; cDNA cloning; gene expression; proteinase inhibitors; hirudin; antileukoprotease.

INTRODUCTION

In recent years oligonucleotide synthesis has attained a high level of automation and has been improved as a fast and reliable method (Hunkapillar and co-workers, 1984). Oligonucleotides are commonly used for gene synthesis, priming and screening procedures, for information manipulation at the DNA level by site directed mutagenesis, or for optimizing cloning and expression constructions by combining for example proper regulatory signals with structural genes. Gene synthesis depends on the fast and reliable determination of entire protein primary structures. However, it is not always possible to obtain this information since often the miniscual amounts of specific available for sequencing is a limitation. With the new automated Edman degradation techniques though, determination of the first 50 - 90 N-terminal amino acids of subnanomole quantities of

© 1987 Pergamon Journals Ltd.
Printed in Great Britain

protein an easier task (Hunkapillar and Hood, 1983; Hunkapillar and coworkers, 1984) and thus, synthesizing gene probes for screening DNA libraries based on partial sequences may be the best strategy to find the desired gene.

In order to give an expression of the work for both the chemical gene synthesis and the cDNA cloning we will present two selected examples from our laboratory namely; The cloning and expression of the genes for two proteinase inhibitors of medical interest (Fritz, 1980). These are hirudin, a polypeptide of 65 amino acid residues, and the most potent natural occuring thrombin inhibitor, consequently a promising candidate for therapeutic application in consumption coagulopathy; and antileukoprotease (ALP), a protein of 107 amino acid residues, which may prove a valuable therapeutic agent in the blocking of excess proteases, elastase for example, in inflammatory processes.

SYNTHESIS AND EXPRESSION OF A HIRUDIN GENE

Synthesis of a Gene Coding for the Thrombin Inhibitor Hirudin

The oligonucleotides for the hirudin gene were designed on the basis of the known amino acid sequence of the protein (Dodt and co-workers, 1984) using codons normally found for highly expressed *E. coli* proteins (Ikemura, 1981) except of the codons for amino acid 9 and 45, these were chosen accordingly to generate *Hpa*I and *Kpn*I sites. The gene including a synthetic Shine-Dalgarno sequence is flanked by an *Eco*RI site at the 5'-terminus and a *Sal*I site at the 3'-terminus. To facilitate total synthesis the gene was subdivided into the three blocks A, B and C equipped with appropriate restriction sites. The coding sequence is flanked by an ATG at the 5'-terminus and TAA and TAG at the 3'-terminus. The Shine-Dalgarno sequence was designed according to given rules (Scherer and co-workers, 1980; Stormo, Schneider and Gold, 1982). This construction allows the expression of the gene as a non-fusion protein as well as a fusion protein, after elimination of the S.-D. sequence by *Hind*III digestion.

Construction of the hirudin gene. The 102 bp fragment A was synthesized from only 2 oligonucleotides by the fill-in method (Rossi, 1982), whereas fragments B (70 bp) and C (84 bp) were prepared from completely overlapping oligonucleotides using the ligation method (Brown and co-workers, 1979). All three fragments were designed with blunt ends containing 3 - 4 bp overhangs beyond the terminal restriction sites to ensure digestion with the corresponding enzymes. Due to the different degrees of phosphorylation, purification and digestion of the ligation products an overall yield of only about 6 % was achieved for fragments B and C. Using the in vitro polymerization method the incorporation of deoxyribonucleoside triphosphates was determined to be a maximum of 70 - 80 % with an overall yield after purification of about 25 %. After subcloning the fragments were joined together, generating the complete gene, and via a three part ligation inserted into an *Eco*RI/*Sal*I digested, dephosphorylated pUC 18 vector. The DNA sequences of both strands were found by Maxam and Gilbert sequencing (1980) to be in complete agreement with the theoretically designed gene.

Expression of the Hirudin Gene in E. coli

For the efficient production of a small polypeptide like hirudin protein folding, disulfide bonding and protection against proteolytic degradation have to be guaranteed. This should be accomplished by co-translational transport of the nascent polypeptide chain

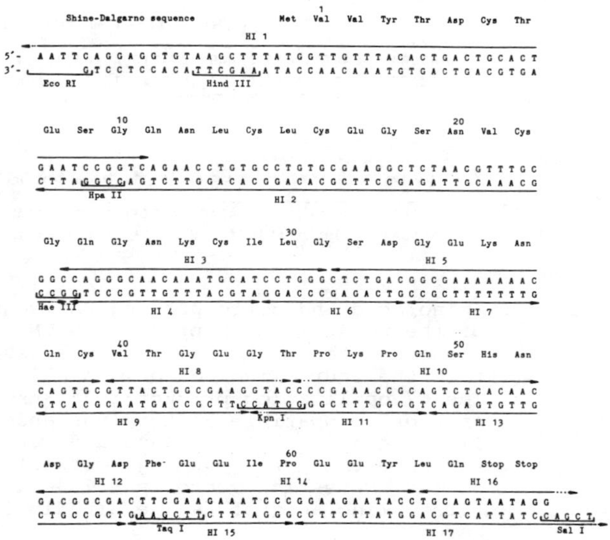

Fig. 1. Amino acid sequence of hirudin and the construction of the synthetic gene. Chemically synthesized oligonucleotides are arrowed; dotted ends indicate oligonucleotides originally carrying a short noncoding sequence adjacent to the restriction site. Fragment A was synthesized from HI 1 and HI 2 by the fill-in method; Fragment B, including oligonucleotides HI 3 - 9, and fragment C including oligonucleotides HI 10 - HI 16, were obtained by the ligation method.

through the cytoplasmic membrane to the periplasmic space using bacterial or eukaryotic signal sequences (Talmadge, Stahl and Gilbert, 1980; Chan and co-workers, 1981). One of periplasmic proteins of E. coli is alkaline phosphatase (phoA) which carries in its precursor form a signal sequence of 21 amino acids (Kikuchi and co-workers, 1981). A DNA coding for the phoA signal peptide was assembled from four chemically synthesized oligonucleotides (Fig. 2) and ligated into the EcoRI/HindIII restriction sites of the expression vector pKK 223-3. A derivative of this vector with the hirudin gene containing HindIII fragment in its correct orientation is designated pHIR21. This plasmid directed a tac-promoter controlled synthesis of a fusion protein starting with the phoA signal peptide at the N-terminus followed by hirudin.

The time course of the expression of hirudin from this construction was followed by a thrombin inhibiting assay. 21 h after induction hirudin activity reached a level of about 350 mIU/1 A 578 unit of cells in total cell extracts and periplasmic fractions (Fig. 3). Proteins and peptides of periplasmic fractions were separated by ion exchange chromatography followed by HPLC. This resulted into two protein peaks (I and II), which were analysed by amino acid and thrombin inhibiting determinations. Protein II indicates that the

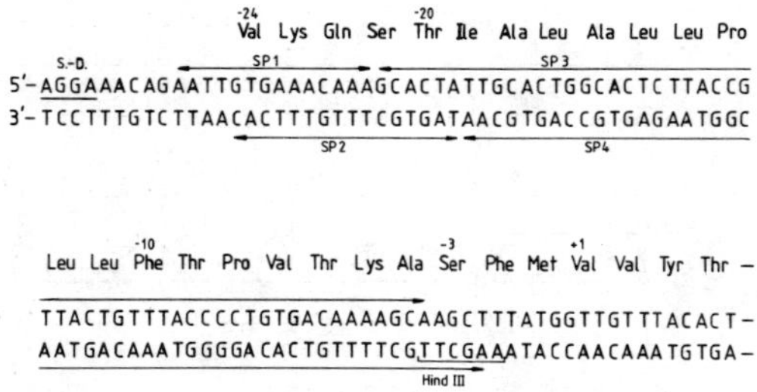

Fig. 2. Amino acid and corresponding DNA sequence of the phoA signal peptide. The DNA sequence of the signal peptide was assembled from the four oligonucleotides SP1 - SP4. The 67 bp ligation product is flanked by an EcoRI compatible 5'-terminus and a Hind III site at the 3'-terminus and was inserted into the corresponding restriction sites of the expression vector pKK 233-3. The first amino acid of native hirudin is numbered 1.

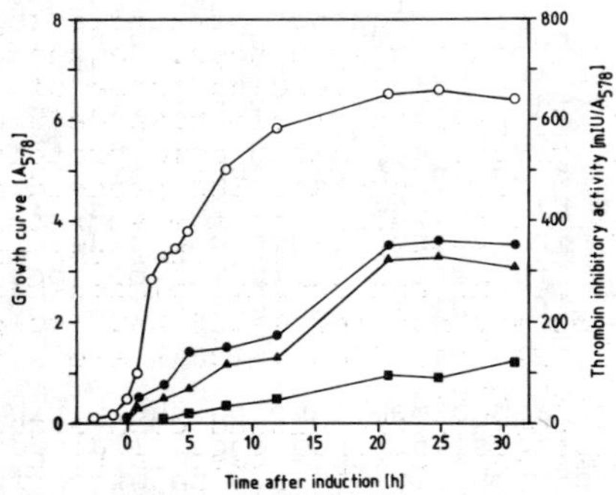

Fig. 3. Thrombin inhibiting activity determinations of E. coli BMH 71-18/pHIR21 after induction of gene expression by IPTG. (○—○) growth curve; Thrombin inhibiting activities in: (●—●) cell extracts, (▲—▲) periplasmic fractions, and (■—■) culture supernatants are given.

signal peptide was correctly processed by the leader peptidase at positions Ala(-4)-Ser(-3) during secretion into the periplasm. Surprisingly, peptide I showed the same amino acid sequence as determined for hirudin except that a sulfated tyrosine residue in position 63 is absent. The specific activity of protein I (550-570 mIU/µg) is identical to the activity of desulfato-hirudin obtained by arylsulfatase treatment of native hirudin from leeches. Thrombin inhibiting activity of protein II was 500 fold less than that for desulfated hirudin and therefore negligible under assay conditions.

Recalculation of thrombin inhibitory activity in periplasmic fractions yielded hence 4 mg desulfated hirudin (protein I) per liter culture. As hirudin is a secretory protein from the salivary glands of the leech the expression of the synthetic hirudin gene fused to the leader sequence may mimic segregation under natural conditions. This indication is supported by the sequence of a cDNA clone of a natural occuring hirudin form which presumably is translated as a precursor (Harvey, 1986).

ISOLATION AND CLONING OF A cDNA CODING FOR ANTILEUKOPROTEASE

From preliminary protein sequence data of the inhibitor two regions for the synthesis of corresponding oligonucleotides were selected. Two oligonucleotide mixtures were synthesized complementary to the mRNA coding for amino acids 79 - 84 (RH1) and 92 - 97 (RH2). The third base degeneracies in amino acid codons were considered and thus the probes contained 16 different species of DNA sequences to ensure that the correct one would be present. RNA isolated from human cervix uterus, (immuno fluorescence investigations of Schill and co-workers (1978) indicate a high concentration of the inhibitor in cervical mucous) was subjected to Northern Blot analysis with RH1 and RH2. The former probe produced a strong signal indicating a mRNA 700 - 800 bases long. Consequently a cDNA library from poly(A) enriched cervix uterus RNA was constructed in pBR322. Colony hybridization screening of the library, consisting of 10000 recombinant clones, revealed several positive clones. One of these containing the largest insert was isolated and designated pRH34. As shown in Fig. 4 it contains the entire coding sequence of antileukoprotease mRNA and additional 5'- and 3'-flanking sequences. Sequence data showed a cDNA of 596 base pairs with an open reading frame of 396 base pairs encoding 132 amino acids. Comparison of the cDNA deduced amino acid sequence with that of the mature 107 amino acid protein from seminal plasma indicate that they are the same. The preceeding 25 N-terminal amino acids are postulated to belong to the secretory part of the protein (amino acids -25 to -1 in Fig. 4). Thus the mature inhibitor molecule of cervical mucous and seminal plasma has 107 amino acids with a calculated molecular mass of 11726 Da. Two possible polyadenylation signals, AATAAA (Proudfoot, 1984) are situated downstream the protein coding sequence.

To investigate whether the cloned cDNA sequence codes for biologically active human antileukoprotease in <u>E. coli</u> the cDNA was subcloned into the expression plasmid pWH701, a derivative of the plasmid pPLc 236 (Remault and co-workers, 1981). By combining synthetic oligonucleotides with antileukoprotease cDNA the coding sequence was cloned behind the P_L-promoter of bacteriophage lambda (Heinzel and co-workers, 1986). The recombinant expression plasmid was propagated in <u>E. coli</u> MC 1061/pRK 248 cIts (Cassadaban and co-workers, 1980; Bernard and co-workers, 1979), a strain that contains the plasmid encoded temperature-sensitive cI-repressor. Thus expression of antileukoprotease cDNA sequence could be regulated by appropriate choice of incubation temperature. Western blot analysis of induced <u>E. coli</u> cell lysates carrying the recombinant expression plasmid revealed expression of antileukoprotease 2 - 6 h after induction. In comparative inhibition assays with bovine chymotrypsin using crude extracts of induced and non-induced <u>E. coli</u> cells harbouring the recombinant expression plasmid only the former was found to exhibit significant inhibition of protease activity.

Comparing the N-terminal antileukoprotease protein sequence (amino acids 1 to 54) with the C-terminal amino acids (55 to 107) an internal homology was found indicating that the protein contains two con-

```
                              -25
                              Met Lys Ser Ser Gly Leu Phe Pro Phe Leu Val Leu Leu Ala
GTC ACT CCT GCC TTC ACC ATG AAG TCC AGC GGC CTC TTC CCC TTC CTG GTG CTG CTT GCC   60

                                        -1  1                                   9
Leu Gly Thr Leu Ala Pro Trp Ala Val Glu Gly Ser Gly Lys Ser Phe Lys Ala Gly Val
CTG GGA ACT CTG GCA CCT TGG GCT GTG GAA GGC TCT GGA AAG TCC TTC AAA GCT GGA GTC  120

                                                                               29
Cys Pro Pro Lys Lys Ser Ala Gln Cys Leu Arg Tyr Lys Lys Pro Glu Cys Gln Ser Asp
TGT CCT CCT AAG AAA TCT GCC CAG TGC CTT AGA TAC AAG AAA CCT GAG TGC CAG AGT GAC  180
                                        ---------------------------
                                                   RH5

                                                                               49
Trp Gln Cys Pro Gly Lys Lys Arg Cys Cys Pro Asp Thr Cys Gly Ile Lys Cys Leu Asp
TGG CAG TGT CCA GGG AAG AAG AGA TGT TGT CCT GAC ACT TGT GGC ATC AAA TGC CTG GAT  240

                                                                               69
Pro Val Asp Thr Pro Asn Pro Thr Arg Arg Lys Pro Gly Lys Cys Pro Val Thr Tyr Gly
CCT GTT GAC ACC CCA AAC CCA ACA AGG AGG AAG CCT GGG AAG TGC CCA GTG ACT TAT GGC  300

                                                                               89
Gln Cys Leu Met Leu Asn Pro Pro Asn Phe Cys Glu Met Asp Gly Gln Cys Lys Arg Asp
CAA TGT TTG ATG CTT AAC CCC CCC AAT TTC TGT GAG ATG GAT GGC CAG TGC AAG CGT GAC  360
                                    -----------------------
                                              RH1

                                                                    107
Leu Lys Cys Cys Met Gly Met Cys Gly Lys Ser Cys Val Ser Pro Val Lys Ala Stop
TTG AAG TGT TGC ATG GGC ATG TGT GGG AAA TCC TGC GTT TCC CCT GTG AAA GCT TGA TTC  420
        -----------------------
                  RH2

CTG CCA TAT GGA GGA GGC TCT GGA GTC CTG CTC TGT GTG GTC CAG GTC CTT TCC ACC CTG  480

AGA CTT GGC TCC ACC ACT GAT ATC CTC CTT TGG GGA AAG GCT TGG CAC ACA GCA GGC TTT  540

CAA GAA GTG CCA GTT GAT CAA TGA ATA AAT AAA CGA GCC TAT TTC TCT TTG CAC          594
                                ---------
```

Fig. 4. Nucleotide and deduced amino acid sequence of human ALP cDNA. The positions of the three oligonucleotide screening probes and the polyadenylation sites are underlined. The putative signal sequence is represented by residues -25 to -1.

secutive domains of equal length. Both domains can form three disulfide bridges and show a high degree of sequence homology (Fig. 5). Furthermore they display homologies with mouse and rat yolk proteins and a proteinase inhibitor protein from red sea turtle (Henninghausen and Sippel, 1982; Henninghausen and co-workers, 1982; Dandekar and co-workers, 1982; Kato and Tominaga, 1980). Detailed biochemical information as to possible functions may be elucidated by the selective expression of the two domains.

```
1                                                   27
S G K S F K A G V C P P K K S A Q C L R Y K K P E - C Q *   A L P  N-terminus
          |   |   | |           | | |         |   |
55                                                  81
N P T R R K P G K C P - V T Y G Q C L M L N P P N F C E >   A L P  C-terminus


  28                                                54
* S D W Q C P G K K R C C P D T C G I K C L D P V D T P     A L P  N-terminus
    |   | |           | |       | |     |     | |
  82                                              107
> M D G Q C K R D L K C C M G M C G K S C V S P V K A.      A L P  C-terminus
```

Fig. 5. Amino acid sequence homologies between the two domains of ALP. The N-terminal amino acids (1 - 54) are projected above the C-terminal amino acids (55 - 107). Identical residues among the N- and C-terminal repeats are indicated by vertical bars.

DISCUSSION

There are in principle two possibilities to isolate a desired protein coding nucleic acid sequence; one can make use of chemical DNA synthesis or, cDNA cloning, as described above for hirudin and antileukoprotease, respectively. The choice of method, however, depends on ones objectives and the information already at hand. If the structural gene is required for expression studies or to perform fine manipulations at the DNA level, (site directed mutagenesis, for example,) to study protein-protein interactions or structure/function relationships (as is our intention with hirudin), then the chemical synthesis of oligonucleotides provides a fast and reliable method for obtaining the desired gene. However, to construct a sequence of 300 - 400 base pairs one must envisage a period of 3 - 4 months work with a prerequisite knowledge, of course, of the entire protein primary structure; The constraints imposed by size and amounts of oligonucleotides producable (nowdays, picomole quantities in the range of 120 - 150 nucleotides long,) is also a point of deliberation since it limits the length of a coding region considerable for synthesis. Nevertheless, synthetic constructions offer the advantage of being able to optimize codon usage for specific host systems or to generate restriction sites whilst also allowing the inclusion of control/regulatory sequences, translation initiation signals, for example; c.f. hirudin.

In the cases where only partial protein coding sequences are available then a different cloning strategy has to be adopted; here one would opt for the construction of a cDNA library and the screening there of with specific oligonucleotide probes synthesized from available sequence data and with the minimal possible third base degeneracies. The identification/isolation of such a cDNA clone from a representative library is rather subject to the frequency of the respective mRNA in the total mRNA population; This means of course, screening large numbers of clones if the sequence desired belongs to a low abundancy class of mRNAs. However, once the cDNA is identified, its subsequent sequence data can complement the prior protein sequence data, as demonstrated for human antileukoprotease. Moreover, pre/pro amino acid regions cannot, in general, be identified by proteinsequencing, cDNA cloning, however, offers this facility.

ACKNOWLEDGEMENT

The authors would like to express their gratitude to Mrs. A. Thornton for reading and Mrs. C. Egner for typing the manuscript.

REFERENCES

Bernard, H. U., and Helinski, D. R. (1979). Methods Enzymol., 68, 482-492.

Brown, E. L., Belagaje, R., Ryan, M. J., and Khorana, H. G. (1979). Meth. Enzymol., 68, 109-151.

Casadaban, M. J., and Cohen, S. N. (1980). J. Mol. Biol., 138, 179-207.

Chan, S. J., Weiss, J., Konrad, M., White, T., Bahl, S.-D., Marks, D., and Steiner, F. (1981). Proc. Natl. Acad. Sci. USA, 78, 5401-5405.

Dandekar, A. M., Robinson, E. A., Appella, E., and Qasba, P. K. (1982). Proc. Natl. Acad. Sci. USA, 79, 3987-3991.

Dodt, J., Müller, H.-P., Seemüller, U., and Chang, J.-Y. (1984). FEBS, 165, 180-183.

Fritz, H. (1980). Protein degradation in health and desease. *Ciba foundation Symposium, 75*, 351-379.
Harvey, R. P., Degryse, E., Stefani, L., Schamber, F., Cazenave, J.-P., Courtney, M., Tolstoshev, P., and Lecocq, J.-P. (1986). *Proc. Natl. Acad. Sci. USA, 83*, 1084-1088.
Heinzel, R., Appelhans, H., Gassen, H. G., Seemüller, U., Machleidt, W., Fritz, H., and Steffens, G. (1986). *Eur. J. Biochem., 160*, 61-67.
Henninghausen, L. G., and Sippel, A. E. (1982). *Nucleic Acids Res., 10*, 2677-2684.
Henninghausen, L. G., Sippel, A. E., Hobbs, A. A., and Rosen, J. M. (1982). *Nucleic Acids Res., 10*, 3733-3744.
Hunkapillar, M. W., and Hood, L. E. (1983). *Science, 219*, 650-659.
Hunkapillar, M. W., Kent, S., Caruthers, M., Dreyer, W., Firea, J., Giffin, C., Horvath, S., Hunkapillar, T., Tempst, P., and Hood, L. (1984). *Nature, 310*, 105-111.
Ikemura, T. (1981). *J. Mol. Biol., 151*, 389-409.
Kato, J., and Tominaga, N. (1979). *Fed. Proc., 38*, 832.
Kikuchi, Y., Yoda, K., Yamasaki, M., and Tamura, G. (1981). *Nucl. Acids Res., 9*, 5671-5678.
Proudfoot, N. J. (1984). *Nature, 307*, 412-413.
Remault, E., Stasseus, P., and Fiers, W. (1981). *Gene, 15*, 81-93.
Rossi, J. J., Kierzek, R., Huang, T., Walker, P. A., and Itakura, K. (1982). *J. Biol. Chem., 257*, 9226-9229.
Scherer, G. F. E., Walkinshaw, M. D., Arnott, S., and Morre, D. J. (1980). *Nucleic Acids Res., 8*, 3895-3907.
Schill, W.-B., Wallner, O., Schiessler, H., and Fritz, H. (1978). *Experienta, 34*, 509-510.
Stormo, G. D., Schneider, T. D., and Gold, L. M. (1982). *Nucleic Acids Res., 10*, 2971-2995.
Talmadge, K., Stahl, S., and Gilbert, W. (1980). *Proc. Natl. Acad. Sci. USA, 77*, 3369-3373.

Animal Collagenase-specific Inhibitor: Structural and Functional Characterizations and Localization

T. Hayakawa*, J. Kishi*, T. Kawai*,
T. Hoshino[†] and K. Kobayashi[†]
*Department of Biochemistry, School of Dentistry, Aichi-Gakuin University, Chikusa-ku, Nagoya 464;
[†]Department of Anatomy, Nagoya University School of Medicine, Showa-ku, Nagoya 466, Japan

ABSTRACT

Root pulps from bovine unerupted wisdom teeth produce a potent collagenase inhibitor together with latent collagenase in Eagle's minimal essential medium. The inhibitor was purified more than 700-fold from the culture medium, and it showed a single band by SDS-polyacrylamide gel electrophoresis, but showed multiple bands by basic polyacrylamide gel electrophoresis and electro-focusing. The inhibitor is a sialoglycoprotein containing approx. 20 % carbohydrate by weight and its composition suggests that it bears complex-type oligosaccharides. A rabbit antiserum was prepared against the inhibitor, and IgG purified from the antiserum can completely abolish the inhibitory activity of the inhibitor. The anti-inhibitor serum permitted localization of the inhibitor protein in sections of bovine dental pulps together with several other tissues like dental sac, gingiva, predentin, dentin, skin, aorta, liver and cornea from the same animal revealing the universal distribution of the inhibitor protein in connective tissues. The localization of the inhibitor in bovine dentin was further investigated using an immunoelectron microscopy. The inhibitor was mainly localized as an amorphous or granular accumulation along the wall of dentinal tubules in decalcified bovine dentin. A part of the immunoreactive inhibitor appeared as two periodic bands on the collagen fibrils surrounding the dentinal tubules.

KEYWORDS

Collagenase inhibitor; bovine dental pulp; glycoprotein; anti-inhibitor rabbit serum; immunohistochemical localization; immunoelectron-microscopic localization; bovine dentin.

Animal collagenase plays a crucial role in connective tissue metabolism by catalyzing the first major step in the degradation of native collagen. Specific collagenase activities have been demonstrated in numerous tissues from a wide variety of species, but the mechanisms which regulate collagen degradation in response

© 1987 Pergamon Journals Ltd.
Printed in Great Britain

to physiological or pathological stimuli have remained unclear. As one of the regulatory factors, tissue-derived collagenase inhibitors have been demonstrated in various tissue extracts or media of cell cultures or tissue explants (Sellers & Murphy, 1981).

Root pulps from bovine unerupted wisdom teeth produce a potent collagenase inhibitor together with latent collagenase when cultured in Eagle's minimal essential medium (MEM) (Kishi & Hayakawa, 1982).

The inhibitor was purified more than 700-fold from the culture medium using ConA-Sepharose, Ultrogel AcA 44, and DE-52 cellulose columns (TABLE 1). It showed a single band (MW=36K) by SDS-polyacrylamide gel electrophoresis, but showed multiple bands by basic (pH 8.3) polyacrylamide gel electrophoresis and electrofocusing (Fig. 1). The inhibitor is a sialo-glycoprotein containing approx. 20 % carbohydrate by weight, and its composition suggests that it bears complex-type oligosaccharides. The electrophoretic heterogeneity of the inhibitor was proved to be due to the attachment of different numbers of sialic acid residues. All the SH groups were demonstrated to exist as six disulfide linkages. The inhibitor seems to be specific for animal collagenases obtained from sources as diverse as amphibians and man (TABLE 2 and 3).

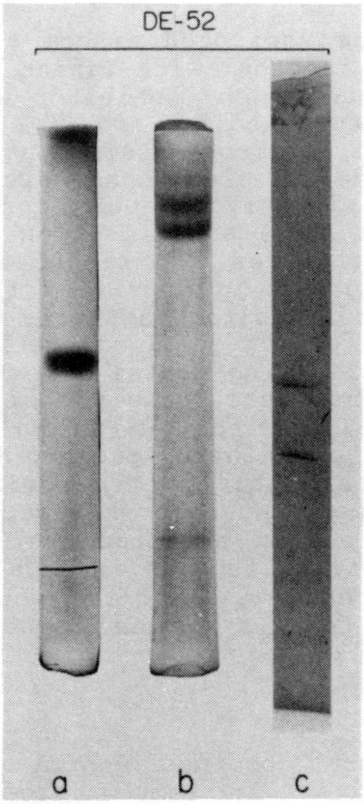

Fig. 1. Electrophoresis and electrofocusing of purified collagenase inhibitor. Polyacrylamide gel electrophoresis patterns of DE-52 cellulose preparation in the presence of SDS (Gel a) and at pH 8.3 (Gel b), and electrofocusing (pH 3.5-9.5) patterns of DE-52 cellulose preparation (Gel c).

Bovine pulp inhibitor does not combine with collagen. The addition of the inhibitor to activated collagenase resulted in a dose-dependent inhibition of the enzyme activity, but the interaction between the inhibitor and activated collagenase is not tight enough to keep the complex intact during gel filtration column chromatography. A rabbit antiserum was prepared against the inhibitor, and IgG purified from the antiserum can completely abolish the inhibitory activity of the inhibitor (Kishi & Hayakawa, 1984).

TABLE 1 Purification and Recovery of Collagenase Inhibitor from Eagle's MEM Cultures of Unerupted Wisdom Tooth Pulps

Fractionation	Total volume	Total protein	Collagenase inhibitory activity			
			Units	Specific activity (units/mg)	Purification factor	Recovery (%)
Explant medium*	360.0	3,096.0	9,648.0	3.1	1	100.0
ConA-Sepharose	166.0	68.7	10,823.2	157.5	50.8	112.2
Ultrogel AcA 44	35.0	4.0	5,145.0	1,286.3	414.9	53.3
DE-52 cellulose	21.5	1.1	2,472.5	2,247.7	725.1	25.6

*Pooled after lyophilization.

TABLE 2 Effect of Inhibitor on Several Collagenases

Source	Enzyme activity (mU)		Inhibition (%)
	Control	+Inhibitor*	
Tadpole back skin	310	0	100
Chick skin	434	0	100
Bovine dental sac	349	0	100
Bovine dental pulp	416	0	100
Human leukocytes	197	0	100
Human post-burn granulation tissue	308	0	100
Cl. histolyticum (30 ng)	391	412	0

Soluble ^{14}C-collagen was used as substrate with the exception that ^{14}C-collagen fibrils were used for tadpole collagenase.

*2 μg of DE-52 cellulose preparation.

TABLE 3 Effect of Inhibitor on Several Proteinases

Enzyme	Substrate	Enzyme activity		Inhibition(%)
		Control	+Inhibitor	
Trypsin (0.2 µg)	Casein (A_{280})	0.367	0.365	0.5
Thermolysin (0.01 µg)	Azocasein (A_{366})	0.238	0.242	0
Elastase, pancreatic (0.1 µg)	Elastin-Congo red (A_{495})	0.035	0.039	0
(0.06 µg)	Suc-$(Ala)_3$-pNA (A_{410})	0.181	0.180	0.6
Elastase, post-burn granulation tissue	Elastin-Congo red (A_{495})	0.040	0.042	0
	Suc-$(Ala)_3$-pNA (A_{410})	0.494	0.502	0
Gelatinase, post-burn granulation tissue	Heat-denatured ^{14}C-collagen (mU)	279	252	9.7

*5 µg of DE-52 cellulose preparation.

The monospecific anti-inhibitor serum did not cross-react with either intact latent collagenase or with the one pretreated by p-aminophenylmercuric acetate. Furthermore, IgG purified from the antiserum quantitatively inhibited the anti-collagenolytic activity of the inhibitor in either the presence or absence of p-aminophenylmercuric acetate. But the IgG did not affect latent collagenase at all. These facts strongly support the possibility that the inhibitor may not be responsible for the latency of collagenase and allow us to propose a dual regulatory mechanism of collagenase activity; that is, an inactive form, *per se*, may thus be additionally kept in the inactive state by the existence of an inhibitor which is synthesized in the same pulp tissues (Fig. 2)(Kishi, Kawai & Hayakawa, 1984).

The anti-inhibitor serum permitted localization of the inhibitor protein in sections of bovine dental pulps together with several other tissues using both immunofluorescence and immunoperoxidase staining techniques. In bovine dental pulps, the inhibitor is primarily localized both in cells and extracellular matrices (Kishi, Kawai & Hayakawa, 1985). Cytoplasmic staining was also seen in both explanted root pulp tissues and immature fibroblast-like pulp cells from primary cell cultures of bovine root pulps (Kishi & Hayakawa, 1985). These findings suggest that the pulp cells are responsible for inhibitor production. Sections of other tissues, like dental sac, gingiva, predentin, dentin, skin, aorta, liver, and cornea from the same animal revealed the universal distribution of the inhibitor protein in connective tissues. As a tissue distinct from dental pulps,

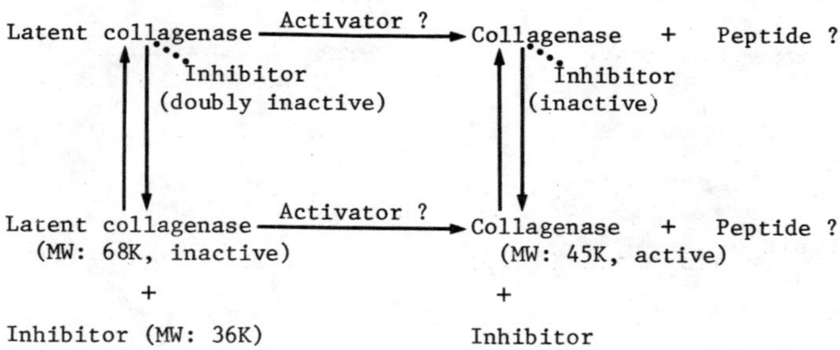

Fig. 2. Proposed mechanism for the dual regulation of collagenase activity in bovine dental pulps.

bovine aorta media were cultured in Eagle's MEM; it was proved that the collagenase inhibitor isolated from the medium using ConA-Sepharose, Ultrogel AcA 54, and DE-52 Cellulose columns was functionally and immunologically equivalent to that from dental pulps.

The localization of collagenase inhibitor in bovine dentin was investigated by immunohistochemistry and immunoelectron microscopy using a specific rabbit antiserum raised against the inhibitor purified from culture medium of bovine unerupted third molar pulps. Immunoreactive collagenase inhibitor was mainly localized as an amorphous or granular accumulation along the wall of dentinal tubules in decalcified bovine dentin (Fig. 3). A part of the immunoreactive inhibitor appeared as two periodic bands on the collagen fibrils surrounding the dentinal tubules (Fig. 4). Those two, a major and a minor band, corresponded to D-periodic fibril bands V-VII and I, respectively (Fig. 5). As those two bands are located on both borders between overlap and hole zones of the collagen fibril, C- and N-terminal non-helical peptides and eight regions which are nD (n=1-4) apart from each non-helical peptide along the triple helix are all equally possible binding sites of the collagenase inhibitor on the collagen molecule. It is interesting that the major band is located not exactly at, but fairly close to, the locus of collagenolytic cleavage, which lies between D-periodic fibril bands IV and V (Fig. 6)(Hoshino and others, 1986).

In conclusion, the inhibitor is a glycoprotein which is likely to be a constitutive part of extracellular matrices and may function in the regulation of collagen degradation.

ACKNOWLEDGEMENT

This work was supported in part by Scientific Research Grants (No. 59570800, 59771380) from the Ministry of Education, Science, and Culture, Japan.

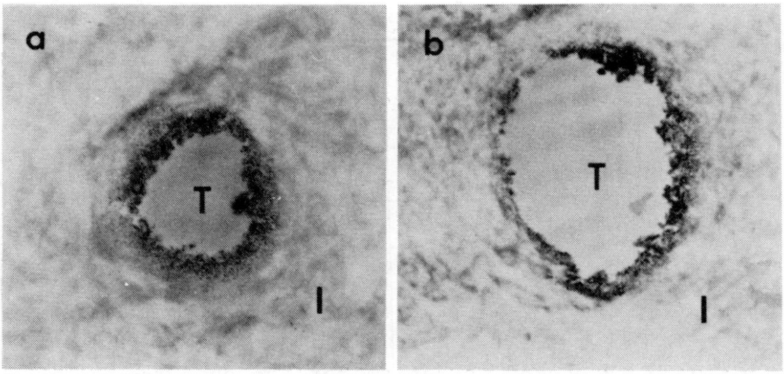

Fig. 3. Cross-sections of dentinal tubules in outer (a) and inner (b) dentin. Immunoreactive collagenase inhibitor is localized along the tubular wall, but no labeling is seen in the intertubular matrix (I). T, lumen of dentinal tubule (unstained, x 9,000).

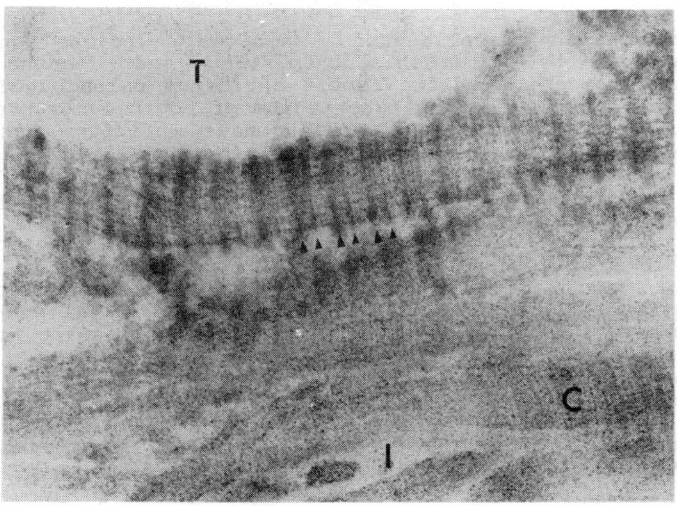

Fig. 4. A portion of a cross-section of dentinal tubular wall. Immunoreactive inhibitor localization corresponds with bands V-VII (large arrowheads) and band I (small arrowheads) on collagen fibrils surrounding the dentinal tubule (T). Collagen fibrils (C) in the intertubular matrix (I) were not labeled with anti-inhibitor serum. (stained further with uranyl acetate and lead citrate, x 90,000).

REFERENCES

Hoshino, T., J. Kishi, T. Kawai, K. Kobayashi, and T. Hayakawa (1986). Collagen Rel. Res. in press.
Kishi, J., and T. Hayakawa (1982). Biochem. Int. 5, 763-770.
Kishi, J., and T. Hayakawa (1984). J. Biochem. 96, 395-404.
Kishi, J., T. Kawai, and T. Hayakawa (1984). Biochem. Int. 9, 343-350.
Kishi, J., T. Kawai, and T. Hayakawa (1985). Archs oral Biol. 30, 571-576.
Kishi, J., and T. Hayakawa (1985). Biomed. Res. 6, 35-40.
Sellers, A., and G. Murphy (1981). In D. A. Hall and D. S. Jackson(Ed.), International Review of Connective Tissue Research, Vol. 9, Academic Press, New York. pp. 151-190.

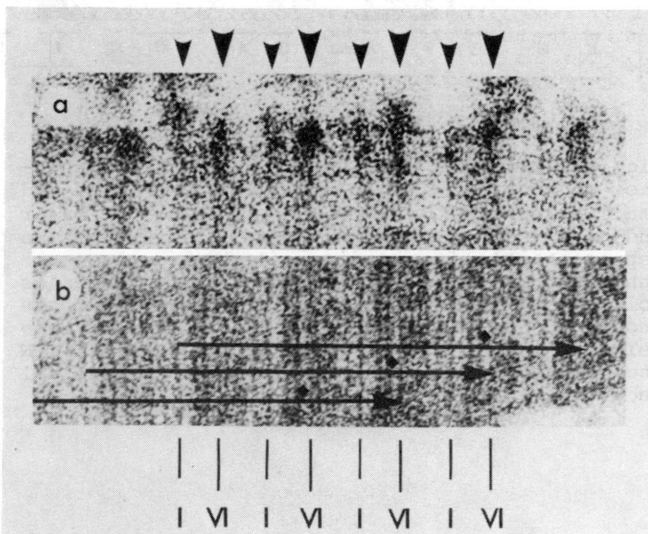

Fig. 5. Higher magnification of collagen fibril surrounding the dentinal tubule which reacted with rabbit anti-inhibitor serum (a) and that in the intertubular matrix which did not react with the antiserum (b). Large and small arrowheads represent the two locations of collagenase inhibitor. The same D-periodic bands of both fibrils are matched. The arrows, pointing in the C-terminal direction, represent collagen molecules within the fibril in their approximate relationship to the band pattern. ◆ , collagenase cleavage site. (x 22,000)

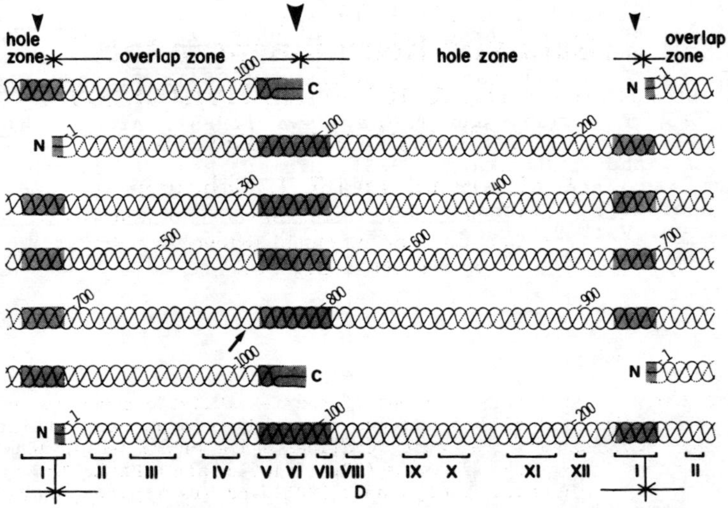

Fig. 6. Schematic drawing of collagenase inhibitor binding sites in collagen fibril. Large and small arrowheads indicate the positions of major and minor immunoreactive bands. Shaded areas represent putative binding sites of collagenase inhibitor on collagen molecules. Arrow indicates collagenase cleavage site. Sloping numbers indicate residue numbers of α-chains along triple helix. I-XII, collagen fibril band numbers; N and C, the respective amino and carboxy terminal ends.

Catalytic Recruitment in the Inactivation of Serine Hydrolases by Phosphonate Esters

I. M. Kovach and R. L. Schowen
Center of Biomedical Research, University of Kansas,
2099 Constant Avenue, Lawrence,
Kansas 66045–2122, USA

ABSTRACT

Molecular details for the recruitment of enzyme catalytic power in the inhibition of six serine hydrolases and two organophosphorus reagents have been investigated. The enzymes are: acetylcholinesterase (AChE) (eel); AChE (RBC); α-chymotrypsin; β-trypsin; elastase (PPE); subtilisin (BPN'). The agents are methyl 4-nitrophenyl propylphosphonate (MPN) and bis-4-nitrophenyl propylphosphonate (NPN). The proton inventory ressults for serine proteases have been reported earlier (Kovach, Larson, and Schowen, 1986b).

KEYWORDS

Serine hydrolase mechanism; irreversible inhibition; solvent isotope effects; proton inventories; molecular modeling by computer-graphics.

INTRODUCTION

The inhibition of serine hydrolases, including acetylcholinesterase (AChE), by organophosphorus inhibitors, has been investigated to characterize the interactions that underlie specificity, productive distortions, and the involvement of acid-base catalysis. The proton-inventory method (Venkatasubban and Schowen, 1984; Kovach, Larson, and Schowen, 1986a) has been used to determine the importance of protonic reorganization in the recruitment of enzyme catalytic power by these selective toxins. This recruitment corresponds to the activation of the evolutionarily developed catalytic machinery of the enzyme in the activated complex for the phosphorylation step (Kovach, Larson, and Schowen, 1986b); the absence or damping of the recruitment in the dephosphorylation step then leads to irreversible inhibition and the toxic properties of the agents.

In a proton-inventory study, rates of reaction (e.g., inactivation) are measured in protium oxide (HOH), deuterium oxide (DOD), and binary mixtures of the two solvents. The overall solvent isotope effect, k_{HOH}/k_{DOD}, is a measure of the general importance of state changes at exchangeable hydrogenic sites in the process of phosphorylation: values larger than unity are observed in enzymic processes involving general acid-base catalysis. If a single protonic site is important, as in simple general acid-base catalysis, then **k** is linear in **n**, the atom fraction of deuterium in binary mixtures. If two or three sites operate, as expected for "charge-relay catalysis", then a quadratic or cubic relationship should result. If a large number of sites each contribute a slight effect, to

© 1987 Pergamon Journals Ltd.
Printed in Great Britain

produce a considerable overall magnitude, as in a protein conformational change, than an exponential dependence of **k** on **n** is seen. The object of the studies reported here is to discover the degree to which highly active toxic agents have activated the catalytic machinery of the enzyme, as reflected in the magnitude of the overall solvent isotope effect, the character of the proton inventory and in the types of interactions that mark specificity and productive distortions.

RESULTS AND DISCUSSION

Three phosphonate esters with 4-nitrophenyl leaving groups were synthesized in our laboratory by coupling of the potassium salt of 4-nitrophenol with the corresponding alkoxy alkylphosphonochloridate. The synthesis of this latter was effected by reactions between an alkylphosphonodichloridate and the desired alcohol in the presence of triethylamine as catalyst. The abbreviations developed for the compounds are meant to be general for compounds of the formula X-R(P=O)-Y, where Y is a leaving group, R an alkyl substituent attached to phosphorus through a P-C bond, and X a second substituent, possibly also a leaving group. The abbreviations are of the form XRY, so that N = 4-nitrophenyl, P = propyl, and M = methoxy.

Kinetics of Inhibition

All reaction rates were monitored spectrophotometrically either by the direct release of 4-nitrophenol at 400 nm under zero or first-order conditions or, for fast inhibition, in the presence of a good, competitive, chromogenic substrate. The substrates were: phenyl acetate with AChE, Glu-Leu-Phe-4-nitroanilide with chymotrypsin and subtilisin, and N-succinyl-Ala-Ala-Ala-4-nitroanilide with elastase. Both inhibitor and substrate solutions were freshly prepared in methanol acetonitrile or their aqueous mixtures for each experiment and they were analyzed spectroscopically to obtain accurate values for the concentrations used in data reduction. The kinetic rate law for irreversible inhibition in the presence of a substrate modifier has been developed and used lately by others (Hart and O'Brien, 1973; Tian and Tsou 1982) and us. The observed rate constant from such experiments is a function of the microconstants for inhibition and the Michaelis constant and concentration for the substrate as shown below:

$$1/k_{obs} = 1/k_2 + K_i/k_2[I_o][1 + [S]K_m] \quad (1)$$

Linear least squares fit of the first-order rate constants to the concentration of the substrate gave the second-order rate constant, k_2/K_i, since the $1/k_2$ term was negligible at inhibitor concentrations $[I_o]$ below K_i, under our conditions. In the inactivation studies we report here, no departure from second-order kinetics was observed. This suggests that enzyme-reagent complexes are probably not formed below 20-100 μM in the case of serine proteases and below 10 μM in the case of AChE.

Inactivation of Serine Proteases

Stochiometry of inhibition

MPN and NPN both react at 25 °C and pH 7-8 with chymotrypsin, elastase, and subtilisin to generate 1 mol of 4-nitrophenol per mol of enzyme active sites. NPN also reacts similarly with trypsin; MPN shows no reaction with trypsin within 30 min.

The reactions of NPN, when either NPN or the enzyme is in substantial excess, are strictly first-order. Liberation of a second mol of 4-nitrophenol, after the initial phosphonylation, occurs realtively slowly. MPN which is a racemic mixture, shows biphasic kinetics with the first 0.5 equiv of 4-nitrophenol appearing several times faster than the second 0.5 equiv. This is consistent with the expected enanatioselectivity. In the solvent isotope effect and proton -inventory studies for MPN, initial rates were used. Consequently, the results for MPN refer to the more reactive enantiomer.

Specificity.

The second-order rate constants for inactivation of serine proteases are given in Table 1 in the first column for water and in the second column for heavy water. Some specificity is observed in these reactions: each reagent

TABLE 1 Second-Order Rate Constants and Solvent Isotope -Effects for Inactivation of Serine Proteases by NPN (A) and MPN (B) at 25° and Appropriate[a] pL

Enzyme (pH), k_{HOH}/k_{DOD}	$(k_i/K_i)_{HOH}$, $M^{-1}s^{-1}$(SD)	$(k_i/K_i)_{DOD}$, $M^{-1}s^{-1}$(SD)	ϕ_1 (SD)	ϕ_2 (SD)
		A		
Chymotrypsin (7.45)	2260(59)	1178(17)	0.57	0.93
1.92 ± 0.06			(0.06)	(0.09)
Elastase (8.45)	1840(50)	1120(20)	0.77	0.66
1.65 ± 0.05			(1.03)	(1.15)
Trypsin (8.48)	264(8)	145(3)	0.58	0.93
1.83 ± 0.06			(0.04)	(0.05)
Subtilisin (8.40)	1330(49)	721(20)	0.50	1.07
1.85 ± 0.07			(0.03)	(0.05)
		B		
Chymotrypsin (6.90),	0.870	0.410	0.40	1.17
2.12 ± 0.05	(0.016)	(0.005)	(0.02)	(0.05)
Elastase (8.43),	3.19	1.59	0.51	0.97
2.01 ± 0.15	(0.20)	(0.08)	(0.07)	(0.10)
Trypsin (8.50)		no reaction observed		
--		($k_i/K_i < 0.1\ M^{-1}s^{-1}$)		
Subtilisin (8.40),	2.57	1.24	0.50	0.96
2.06 ± 0.05	(0.01)	(0.02)	(0.08)	(0.13)

[a]For chmotrypsin, a value approximating the pH of maximum rate was used. For other enzymes, a pH in the pH-independent region was chosen. Corresponding pL in mixtures of HOH and DOD was achieved by employment of constant buffer rations. Taken from Kovach, Larson, and Schowen, 1986b. With permission.

discriminates among the four enzymes, and each enzyme reacts differentially with the reagents. For example, the presence of the second aryloxy group in NPN (as opposed to the methoxy group in MPN) is obviously advantageous: NPN reacts with every enzyme at least 500-fold more rapidly than MPN. This is not a simple electronic effect. In nonenzymic hydrolysis, NPN shows a rate constant only of 10-fold larger than for MPN (Kovach, Larson, and Schowen, 1986b). With chymotrypsin, trypsin and subtilisin, a logical hypothesis would be that this aromatic group inserts itself into the "pocket" that accomodates a relatively large C-alpha side chain of the natural substrate for each of these enzymes (Fersht, 1985). Molecular modeling with computer-graphics gives support to this idea. In Fig. 1-2, probable interactions between residues that comprise of the catalytic triad, oxyanion hole, some of the N-acyl binding-site, and the specificity pocket of trypsin and the constituents of the 4-nitrophenyl propylphosphono (from NPN) and methyl propylphosphono (from MPN) moieties are shown as photographed from the screen of an Evans & Sutherland PS330 terminal. The X-ray

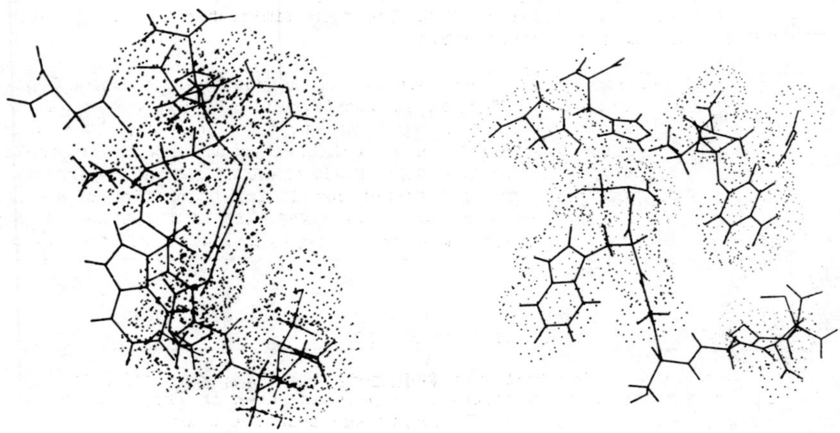

P-(R) P-(S)

Fig. 1 Computer-graphics images of trypsin inhibited by NPN

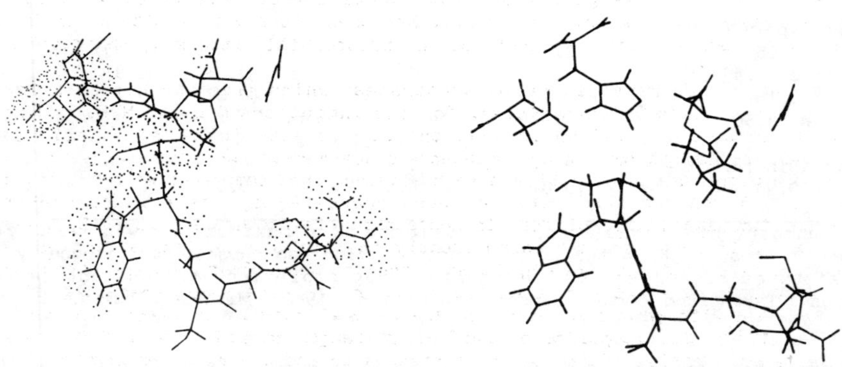

P-(R) P-(S)

Fig. 2 Computer-graphics images for trypsin inhibited by MPN

crystallographic coordinates for a 5 A° sphere around the oxygen of the active-site serine of trypsin were retrieved from the Brookhaven data-base. The phosphorus and its ligands were attached so that the P-O oxygen is oriented into the oxyanion hole for hydrogen-bonding interations. The alkyl and alkoxy ligands were placed into the specificity pocket in each of the possible sequences that correspond to the P-(R) and P-(S) enantiomers. Models of trypsin after inhibition by NPN are shown (the van der Waals volumes are indicated by the use of dot structures) for the P-(R) and P-(S) configurations in Fig. 1. The latter is at a 90° rotation from the former. Both ligands can essentially be accomodated by the specificity pocket. Fig. 2 shows the corresponding pictures for the P-(R) and P-(S) isomers of trypsin inhibited by MPN. The much diminished level of interaction in these cases is quite discernable.

The hypothesis developed above is excluded for elastase: only a small "pocket" is present in the active site of this enzyme (Fersht, 1985), not capable of accepting a 4-nitrophenyl group. The spesificity for these highly reactive organophosphorus inhibitors may not be as restrictive as it is with substrates; this is suggested by Fig. 1-2. In the case of elastase or perhaps with some of the other proteases, the stabilizing interactions at the transition state might occur between the remaining 4-nitrophenyl ligand and part of the <u>N</u>-acyl binding site, which serves to accomodate the "poly-peptide tail" of the natural substrates (Fersht, 1985).

Solvent isotope effects and proton inventories

The reactions of all agents with all proteases show k_{HOH}/k_{DOD} = 1.7 to 2.1 and in all cases the proton inventory is linear as shown in Table 1 and in Fig. 3. When proton inventory data were fitted to the equation below;

$$k_n/k_o = (1-n-n\phi_1)(1-n-n\phi_2) \qquad (2)$$

the value for one of the ϕ was unity within 10% and mostly within 3%. These fractionation factors (ϕ) are listed in the third and fourth columns of Table 1. The most straightforward interpretation of this finding is that a single site is acting catalytically. Presumably the active-site histidine is acting as a general base to promote serine nucleophilic attack at phosphorus.

As with truncated peptide-analog substrates, which give one-site catalysis (Venkatasubban and Schowen, 1984), for the inhibitors NPN and MPN, the most straightforward assumption is that the single site involved in acid-base catalysis is the hydrogen bridge between the active-site serine hydroxyl and the imidazole nitrogen of the active-site histidine. This component of the catalytic power of the enzyme has therefore been recruited by the organophosphorus reagents. The likelihood of this interpretation is further increased by the dependence of the rate on what is apparently the histidinium ionization, as seen in the pH/rate profiles (Kovach, Larson, and Schowen, 1986b). On the other hand, the inhibitors have not recruited the multi-proton catalytic capability of the enzyme, seen with good simulators of the natural substrates (Venkatasubban and Schowen, 1984). Oligopeptides of sufficient length usually "couple" the acid-base catalytic entity of the serine proteases to produce multi-proton catalysis and therewith non-linear ("bowl-shaped") proton inventories. Since the inhibitors, even in the cases with the greatest rates, give linear proton inventories, the multi-proton component of the catalytic apparatus remains unrecruited.

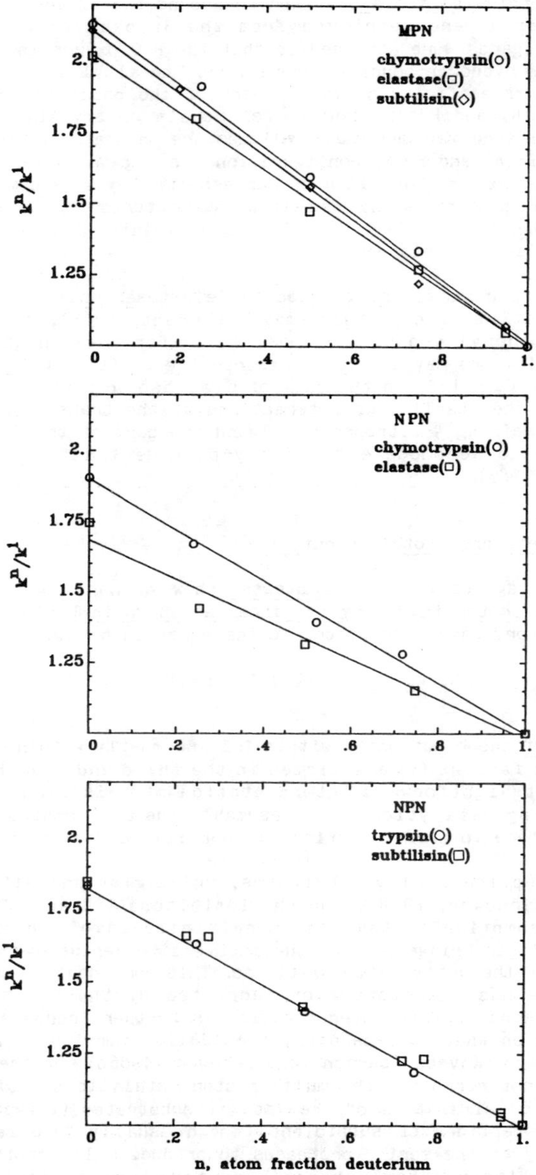

Fig. 3 Proton inventories for the inactivation of serine proteses by MPN and NPN (Taken from Kovach, Larson, and Schowen, 1986b. With permission.)

Inactivation of Acetylcholinesterases

Specificity.

Table 2 lists the second-order rate constants for inactivation of AChE-s by phosphonate esters. MPN shows moderate rates and distinguishes between the two types of AChE-s, while NPN reacts very poorly. An attractive explanation for this finding is a steric hindrance preventing efficient binding of the rigid 4-nitrophenyl substituent on the apparently restricted active-site of AChE (Rosenberry, 1975).

TABLE 2 Second-Order Rate Constants and Solvent Isotope -Effects for Inactivation of AChE eel and RBC at pH 7.70, pD 8.20, and 25°

Inhibitor	AChE	$(k_i/K_i)_{HOH}$, $M^{-1}s^{-1}$(SD)	$(k_i/K_i)_{DOD}$, $M^{-1}s^{-1}$(SD)	k_{HOH}/k_{DOD}
MPN	RBC	922 ± 50	608 ± 3.6	1.50 ± 0.10
MPN	eel	354 ± 26	236 ± 21	1.50 ± 0.17
NPN	eel	7.32 ± 0.01	5.77	1.3 ± 0.1

Solvent isotope effects

The reactions of the two agents with both AChE (eel) and AChE (RBC) show k_{HOH}/k_{DOD} = 1.3 to 1.6 which suggests that general base catalysis by the imidazole of the histidine residue in the active-site of AChE is partially rate determining. Obviously in these cases of central importance with regard to toxicity, the solvent isotope effect is generally not consistent with simple general-base catalysis in a single rate determining step for inactivation, but rather with the incursion of other rate determining steps. Other mechanistic probes and solvent isotope effects with other organophosphorus inhibitors give results (Kovach, Huber, and Schowen, in preparation) pointing in this direction. At least partially rate determining conformational induction has been proposed by Rosenberry (1975) for acetylation of AChE by several good substrates. Solvent isotope effect and proton inventory results for the reaction of phenyl acetate with AChE-s obtained by us (Kovach, Larson, and Schowen, 1986a) show a similarity between acetylation and phosphonylation. In the (rate limiting) physiological reaction, deacetylation of acetyl-AChE, participation of only a single-proton catalysis, supposedly by the imidazole of the histidine residue, has been established with a solvent isotope effect of 2.42 ± 0.03 for the eel enzyme and 2.28 ± 0.11 for the erythrocyte enzyme. Although there is some likeness between the dynamics of acetylation and phosphonylation of AChE, since general acid-base catalysis of the attack by serine can be similar, the heavy atom rearrangements for acyl transfer and phosphoryl transfer are expected to follow substantially different geometrical courses. This is particularly obvious if addition intermediates form, which would be tetrahedral for acyl transfer and bipyramidal for phosphonyl transfer. Even if the reactions were concerted, however, the transition states should have correspondingly different geometries. If intermediates form, there may be a difference in the balance between rate limiting steps and the importance of base catalysis.

CONCLUSIONS

Acceleration of the hydrolytic decomposition of the phosphonate esters by each enzyme might be compared to enzymic acceleration factors for hydrolysis of the natural substrate, in each case, to reveal enzyme efficiency. AChE, a remarkably efficient enzyme, accelerates the hydrolysis of MPN by a factor of 10^{10} and the hydrolysis of soman, one of the most potent of organophosphorus inhibitors, by a factor of 10^{13}. The latter amounts to 67% (12/18) mobilization of the catalytic power of the enzyme with respect to acetylcholine (acceleration factor of 10^{18}). Similar recruitment of the catalytic power of chymotrypsin is demonstrated by NPN, with an acceleration of its hydrolysis by ~10^{9} fold to be compared with 10^{12} fold acceleration of the hydrolysis of natural substrates.

These inhibitors show very **different potencies** against the AChE-s from what is observed with the serine proteases. The reaction of MPN with AChE occurs about 350 fold faster than its reaction with chymotrypsin. NPN is a very poor inhibitor of AChE, one to 500 times worse than with serine proteases. The efficient inactivation of serine proteases by NPN appears to stem from strong hydrophobic interactions between the aromatic ligand and residues in the specificity pocket and the *N*-acyl binding-site.

It is a salient feature of our observations with both group of enzymes, the AChE-s and the serine proteases, that the solvent isotope effects and proton inventories for inactivation are all similar for each group and similar to what has been found for unnatural substrates of the respective enzymes. This suggests that the balance of contribution of all limiting steps in phosphorylation does not change significantly for a wide range of reactivity. Phosphorylation takes place with rate limiting attack of the serine assisted by general base catalysis by the imidazole residue for the serine proteases and with the rate limiting contribution of an additional step, supposedly the departure of the leaving group in case of the AChE-s.

ACKNOWLEDGEMENT

This research was supported by the U.S. Army Medical Research and Development Command, Contract No. DAMD 17-83-C-3199.

REFERENCES

Fersht, A. (1985.) *Enzyme Structure and Mechanism*. 2nd ed., Freeman, San Francisco; Chapter 1.

Hart, G. J., and R. D. O'Brien (1973). *Biochemistry, 12*, 2940.

Kovach, I. M., M. Larson, and R. L. Schowen (1986). *J. Am. Chem. Soc., 108*, a. 3054, b. 5490.

Kovach, I. M., J. A. Huber, and R. L. Schowen. Manuscript in preparation.

Rosenberry, T. L. (1975). Acetylcholinesterase *Adv. Enzymol. Relat. Areas Mol. Biol., 43*, 103.

Tian, W. X., and C. L. Tsou (1982). *Biochemistry, 21*, 1028.

Venkatasubban, K. S. and R. L. Schowen (1984). *CRC Crit. Rev. Biochem., 17*, 1.

Design of Alternate Substrate Inhibitors of Serine Proteases: Tactics for Obstructing Deacylation Pathways[1]

Allen Krantz, Leslie J. Copp and Robin W. Spencer
Syntex Research (Canada), 2100 Syntex Court, Mississauga, Ontario L5N 3X4, Canada

ABSTRACT

4H-3,1-Benzoxazin-4-ones are alternate substrate inhibitors of serine proteases that function by forming stable acyl enzymes. 2-NHR benzoxazinones are acceptably stable compounds that are good inhibitors of human leukocyte elastase. The acyl enzymes formed during inhibition have three routes of deacylation which must be slowed to achieve low K_i values. Deacylation via hydrolysis to a 2-ureidobenzoate is effectively blocked by 5-alkyl substitution, and deacylation via cyclization to a 3-R-2,4-(1H,3H)-quinazolindione is inhibited by branched R. The combined effects of these substitutions result in 10,000-fold improvements over lead compounds.

KEYWORDS

serine proteases, human leukocyte elastase, mechanism-based inhibition, alternate substrate inhibition, steric hindrance, proton inventory, benzoxazinones

INTRODUCTION

Serine proteases are attractive targets for medicinal chemists engaged in the design of mechanism-based inhibitors. Not only are there a number of enzymes in this class that are implicated in the pathophysiology of human disease, but also the design of inhibitors of enzymes that catalyze their reactions by acyl transfer mechanisms presents an interesting challenge.

There are three distinct intermediates (1-3) during serine protease catalyzed hydrolyses in which the enzyme is bound covalently to its substrate (or to a fragment of its substrate). These intermediates offer intriguing possibilities for curbing enzyme activity. In principle, an alternate substrate inhibitor could be designed to block enzyme

[1] Contribution number 256 from the Institute of Bioorganic Chemistry

© 1987 Pergamon Journals Ltd.
Printed in Great Britain

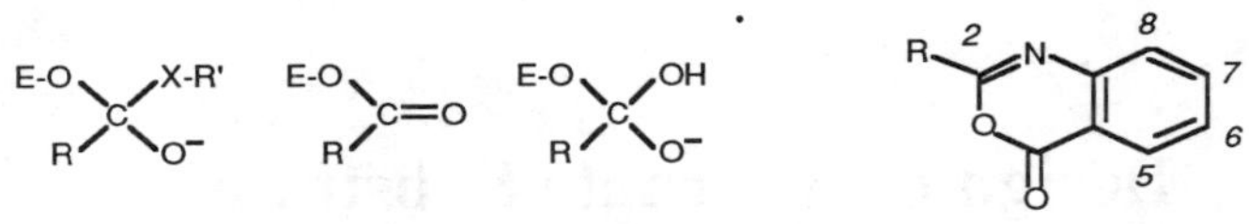

1 2 3 4

catalyzed deacylation by giving rise, during the course of catalysis, to an extremely long lived species for which the machinery of catalysis has been severely disrupted. This is an especially appealing strategy for the design of stable acyl enzymes of serine proteases, since so much is known about hydrolytic mechanisms and the means of controlling ester reactivity.

An additional challenge is to design specific inhibitors of high potency without using peptidyl functionality (Ringrose, 1983) to mimic the enzymes' natural substrates or inhibitors. 4H-3,1-Benzoxazin-4-one 4 represents an attractive lead structure for accomplishing this task (Alazard and colleagues, 1973; Hedstrom and colleagues, 1984; Teshima, Griffin, and Powers, 1982), provided that R is an electron donating group to counter the lability of the oxazinone ring to nucleophiles.

Hedstrom and colleagues (1984) reported that 2-aminobenzoxazinone 5 (R = H, R_5 = H, Scheme 1) is rapidly isomerized to the 2,4-(1H,3H)-quinazolindione 7 by chymotrypsin. Apparently, deacylation to regenerate free enzyme with concomitant formation of 7 occurs via cyclization of the acyl enzyme 6 and is considerably faster than the "normal" enzyme catalyzed hydrolysis observed for 2-oxa- and 2-carba- benzoxazinones. We have observed similar results with 5 (R=H, R_5=H) and human leukocyte elastase (HLE), an enzyme of potential clinical interest (Powers and others, 1986; Schnebli, 1985). We show here that the lifetime of acyl elastases derived from benzoxazinones can be enormously increased using tactics that are intended to obstruct deacylation pathways 6 → 7 and 6 → 8. Since $K_i = k_{off}/k_{on}$ for alternate substrate inhibitors (Segel, 1975; Spencer and colleagues, 1986), decreasing k_{off} ($= k_q + k_u + k_b$) will result directly in lower inhibition constants.

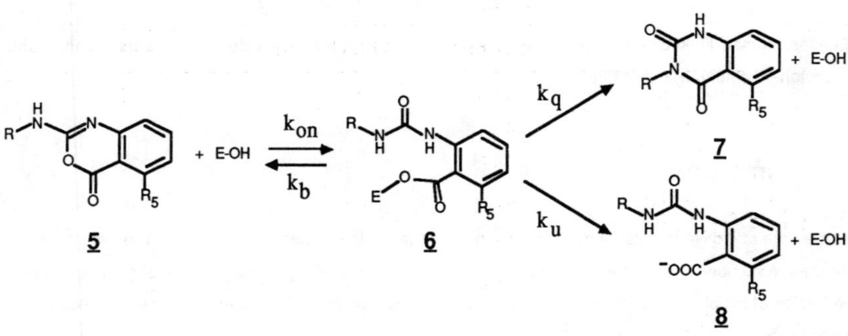

Scheme 1. The mechanism of protease inhibition by benzoxazinones.

EXPERIMENTAL

2-Alkylamino-4H-3,1-benzoxazin-4-ones were prepared by published procedures (Butula, Vela, and Zorc, 1981; Hamprecht and Weierzer; Karl and Wolfgang; Krantz, Tam, and Spencer, 1985; Lempert and Doleschall, 1963a,b; Papodopoulos, 1980, 1981; Papodopoulos and Petridou-Fischer, 1982; Papodopoulos and Torres, 1982; Sayigh, Haven, and Ulrich). Human leukocyte elastase and human sputum elastase were obtained and assayed by the progress curve method as previously described (Copp, Krantz, and Spencer, 1986; Spencer and colleagues, 1986). Proton inventory measurements on the rates of deacylation were measured and analyzed as in Schowen (1978) and Schowen and Schowen (1982).

RESULTS AND DISCUSSION

Inhibition by Benzoxazinones with Linear 2-Alkylamino Substitution

The inhibition of HLE by various 2-alkylamino benzoxazinones **5** is shown in Fig. 1 and Table 1. The acylation rate k_{on} is maximal for R = ethyl, n-propyl, and n-butyl, which results in large increases in potency (ca. 1000-fold decreases in K_i) over the parent compound (R = H, R_5 = H). This effect on k_{on} cannot be due to electronic effects (as opposed to other series of benzoxazinones in which k_{on} is enhanced by electron withdrawal at C_2 : Spencer and colleagues, 1986), and must be due to favorable interactions of R at the active site. The deacylation rates (k_{off}) are not affected by the size of R until R is larger than n-propyl (when R_5 = H); evidently for larger R there is an interaction that distorts the acyl enzyme to a conformation that slows deacylation. Note the interactive effect of bulk at R_5 and R; when R_5 is methyl there is a very steep decline in k_{off} for R larger than methyl (Fig. 1, right, open symbols).

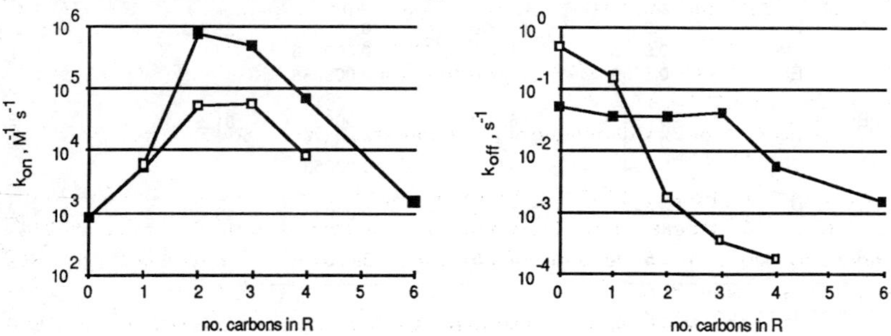

Fig. 1. Inhibition of HLE by 2-(linear alkylamino) benzoxazinones.
filled symbols : R_5 = H; open symbols : R_5 = CH_3.

The Effect of Branched R on Product Composition

Table 1 and Fig. 2 show how the ultimate product composition varies with the structure of R. The nature of the alkyl group is clearly a determinant of product, as 2-NHR-benzoxazinones with linear R give predominantly

AB VOL65-O

quinazolindiones **7**, and those with branched R form primarily 2-ureidobenzoates **8**. We have also observed significant reductions in the rate of a model reaction, the base-catalyzed cyclization of methyl esters of 2-(R-NH-CO-NH)-benzoic acids (**9**, not shown) to 3-R-quinazolin-2,4-diones (**7**), by increasing the size of R. Compared to the cyclization rate of the N-methyl derivative (**9**, R = Me) the values corresponding to the linear (R = nBu) and branched (R = iPr) groups are lower by factors of 8 and 640, respectively. It seems reasonable to assume that steric interactions that develop in the transition state (between R and vicinal oxy functions) will intensify as R increases in bulk, and that such steric effects are at least partly responsible for the diminished deacylation rates and altered product partition observed in the enzyme catalyzed reaction.

TABLE 1 HL Elastase Inhibition by Benzoxazinones **5**

R	R_5	K_i (μM)	k_{on} ($M^{-1}s^{-1}$)	k_{off} (s^{-1})[a]	% **7** [b]
H	H	63.	810.[c]	0.051[c]	91 ± 1
	Me			≈ 0.47	≈ 100[d]
	Et	5.1	< 15.		
Me	H	6.9	5,500.[c]	0.038[c]	98 ± 2
	Me	31.	4,800.[c]	0.15 [c]	95 ± 5
Et	H	0.051	730,000	0.037	81 ± 1
	Me	0.034	52,000	0.0018	
nPr	H	0.091	450,000	0.041	94 ± 3
	Me	0.0060	56,000	0.00034	
iPr	H	0.052	59,000	0.0030	28 ± 9
	Me	0.0015	11,800	0.000018	
	Et	0.00094	70,000	0.000066	
nBu	H	0.084	66,000	0.0055	81 ± 1
	Me	0.023	7,600	0.00017	
	Et	0.0042	130,000	0.00053	
sBu	H	0.043	210,000	0.0091	18 ± 9
nHex	H	3.6	1,500	0.0053	

[a] Except as noted[c], $k_{off} = k_{on}K_i$

[b] = 100 [**7**]/ ([**7**] + [**8**]), by analysis of fluorescence emission spectra after exhaustive incubation of HLE and **5**. Fluorescence standards used were authentic **7** and **8** (R = nBu and iPr, R_5 = H).

[c] k_{on} calculated as k_{off}/K_i, with k_{off} = measured rate of recovery of enzyme activity from the isolated acyl enzyme.

[d] From a comparison of ultraviolet difference spectra to standards.

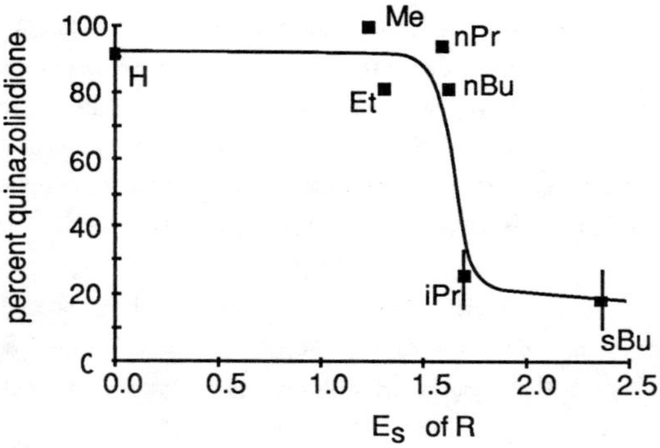

Fig. 2. Final product partition vs. steric bulk of R. E_s is Taft's steric parameter.

The data of Fig. 2 represent the products observed after exhaustive turnover of the given benzoxazinones by elastase. We have recently found that cyclization of the acyl enzyme to benzoxazinone (6 → 5, Scheme 1) can be observed if deacylation occurs in the presence of a "trap" for the regenerated active enzyme. Acyl elastases 6 (R = nBu or iPr, R_5 = H) were separated from excess benzoxazinone by rapid, low pH gel filtration (Copp, Krantz, and Spencer, 1986), and then allowed to deacylate at pH 7.8 in the presence of excess soybean trypsin inhibitor. Products 5 and 7 were assayed by HPLC. Combining these preliminary results with the final partition data (Fig. 2), we can estimate the overall partitioning of 6 as follows. For linear R (=nBu), 5 ≈ 4%, 7 ≈ 76%, and 8 ≈ 20%, and for branched R (=iPr), 5 ≥25%, 7 ≤15%, and 8 ≤60%. Thus for linear R, deacylation by N-cyclization to quinazolindione is the major pathway, but when branched R curtail the formation of quinazolindione, O-cyclization to benzoxazinone becomes significant and may, in fact, predominate. The cyclization of esters of 2-(H_2N-CO-NH) benzoic acid to benzoxazinones has chemical precedent when the leaving alcohol is sufficiently acidic (Hegarty and Bruice, 1970). The effects of other ring substituents on this three-way enzyme partition are under continuing study.

The Effect of 5-Alkyl Substitution on Deacylation Rates

5-Alkyl substituted benzoxazinones give rise to acyl enzymes which are di-ortho- substituted benzoic acid esters (6, $R_5 \neq$ H). Bulky R_5 should have a profound effect on the hydrolytic rate (k_u) at the very least, since it is well known that $B_{AC}2$ hydrolysis of benzoic acid esters is blocked by flanking ortho substituents which shield the carbonyl from nucleophilic attack (Goering, Rubin, and Newman, 1954).

Table 1 presents the impact of alkyl R_5 substitution on K_i and on the constituent rate constants k_{on} and k_{off}. For the 2-n-butylamino system, K_i is lowered four-fold by replacing the C_5 hydrogen with a methyl group, and by a factor of 20 on changing hydrogen to ethyl. The major effect of these changes is on k_{off}, which is reduced 32-fold for the methyl substitution and 10-fold for the ethyl. Essentially the same story unfolds for all R larger than methyl, suggesting an important synergy between the sizes of R and R_5. When R is methyl or hydrogen, R_5 substitution does not slow deacylation. When R is larger, R_5 substitution slows deacylation dramatically. Note that this synergy is due to combined steric interactions in the active site, since there is only a slight effect of di-ortho substitution in the model reaction for quinazolindione formation (for example, **9** (R = nBu, R_6 = H) → **7** (R = nBu, R_5 = H) and **9** (R = nBu, R_6 = Et) → **7** (R = nBu, R_5 = Et) have relative rates of 1.6:1, aqueous, 25°).

It is especially noteworthy that k_{off} is diminished by alkyl R_5 when either linear or branched alkylamino functions reside at C_2. Since the predominant mode of deacylation is N-cyclization (k_q) for R = nBu and O-cyclization (k_b) or hydrolysis (k_u) for R = iPr, 5-alkyl substitution must significantly depress all component rates of the enzyme catalyzed deacylation.

Details of Deacylation : Effects of pH and Proton Inventory

The apparent three-way partition of the acyl enzyme **6** is novel. It also presents a severe challenge, since tactics designed to slow deacylation must block all three routes to be completely effective. Experiments which shed additional light on the reactivity of **6** are therefore desirable.

Deacylation of **6** shows pH dependence similar to that of benzoyl-chymotrypsins (Caplow and Jencks, 1962) and is thus likely to be general base catalyzed by histidine 57. Specifically, the acyl enzyme formed by HLE and **5** (R = iPr, R_5 = H) has $k_{off,\ max}$ = 0.0052 s^{-1} and pK_a = 7.3. Similarly, for the acyl enzyme from bovine trypsin and **5** (R = nBu, R_5 = H), $k_{off,\ max}$ = 0.013 s^{-1} and pK_a = 8.2 (in 0.14 M NaCl, 20 mM KP_i, 25°), and for trypsin and **5** (R = iPr, R_5 = H), $k_{off,\ max}$ = 0.0027 s^{-1} and pK_a = 7.5. The final product partitions for these acyl trypsins are essentially the same as those observed for HLE (Table 1 and Fig. 2). Neutral histidine 57 is therefore likely to be involved in the rate determining step for deacylation by any of the three routes.

The proton inventory (Schowen, 1978; Schowen and Schowen, 1982) of deacylation was measured for HLE and **5** (R = Me or iPr, R_5 = H), with the results in Fig. 3. The acyl enzymes were isolated by gel filtration (vide supra), kept at pH 3, and deacylation followed by the first order recovery of enzymatic activity upon 50-fold dilution into pH 8.2 buffer (20 mM Tris, 40 mM Tris·HCl, 0.5 M NaCl, 0.05% w/v Brij 35, 2.5 μM substrate (MeO-Succ-Ala-Ala-Pro-Val-7-(4-CF_3-coumarinamide)), 25°, in final H_2O/D_2O mixtures as in Fig. 3)[2]. For R = Me, ${}^{H}k_{off}/{}^{D}k_{off}$ = 2.6, a value similar to that observed for deacylation of simple acyl-chymotrypsins (Schowen and Schowen, 1982).

---

[2] In addition to the deacylation data, we obtained the proton inventory for HLE turnover of the substrate under V/K conditions ([S] = $0.04K_m$). We find that $({}^{H}V/K)/({}^{D}V/K)$ = 1.3, in agreement with Stein (1983). We also find a pronounced "bulging up" with the best fit $({}^{n}V/K)/({}^{H}V/K) = 1 + 0.18n - 0.435n^2$ (n^2 term significant at $p < 0.002$). Such behavior has also been seen for α-lytic protease and chymotrypsin (Quinn and others, 1980); see also Stein (1985) on the importance of extended HLE substrates.

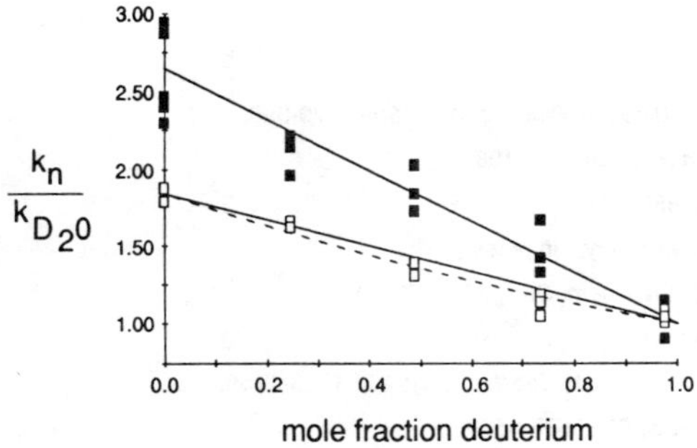

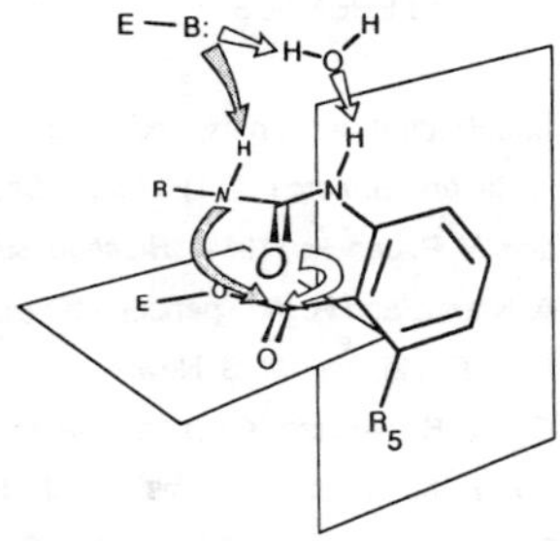

Fig. 3 (left). Proton inventory of the deacylation of **6**. Filled symbols, R = Me; open symbols, R = iPr. Solid lines, best linear fits; dashed line, best quadratic fit for R = iPr.

Fig. 4 (right). Hypothetical cyclization pathways. Dark arrows, **6** → **7** ; light arrows, **6** → **5**.

There is no apparent curvature in the plot, though the quality of the data would allow only a large multiple site effect to be seen. When R = iPr, $^{H}k_{off}/\,^{D}k_{off}$ = 1.8, a difference in the solvent isotope effects that confirms that different structures are involved in the deacylation of linear vs. branched acyl enzymes. In addition, for R = iPr there is a slight but significant (p = 0.065) downward curvature for intermediate H/D mixtures, with a best fit for two-proton catalysis (Stein, 1983) of ϕ^{-1} = 1.37 for each proton.

Figure 4 is a hypothetical view of the acyl enzyme which suggests that for linear alkyl R, histidine 57 (= E-B:) may abstract the ureido proton (distal to the ring)directly in the transition state for quinazolindione formation (dark arrows), while for branched alkyl R, histidine catalysis must proceed by abstraction of the proximal proton, perhaps with an intervening water molecule to account for the apparent proton inventory results (light arrows).

CONCLUSION

Simple alkyl groups at C_5 and in NHR at C_2 enormously enhance the potency of the 2-amino benzoxazinone lead (**5**, R = H, R_5 = H). Typically, a branched (iPr) group in 2-NHR serves to increase k_{on} (70-fold), lower k_{off} (17-fold) and block the N-cyclization mode of deacylation. A methyl or ethyl group at C_5 further diminishes the rate of deacylation 45-160 fold. The combination of 2-NHiPr and 5-Et lowers the deacylation rate 770-fold over the lead structure and increases the potency (lowers K_i) by a factor of 67,000, thereby demonstrating the utility of the mechanistic approach in guiding the design of enzyme inhibitors.

ACKNOWLEDGEMENTS

We are grateful to Tim F. Tam, Everton Thomas, and Roland Billedeau for their syntheses of all of the benzoxazinones and model system esters.

REFERENCES

Alazard, R., J.-J. Béchet, A. Dupaix, and J. Yon (1973). *Biochim. Biophys. Acta, 309,* 379-396.

Butula, I., V. Vela, and B. Zorc (1981). *Croatia Chemica Acta, 54,* 105-108.

Caplow, M. and W. P. Jencks (1962). *Biochemistry, 1,* 883-893.

Copp, L. J., A. Krantz, and R. W. Spencer (1986). *Biochemistry,* in press.

Goering, H. L., T. Rubin, and M. S. Newman (1954) *J. Am. Chem. Soc., 76,* 787-791.

Hamprecht, G. and B. Weierzer *German Patent* 2,914,915.

Hedstrom, L., A. R. Moorman, J. Dobbs, and R. H. Abeles (1984). *Biochemistry, 23,* 1753-1759.

Hegarty, A. F. and T. C. Bruice (1970). *J. Am. Chem. Soc., 92,* 6575-6588.

Karl, D. H. and G. Wolfgang *German Patent* 2,315,303; *Chem. Abstr. 82,* 16851g.

Krantz, A., T.-F. Tam, and R. W. Spencer (1985) *Eur. Pat. Appl.* EP 147,211,03; *Chem. Abstr. 104,* 88568r.

Lempert, K. and G. Doleschall (1963a). *Tetrahedron Lett.,* 781-784

Lempert, K. and G. Doleschall (1963b). *Tetrahedron Lett.,* 1195-1199.

Papodopoulos, E. P. (1980). *J. Heterocyclic Chem., 17,* 1553.

Papodopoulos, E. P. (1981). *J. Heterocyclic Chem., 18,* 515-518.

Papodopoulos, E. P. and J. Petridou-Fischer (1982). *J. Heterocyclic Chem., 19,* 123- 126.

Papodopoulos, E. P. and C. D. Torres (1982). *J. Heterocyclic Chem. , 19,* 269-272.

Powers, J. C., J. W. Harper, K. Hemmi, A. Yasutake, and H. Hori (1986). *3rd SCI-RSC Med. Chem. Symp., 1985,* 241-256.

Quinn, D. M., J. P. Elrod, R. Ardis, P. Friesen, and R. L. Schowen (1980). *J. Am. Chem. Soc., 102,* 5358-5365.

Ringrose, P. S. (1983). *Biochem. Soc. Trans. ,11,* 804-808.

Sayigh, A. A. R., N. Haven, and H. Ulrich *U.S. Patent* 3,450,700.

Schowen, K. B. J. (1978). Solvent Hydrogen Isotope Effects. In R. D. Gandour and R. L. Schowen (Eds.), *Transition States of Biochemical Processes.* Plenum Press, New York. Chap. 6, pp. 225-283.

Schowen, K. B. and R. L. Schowen (1982). Solvent Isotope Effects on Enzyme Systems. *Meth. Enzymology, 87C,* 551-606.

Schnebli, H. P. (1985). In I. L. Bonta, M. A. Bray, M. J. Parnham (Eds.), *Handbook of Inflammation,* vol 5. Elsevier, New York. pp. 321-333.

Segel, I. (1975) *Enzyme Kinetics,* Wiley-Interscience, New York.

Spencer, R. W., L. J. Copp, B. Bonaventura, T.-F. Tam, T. J. Liak, R. Billedeau, and A. Krantz (1986). *Biochem. Biophys. Res. Commun., 140,* 928-933.

Stein, R. L. (1983) *J. Am. Chem. Soc. , 105,* 5111-5116.

Stein, R. L. (1985) *J. Am. Chem. Soc. , 107,* 5767-5775.

Teshima, T., J. C. Griffin, and J. C. Powers (1982). *J. Biol. Chem., 257,* 5085-5091.

MECHANISMS OF ACTION OF PROTEASES AND RELATED ENZYMES

Probing Enzyme Dynamics and Regulation by Cryo-studies

P. Douzou and G. Hui Bon Hoa
INSERM U. 310, Institut de Biologie Physico-Chimique,
13, rue Pierre et Marie Curie, 75005 Paris, France

ABSTRACT

The mechanisms of enzyme dynamics and regulation can be probed by cosolvent, low temperature and high pressure studies. These investigations reveal some of the most intimate processes of enzyme-specific activity and open the way to further analysis.

KEYWORDS

Solvent effect; low temperature studies; high pressure studies; protein dynamics; enzyme regulation.

INTRODUCTION

Enzymes are not only extremely effective catalysts but also the operating components of very fine regulatory mechanisms. Binding specific substrates, they must change configuration in such a way as to bring reactants and participating functional groups into a reactive configuration, and, in most cases, move parts of the protein aside to allow dissociation of the products.

Unfortunately most of available fast kinetic techniques are unable to give sufficient energetic and structural information about these dynamic processes, too fast to be analyzed, and even when work at subzero temperatures permit to stabilize kinetically significant enzyme-substrate and enzyme-product intermediates, no dynamic information is obtained.

Thus, one may sometimes obtain a series of "stop-action" pictures by X ray analysis of these intermediates in the crystalline state (Douzou, 1984) but activated complex governing their interconversion remain virtual.

© 1987 Pergamon Journals Ltd.
Printed in Great Britain

However, in most cases, cosolvents used as "antifreeze" to carry-out investigations at sub-zero temperatures can be exploited as perturbing agents : they influence reversibly and differentially most of the elementary steps of the sequential reaction in such a way as to slow down dramatically the interconversion of intermediates, and then to obtain a temporal resolution of the most kinetically significants and therefore of those involving conformational changes.

RESULTS AND DISCUSSION

In recent years, we carried-out the temporal resolution of enzyme reactions involving flavoproteins (Yagi, 1980), hemoproteins (Douzou, 1977), hydrolytic enzymes (Barman, 1986), and focused on the steps involving triggered conformational changes after binding of the substrate and before the dissociation of the product. Such results can be summarized as follows :

Two step binding of a substrate :

$$E + S \underset{}{\overset{K1}{\rightleftharpoons}} E - S \quad \text{(binding complex)} \qquad (1)$$

$$E - S \underset{k-2}{\overset{k2}{\rightleftharpoons}} E^{*}- S \quad \text{(induced fit complex)} \qquad (2)$$

Dissociation of a product :

$$E^{*}- P \underset{k-n}{\overset{k_n}{\rightleftharpoons}} E - P \qquad (3)$$

$$E - P \overset{K_{n+1}}{\rightleftharpoons} E + P \qquad (4)$$

Equilibrium (K1) and rate-constants (k+, k-) can be analyzed in isolation when conditions for a good resolution have been found (solvent composition, substrate concentration, pH, ionic strength, temperature) and then single step kinetic studies permit to obtain the rates of interconversion and thermodynamic data (enthalpy, entropy, free energy of activation, as well as activation volume by varying the pressure).

Such resolution in single steps obtained by a cosolvent and often with the help of subzero temperatures did permit to confirm the involvement of steps just postulated by fast kinetic recordings in normal conditions of medium and temperature, and also to analyze their responses to changes in "physiological" parameters such as pH and ionic strength. These appeared to act as modulators of the rate of interconversion of many intermediate, and then as "pace-setters" of a reaction pathway.

The most illustrative example was shown by myosin ATPase activity ($ATP \xrightarrow{MYOSIN} ADP + Pi$) : such reaction has been postulated as a seven steps process involving three triggered "isomerisations" (conformational changes) governing the activation of the initial enzyme-substrate complex and of the enzyme-products complexes releasing the products sequentially (Pi and ADP, respectively). Through conducting experiments in mixed solvents by suitable fast kinetic techniques (Barman,

1983), it was possible to confirm such assumption and to resolve the reaction step by step :

$$M + ATP \overset{K1}{\rightleftharpoons} M.ATP \underset{k-2}{\overset{k+2}{\rightleftharpoons}} M^{*}ATP \quad \text{(induced-fit)} \qquad (5)$$

$$M^{*}ATP \underset{k-3}{\overset{k+3}{\rightleftharpoons}} M^{**}ADP.Pi \quad \text{(chemical step)} \qquad (6)$$

$$M^{**}ADP.Pi \underset{k-4}{\overset{k4}{\rightleftharpoons}} M^{*}ADP.Pi \underset{k-5}{\overset{k5}{\rightleftharpoons}} M^{*}ADP + Pi\nearrow \underset{k-6}{\overset{k6}{\rightleftharpoons}} M.ADP \overset{K7}{\rightleftharpoons} M + ADP\nearrow \qquad (7)$$

(Release of Products)

The step $M^{**}ADP.Pi \rightleftharpoons M^{*}ADP.Pi$ (step 4) is rate-limiting and, surprisingly, is not influenced by the cosolvent. Therefore the overall rate is practically unaffected (turn-over constant) while the "pace" (rate of progress) or the reaction is affected. In these conditions, changes in pH or ionic strength appear to modulate K1, k2 and to a less extent k3, as shown in Table I.

Table I Effect of solvent composition on certain of the rate constant of SF1 ATPase at 15°C

Solvent Composition			Kinetic constant					
			K1 10^3M^{-1}	k2	k3	k-3 S^{-1}	k4	kcat
40%	5 mM KCl	pH 8	125	16	11	11	0.12	0.06
	150 mM KCl	pH 8	9	45	14	7	0.09	0.06
ethylene	150 mM KCl	pH 6.4	37	8	–	–	0.06	0.04
glycol	150 mM KCl	pH 9	–	–	–	–	0.12	0.07
water	5 mM KCl	pH8	>5	<160	13	6	0.12	0.08

Thus hydrogen ion and cation concentration act as "pace-setters" of the reaction, a result of importance since this reaction is normally coupled with mechanical effect in muscle contraction. We found other examples of "pace setting" of reactions (Barman, 1986) and such process could represent a fine regulatory mechanism in a number of enzyme-catalyzed processes.

These results clearly confirm that the rate of progress (pace) of a reaction is dependent on how rapidly the protein can undergo a conformational change which result from enzyme-substrate and enzyme-product interactions. These interactions are most often dramatically influenced by external conditions modulating the pace, and in fact the interconversion of intermediates.

Distinction between the rate and pace of a reaction is essential for a better understanding of the mechanism and impact of enzyme reactions studied in isolation, and single step kinetic studies certainly are a way to analyze these essential features.

It is tempting, and useful, to exploit the control of any elementary step to evalutate its thermodynamic parameters, and in fact its "activated" state. Working over temperature range of 80oC (from + 30 to - 50oC), as well as over pressure range of 1,000 to 1,500 atmospheres, it is possible to extract the "activation" parameters $\Delta H^{\neq}$, $\Delta S^{\neq}$ and $\Delta V^{\neq}$ respectively.

Their derivation from the single step kinetic techniques now available is a relatively easy matter, but their interpretation in structural and energetic terms is notoriously difficult. They are of course a reflection of the structural and energetic features of the activated complexes ($E.S^{\neq}$, $E.P^{\neq}$) but they are in themselves rather uninformative about the actual mechanisms involved in an interconversion.

These difficulties prompted us to select simple models such as abortive enzyme-substrate complexes and to use suitable fast perturbations in order to investigate rate-coefficients (k+, k-) through relaxations.

Our favorite model is a monooxygenase, a bacterial cytochrome P450 (extracted from Pseudomonas putida) which binds camphor in the first step of a sequential reaction leading to the hydroxylation of this specific substrate (Gunsalus, 1974). In the absence of a reducing agent of the iron-heme, the complex cytochrome P450-Cam is stable. Its formation involves a low spin to high spin transition (LS → HS) of the ferric iron (the cytochrome) is termed m, after monooxygenase, and in fact m^{o}, o for oxidized, in its ferric form). The reaction is expressed in the following way :

$$m^{o}_{LS} + S \rightleftharpoons m^{os}_{HS} \qquad (8)$$

In fact, there are strong arguments favoring a two-step process, in which the cytochrome binds camphor and remains in the low spin (LS) state, and next in which the complex m^{os}_{HS} undergoes a spin transition LS ⇌ HS. Such electronic transition is not rate-limiting and is presumably coupled with a rate-limiting conformational change (Sligar, 1976; Lange, 1979). These steps can be schematized as follows :

$$m^{o}_{LS} + S \underset{}{\overset{K1}{\rightleftharpoons}} m^{os}_{LS} \qquad (9)$$

$$m^{os}_{LS} \underset{k-2}{\overset{k2}{\rightleftharpoons}} m^{*os}_{HS} \qquad (10)$$

Working in mixed solvents (such as ethylene glycol-water 50 : 50 v/v) and at various concentrations in camphor, it is possible to resolve these steps, and to analyze the second one in isolation. The first one, at high concentration in camphor (> 500 μM) is too fast (0.4 ms) to be detected, while the second (rate-limiting) can be fairly analyzed, through recording of K equilibrium (k2/k-2) and kobs rate constant (k observed = k2 + k-2).

These equilibrium and rate constants obtained by using static (Hui Bon Hoa, 1982) and kinetic (Hui Bon Hoa (1973) (rapid mixing) technics built in our laboratory, can be studied over a broad range of temperature (+40oC $\rightleftharpoons$ -40oC) and pressure (1 $\rightleftharpoons$ 1,500 atmospheres) to obtain estimates of the thermodynamic parameters of the overall process (ΔH^o, ΔS^o, ΔG^o, ΔV^o) and of the activation process ($\Delta H^{\neq}$, $\Delta S^{\neq}$, $\Delta G^{\neq}$, $\Delta V^{\neq}$), and we added a new possibility by inducing positive and negative pressure jumps perturbing the spin state equilibrium (LS $\rightleftharpoons$ HS) with relaxation rates (individual rate coefficients k2, k-2) varying with the external conditions.

Pressure is an important parameter which modifies the rate of most reactions, but in spite of expectations about its usefulness, it is not commonly used in enzymology. The reason for that is mostly technical, since pressure devices raise a number of difficult problems along with problems dealing with the effects of pressure on physical chemical parameters such as viscosity and pH, not to mention possible denaturing effects. On the other hand, investigators using pressure have to heat complex data and, for instance, the actual physical significance of the "activation volumes" ($\Delta V^{\neq}$) defined as the volume difference $V^{\neq}_{ES} - V_{ES}$ (where $V^{\neq}_{ES}$ is the volume of the activated complex $ES^{\neq}$) is yet uncertain.

Pressure is most often applied in static conditions to alter the equilibrium overall reactions. Its application to the kinetic studies of steps in isolation required the use of flow-stopped devices (Heremans, 1980) and of subzero temperatures (Balny, 1984), and in spite of constant progress in the making of appropriate devices, it is still difficult to practice single step kinetic studies as a function of pressure. Moreover the physical significance of data is not obvious because of unexpected variations of magnitude and sometimes of sign of the activation volumes under the influence of external conditions.

For these reasons, we decided to build an apparatus allowing to induce positive and negative pressure-jumps (30 to 200 bars) in the pressure range 1 - 1,500 bars, and between + 40 and - 40oC. In such conditions one can shift the low spin $\rightleftharpoons$ high spin equilibrium of the $m^{os}_{LS} \rightleftharpoons m^{*os}_{HS}$ step, record relaxation curves, and from K equilibrium and k observed values at same final pressure obtain the rate coefficients k2 (LS → HS) and k-2 (HS → LS).

Positive jumps (200 bars) determine a monoexponential increase of low spin specie (HS → LS), while negative jump determine a decrease of low spin specie (LS → HS). The dead-time of the apparatus is 2.6 msec. Relaxation curves are monoexponential, reversible and symmetric. The rate coefficients k2 and k-2 can be calculated form each one (i.e., after positive or negative P. jump).

Figure 1 shows an example of relaxations obtained after positive and negative jumps (200 bars), with two different final pressures (772 and 976 bars). From kobs and Keq values, k2 and k-2 have been evaluated, and it has been found that k2 (LS → HS) varies significantly with final pressure, a result due to the fact that k2 is pressure sensitive while k-2 is unsensitive.

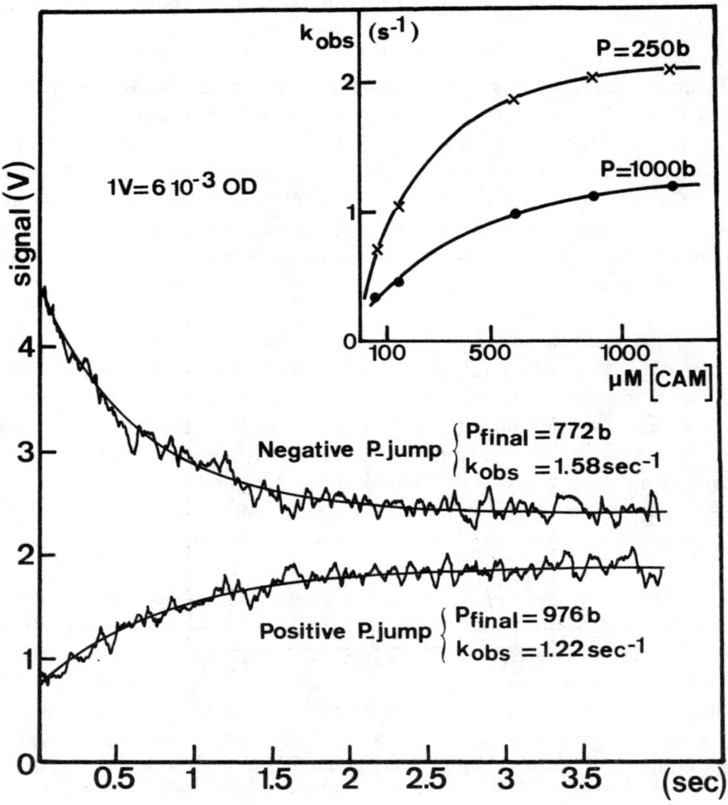

Fig. 1. Relaxation curves after a negative and positive P-jump in 50% (v/v) ethylene glycol-water. Tris buffer 100 mM pH* 7.8, t = 0oC, [P450] = 4.6 μM, [cam] = 400 μM, [KCl] = 60 mM. Insert : relaxation observed rate as a function of concentration of camphor at two pressure.

Recording of the individual rate coefficients k2 and k−2 as a function of pressure (between 1 and 1 200 bars) permits to evaluate the activation volume ($\Delta V^{\neq}$) in varied conditions and then to gather data which may permit to determine the actual physical significance of this parameter. Present results suggest that $\Delta V^{\neq}$ values do not merely express a physical volume change, but rather interactions between the reacting species and their environment, and further comparative investigations are needed to find out a correct interpretation of $\Delta V^{\neq}$. In spite of these difficulties, exploration of the spin-state changes of cytochrome P450 and of other hemoproteins would be rewarding in several aspects since spin-state changes obviously result from conformational changes that are rate-limiting of the spin-transitions. Both are quite temperature sensitive and the significant variations of absorption on the Soret band indeed permit nice and confortable recordings of their rates. Figure 2 shows absorption spectra of the intermediate m^{os} after a pressure jump from 1,400 to 1,700 bars, in 80% (w:w) glycerol at − 20oC.

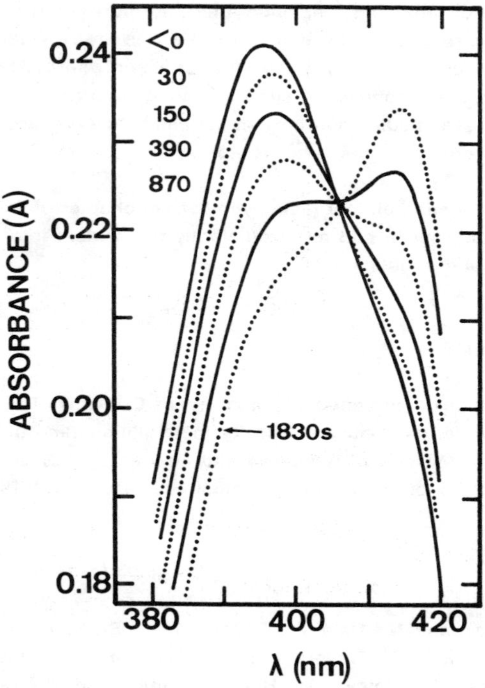

Fig. 2 Absorption spectra at various times (sec) after a P-jump from 1,400 to 1,700 bars . Sample in 80% (w/w) glycerol at −20oC. Tris buffer pH* 7 with 50 mM KCl.

The pressure jump (300 bars) inflicted in less than 5 msec determines a progressive increase in low spin, and such shift in the spin-state equilibrium can be recorded at 391 and 417 nm, with an isobestic point at 404 nm. In these conditions (the viscosity is of about 10^3 cp) the high to low spin transitions lasts almost two thousands of second. The spin transition rates versus viscosity have been measured in various mixed solvents in different volume ratio (Marden, 1982), and these rates have shown a $1/\eta$ dependence a result confirming Kramers assumption according to which the Arrhenius relationship (with its constant pre-exponential term A) is valid, at best, in a limited range of viscosity. On the other hand the dramatic viscosity effect observed clearly indicates the protein flexibility : if the protein was a rigid system, the solvent viscosity would not affect its interior transitions.

In summary, the above result is just an example among others of the usefulness of single step kinetic studies involving temperature and pressure as varying external parameters and suitable fast kinetic recordings, and of simple models with exploitable signals.

Preliminary results obtained on bacterial cytochrome P450 with camphor as specific substrate clearly show that the barriers governing the transitions between low and high spin states are not stationary potential energy barriers postulate by the conventional transition state theory, and suggest that the free energy of binding of camphor may control the free energy of the spin transition and therefore of the conformational relaxation. In spite of such results and assumption, we are in the very early stages of the work and much has to be done to analyze the structural and energetic features of the interconversion of intermediate complexes such as those reported in this lecture and which are just part of intensive studies now carried out almost everywhere to establish the structural dynamics of proteins at rest and at work.

So enzyme dynamics is becoming an increasingly important chapter of the long history of enzymology, now integrating more and more efficiently the protein into mechanisms of enzyme specific activity and regulation.

ACKNOWLEDGEMENT

The authors would like to thank their colleagues and friends C. Balny, Th. Barman and F. Travers for their experimental contributions and for many illuminating discussions. This work was supported by the Ministry of Research and Technology (Grant n° 85-TO717) and by the National Institute of Health and Medical Research (INSERM U. 128).

REFERENCES

Balny, C., J.L. Saldana, and N. Dahan (1984). *Anal. Biochem.*, 139, 178-189.

Barman, T., D. Hillaire, and F. Travers (1983). *Biochem. J.*, 209, 617-626.

Barman, T., F. Travers, C. Balny, G. Hui Bon Hoa, and P. Douzou (1986). *Biochimie*, 68, 1041-1051.

Douzou, P., (1977). In "*Cryobiochemistry*". Academic press, London-New York, pp. 185.

Douzou, P., and G. Petsko (1984). *Advances in Protein Chemistry*, 36, 246-361.

Gunsalus, I.C., J.R. Meeks, J.D. Lipscomb, P. Debrunner, and E. Munck (1974). (O. Hayaishi, ed.) In *Molecular Mechanisms of Oxygen Activation*, pp. 559, Academic Press, New York.

Heremans, K., J. Snauwaert, and J. Rijkenberg (1980). *Rev. Sci. Instrum.*, 51, 806-808.

Hui Bon Hoa, G., and P. Douzou (1973). *Anal. Biochem.*, 51, 127-136.

Hui Bon Hoa, G., P. Douzou, N. Dahan, an d. C Balny (1982). *Anal. Biochem.*, 120, 125-135.

Lange, R., and P. Debey (1979). *Eur. J. Biochem.*, 94, 485-489.

Lange, R., G. Hui Bon Hoa, P. Debey, and I.C. Gunsalus (1979). *Eur. J. Biochem.*, 94, 491-496.

Marden, C.M., and G. Hui Bon Hoa (1982). Eur. J. Biochem., 129, 111-117.

Sligar, S.G. (1976). *Biochemistry*, 15, 5399-5406.

Yagi, K., R. Lange, and P. Douzou (1980). *Biochem. Biophys. Res. Comm.*, 57, 370-374.

Molecular Modeling of the Specificity Requirements of Dipeptidyl Peptidase IV

Jochen Heins, K. Neubert, A. Barth and Richard L. Schowen

Department of Chemistry and Center of Biomedical Research, University of Kansas, Lawrence, Kansas 66045–0046, and Biosciences Section, Department of Biochemistry, Martin Luther University, Domplatz 1, DDR 4020 Halle, German Democratic Republic

ABSTRACT

Computational molecular modeling is used to construct an image which should predict the reactivity of substrates of dipeptidyl peptidase IV at least qualitatively and in some cases quantitatively.

KEYWORDS

Dipeptidyl peptidase IV; substrate reactivity; molecular modeling; molecular mechanics; molecular graphics.

INTRODUCTION

Dipeptidyl peptidase IV (DP IV) [EC 3.4.14-] is an enzyme of considerable interest, both from the biomedical viewpoint and with respect to basic questions of enzyme catalytic power. A membrane-bound enzyme of various mammalian tissues, it also occurs in such pathogenic bacteria as Flavobacterium meningosepticum (Walter et al., 1980). A volume (Barth and Schowen, 1987) now appearing reviews current information on DP IV.

Recent work (Feller et al., 1982) has shown that DP IV is localized in T-cells of the immune system and is in fact a good T-cell marker. For this reason, DP IV is of potential clinical significance. It may also play a role in the regulation of T-cells.

DP IV from pig kidney is a dimeric enzyme with two identical subunits (Kenny et al., 1976) and it is a serine enzyme (Barth et al, 1974). The second order rate constant of the inactivation of DP IV by diisopropyl fluorophosphate is about 25 $M^{-1}s^{-1}$.

This enzyme catalyzes the hydrolytic removal of dipeptide units from the N-terminus of oligopeptides that meet the following structural prerequisites:

1. Pro or Ala must be present at the scissile (P_1) position;

AB VOL65-P

© 1987 Pergamon Journals Ltd.
Printed in Great Britain

2. A protonated and thus positively charged nitrogen must be present at the N-terminus;

3. The residue at the P_1 position n(i.e., at the C-terminal and of the scissile bond must be other than Pro;

4. The peptide bond configuration between P_1 and P_2 positions must be trans (Fischer at al., 1983).

The rate limiting step for the DP IV catalyzed hydrolysis of proline substrates (Pro in P_1) is the deacylation reaction and of alanine substrates (Ala in P_1) is the acylation reaction (Wolf et al., 1978).

The exploration of the interactions between enzyme and substrate in the encounter complex(es), the recognition complex(es), the acylation transition state(s), the acyl enzyme(s) and the deacylation transition state(s), as well as in the various complexes and transition states associated with release of the two products, is a promising approach to understanding the physiological origins of the catalytic power and specificity of DP IV. Ideally such an exploration begins with the X-ray crystallographic determination of the structure of the enzyme and, if possible, of its complexes with substrates and transition state analogs. In the case of DP IV, this step cannot be expected in the near future. At present, even the primary sequence of DP IV remains unknown. Its crystal structure seems still further away. Although DP IV has been crystallized, the available crystals are unsuitable for crystallographic study. Even with good crystals, the large molecular weight and complex, glycoprotein structure of DP IV may present formidable obstacles to structure elucidation.

In the meantime, the computational approach of molecular graphics and modeling offers a method that requires only substrate-structural information. Here we present the results of a modeling study based on rate data for substrates of variable structure. As we shall emphasize, the nature of the model obtained is complex. Most importantly, it is not an image (positive or negative) of the active site of the free enzyme or even of the enzyme in the enzyme-substrate complex. But it is the entity of central significance and greatest importance, whether the problems to be solved have to do with chemical factors in catalysis, the molecular evolution of enzymes or the design of enzyme effectors as pharmacological agents.

COMPUTATIONAL METHODS

Modeling software: The SYBYL system (Tripos Associates, St. Louis, Missouri), versions 3.2 and 3.3 were used with an Evans & Sutherland PS300 graphic interface which permits molecular manipulation (translation, rotation and display) in real time. The following describes the relevant programs of the SYBYL package and how they were used.

Sketch

Substrate molecules were constructed from the standard fragment library of SYBYL. The torsional angles of substrate molecules were then altered to bring the substrate into the closest attainable superposition upon the crystal structure of (S,S)-Ala-Pro-pNA (Reck and Barth, 1981).

Fit

This program uses a least-squares procedure in order to "match" geometrically sets of paired atoms in two molecules i.e., to bring the corresponding atoms in two molecules as closely as possible onto the same point in space. This procedure was employed to bring different substrate molecules into roughly the same orientation as a starting structure for the energy minimization procedure.

Maximin

This energy minimization procedure uses the Molecular Mechanics method (Force Field method) to find minimum-energy structures. The MULTIPLE option means that several substrate molecules were minimized together.

Mvolume

This routine creates a three-dimensional grid where each node has a value which measures the electron density at that point in space. From this information, it generates a surface which, for a particular molecule, encloses a chosen fraction of the total electron density of the molecule. The surface thus generated may be considered to enclose the effective volume of the molecule. The program can also calculate sums and differences of the effective volumes of different molecules. In this way, the combined, or summed, volume of all active substrates was generated, as well as differences between this summed volume of all active substrates and the volume of any compound which was not hydrolyzed by DP IV.

Energy

The potential energy of a molecule is calculated as the weighted sum of the terms for torsional, bond stretching, angle bending, electrostatic and van der Waals interactions. In this work, unit weighting factors were assigned to all energy terms.

DATA BASE

The data upon which we shall rely in constructing the model for DP IV activity are given in Table 1. They consist of the values of k_{cat}/K_m and k_{cat} for a series of dipeptide p-nitroanilides (Barth and Schowen, 1987; Fischer et al., 1983; Wolf et al., 1978).

RESULTS

Template Structure

The structure of Pro-Pro-pNA was fitted to the crystal structure of Ala-Pro-pNA (Reck and Barth, 1981). Thereafter the resulting conformation of Pro-Pro-pNA was minimized with respect to its energy. This minimization process did not lead to a significant alteration of the structure, which means that the structure is in a local energy minimum.

Superposition of Active Substrates

The two substrates employed in establishing the template, Pro-Pro-pNA and Ala-Pro-pNA, are among the most active substrates known for DP IV, with k_{cat}/K_m around $10^6 M^{-1} s^{-1}$ (Table 1). There are a number of other substrates with approximately the same reactivity, but a second category of three substrates is known with k_{cat}/K_M in the range $10^{1-3} M^{-1} s^{-1}$ consisting of the p-nitroanilides of Ala-Ala, (R)-Ala-Ala and Aib-Ala. These five substrates thus span five orders of magnitude in rate. If this range of rates arises solely from differences in transition state stabilization, it represents a range of 29.3 kJ/Mol

Table 1. Rate Constants for DP-IV Catalyzed Hydrolysis of X-Pro-pNA

X	kcat [sec^{-1}]	Km*10^5 [M]	kcat/Km*10^{-6} [sec^{-1} M^{-1}]
Pro	81.99	1.42	5.77
Ala	96.31	2.36	4.08
Val	59.94	1.47	4.08
Ile	43.84	1.36	3.22
Glu	58.81	2.10	2.80
Leu	90.91	3.42	2.66
Phe	107.64	4.43	2.43
Gln	93.18	3.84	2.43
Ser	90.63	3.99	2.27
Tyr	104.55	5.15	2.03
Lys	103.98	6.90	1.51
Gly	130.00	16.80	0.77
Asp	44.60	5.86	0.76
Asn	76.40	11.90	0.64

in enzyme-substrate interaction energy in the transition state. This is a maximum range of interaction energy because of course some or in principle all of the range of free energies of activation could arise from differences in reactant-state free energy. Thus by considering these substrates as a set of "active substrates" we may ask what substrate features are required to admit a substance to the group with activation free energies falling into this 29-30 kJ/Mol range. For this purpose, the five three last-named substrates were superimposed on the Pro-Pro-pNA template and the energy was jointly minimized (MULTIPLE option). The resulting combination of structures is portrayed in Figure 1.

From this, it can readily be seen that a rather narrow range of substrate spatial dispersion is associated with the set of active substrates.

Figure 2 provides two other renditions of the structural characteristics of this class of active substrates. Figure 2a corresponds to application of the logical OR operation to the molecular volumes (MVOLUME) and is thus the sum of the molecular volumes. Figure 2b results from application of the logical AND operation; it thus summarizes features common to all active substrates.

Evaluation of Other Substrates

As a tool for evaluation of potential substrates, Figure 2a should be useful. As an initial hypothesis, it can be supposed that substrates with structure that protrudes beyond the limits of this envelope may encounter repulsive interactions with enzyme structure. The repulsive energy will countervail against transition-state stabilization, raising the free energy of activation outside the 29-30 kJ/Mol range and resulting in a lowered substrate reactivity.

This application is illustrated in Figure 3. Here the total volume for active substrates (Figure 2a) is subtracted from the molecular volume of (S)-Ala-(R)-Ala-pNA (top) and from the molecular volume of (S)-Ala-(R)-Pro-pNA (bottom), after each of these substrates had been superimposed on the template and the energy minimized. Substantial excess volume remains, suggesting a lowered activity. In fact, these materials are not considered substrates and have $k_{cat}/K_m \ll 10^{-3}M^{-1}s^{-1}$.

Related uses of this molecular-volume comparison were made with other compounds which are not substrates for DP IV (results not shown). All such compounds have a considerable excess volume which may be the reason for nonsusceptibility to DP IV. Therefore a similar approach could be used in general to get qualitative information concerning substrate specificity.

In Table 2 are shown the values for torsional angles of all active and inactive compounds after the MAXIMIN procedure.

Ala-Ala-pNA (S,R), Ala-Pro-pNA (R,S), ALa-Pro-pNA (S,R) and Aib-Pro-pNA are not hydrolyzed by DP IV. Comparing the torsional angles of the "best" substrate (Pro-Pro-pNA) with those of Aib-Pro-pNA reveals that there are not significant differences and therefore our results do not explain substrate specificity of DP IV on the basis of the values of the torsional angles.

Quantitative Description of Substrate Specificity

For some of the proline substrates, i.e., Val-Pro-pNA, Ile-Pro-pNA, Leu-Pro-pNA, ALa-Pro-pNA and Pro-Pro-pNA it is possible to construct a structural model where the P_2-part of the N-terminal amino acid residue was as closely fitted to that of Pro-Pro-pNA as possible. The underlying hypothesis for this procedure is that the properties of a given compound with respect to DP IV depend on how

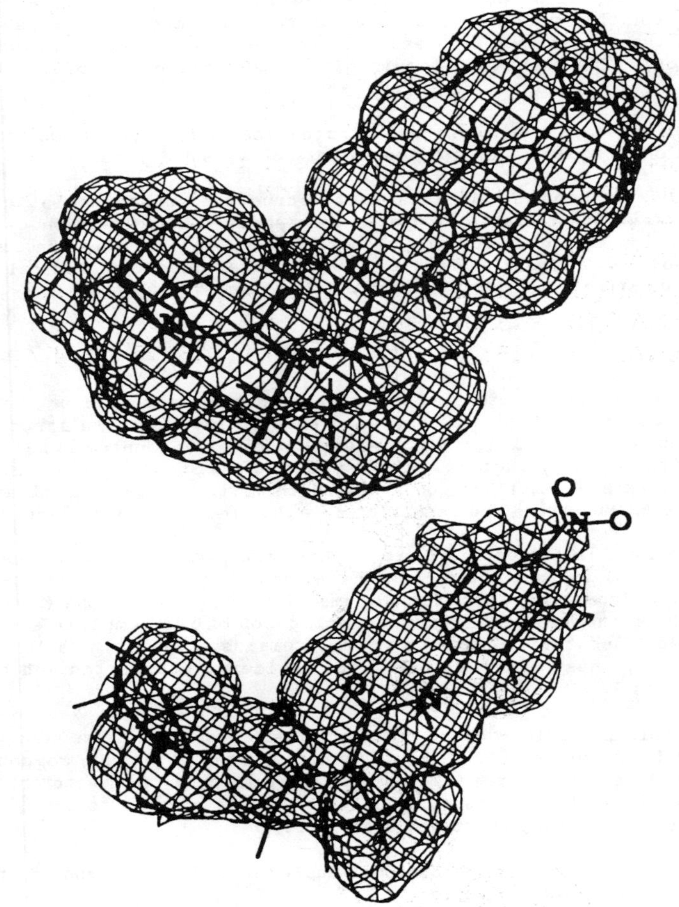

Figure 2. (a) Top: Application of the logical OR to the active structures. (b) Bottom: Application of the logical AND to the active structures.

Table 2
Torsional angles of active substrates and inactive compounds after MAXMIN

Substrate	ψ_1	ω_1	φ_2	ψ_2	ω_2
Pro-Pro-pNA	152.2	174.6	-63.5	153.1	172.3
Ala-Pro-pNA	146.6	178.8	-65.5	153.3	175.2
Ala-Ala-pNA	156.7	169.5	-81.6	150.9	176.2
Aib-Ala-pNA	178.1	-177.1	-124.4	143.0	-173.9
Ala-Ala-pNA(R,S)	130.1	153.6	-77.9	151.3	160.8
Ala-Ala-pNA(S,R)	-132.6	-178.7	-154.1	129.4	170.5
Ala-Pro-pNA(R,S)	92.2	173.7	-59.7	151.7	165.2
Ala-Pro-pNA(S,R)	132.8	162.9	-1.6	115.8	-136.6
Aib-Pro-pNA	151.4	177.6	-69.1	157.2	174.7

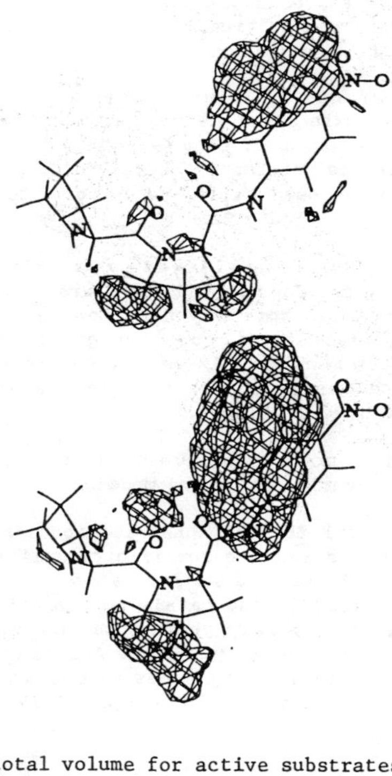

Figure 3. The total volume for active substrates (Figure 2a) is subtracted from the volume of (S)-Ala-(R)-Ala-pNA (top) and from the volume of (S)-Ala-(R)-Pro-pNA (bottom), the outline of (S)-Pro-(S)-Pro-pNA being included in both cases as a guide.

well it can attain a conformation similar to the Pro-Pro-pNA conformation, at least in the P_2 part of the molecule. In Figure 4 is shown a matrix which describes this structural feature.

By relating matrix A (which contains the structural information for a set of substrates) with their reactivity parameters (the activation free energy differences for either k_{cat} (ΔG^{*}_{3}) or k_{cat}/K_m (ΔG^{*}_{2}) for a given compound which may have methyl groups in the lettered positions, relative to those for Gly-Pro-pNA which has no corresponding methyl groups) it should be possible to calculate the contribution of each structural element (Figure 4) to the transition state stabilization of the acylation (ΔG^{*}_{2}) and deacylation (ΔG^{*}_{3}) reactions. The results of these calculations are shown in Table 3.

It is interesting to note the different effect of some structural elements on acylation and deacylation reaction, respectively (e.g., the C-beta position).

DISCUSSION

The Construct Produced by Modeling

The simplest statement about the nature of the model we have reached for the activity of DP IV is that it is represented by the combined-volume illustration of Figure 2a. This is on the one hand true, and on the other hand potentially a dangerous oversimplification.

The danger lies in the appearance of Figure 2a, which seems to represent a molecular volume in three-dimensional space. It is tempting to consider it a "negative image" of the active site of DP IV, and to imagine that when the crystal structure of the enzyme is determined in the future, it will be possible (if our modeling procedure has been valid and correct) to find in the active site the structural complement of Figure 2a.

Perhaps the most confident prediction allowed by our findings is that this structural complement will <u>not</u> be found in the structure of DP IV, as eventually determined by whatever means. Not only does the data base on which our model is constructed have no information about the ground-state active-site structure of DP IV, but in fact much that is well-known about enzyme catalysis tell us that the ground-state enzyme will not contain a region structurally complementary to the model we have developed.

We have noted that our model is not a negative image of the active site of DP IV. There is the question, however: what is the model?

Our model is a mathematical model for the substrate specificity of dipeptidyl peptidase IV, which describes substrate specificity qualitatively (Figure 3) and in some cases quantitatively (Table 3). Its basis is the interaction between enzyme and substrate, mainly in the catalytic transition state. It will be coincidental, therefore, if it should contain much information about the resting, free-state structure of the enzyme. The features of the model that allow it to predict substrate specificity makes it suitable for testing new substrate classes and also possible new inhibitors of DP IV.

REFERENCES

Barth, A. and R. L. Schowen (1987). *Dipeptidyl Peptidase IV*. Deutscher Verlag der Wissenschaften, Berlin

Figure 4

MATRIX A

	C^{β}	$C^{\beta'}$	C^{γ}	$C^{\gamma'}$	C^{δ}
Ala	1	0	0	0	0
Val	1	1	1	0	0
Ile	1	1	1	0	1
Leu	1	0	1	1	1
Pro	1	0	1	0	1

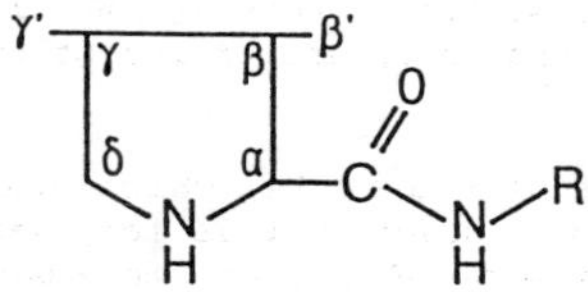

Table 3

Relative contribution of each methyl group to the free energy difference $\Delta\Delta G_2^*$ and $\Delta\Delta G_3^*$, calculated for Ala, Val, Ile, Leu and Pro in P_2 with respect to Gly–Pro–pNA

Position	$\Delta\Delta G_2^*$ [kcal/mole]	$\Delta\Delta G_3^*$ [kcal/mole]
C^{β}	−1.004	0.181
$C^{\beta'}$	0.351	0.376
C^{γ}	−0.351	−0.091
$C^{\gamma'}$	0.466	−0.063
C^{δ}	0.142	0.188

Barth, A., H. Schulz, and K. Neubert (1974). *Acta Biol. Med. Germ.*, 32, 157-174.

Feller, A. C., C.J. van Heijnen, R. E. Ballieux, and M. R. Parwaresch (1982). *Br. J. Haematol.*, 51, 227-234.

Fischer, G., J. Heins, and A. Barth (1983). *Biochim. Biophys. Acta*, 742, 452-462.

Kenny, A. J., A. G. Booth, S. G. George, J. Ingram, D. Kershaw, E. J. Wood, and A. R. Young (1976). *Biochem. J.*, 155, 169-182.

Reck, G., and A. Barth (1981). *Cryst. Struct. Comm.*, 10, 1001-1005.

Walter, R., W. H. Simmons, and T. Yoshimoto (1980). *Molecular & Cellular Biochemistry*, 30, 111-127.

Wolf, B., G. Fischer, and A. Barth (1978). *Acta Biol. Med. Ger.*, 37, 409-420.

β-Lactamases — Normal Peptidases?

M. I. Page and S. Buckwell
Department of Chemical and Physical Sciences, The Polytechnic, Huddersfield, HD1 3DH, UK

ABSTRACT

The mechanism of the β-lactamase catalysed hydrolysis of β-lactam antibiotics is discussed and compared with that of normal serine proteases. There is little evidence for a basic group on the enzyme acting as a general base catalyst. The effect of chain length on the kinetic parameters for the hydrolysis of 6-alkyl-penicillins is described. The expulsion of the leaving group at C3' in cephalosporins is reversible and is not catalysed by β-lactamase.

KEYWORDS

β-lactamase, kinetics, mechanism, β-lactam antibiotics, substituent effects.

Penicillins 1 and cephalosporins 2 are typical β-lactam antibiotics and there are at least two families of enzymes which specifically recognise the β-lactam ring. The β-lactamases (penicillin amido β-lactam hydrolase EC 3.5.2.6 and EC 3.5.2.8) catalyse the hydrolysis of the β-lactam ring yielding biologically inactive amino acid products. Plasmid-mediated formation of β-lactamases by bacteria is largely responsible for the resistance of many bacteria to the normally lethal action of

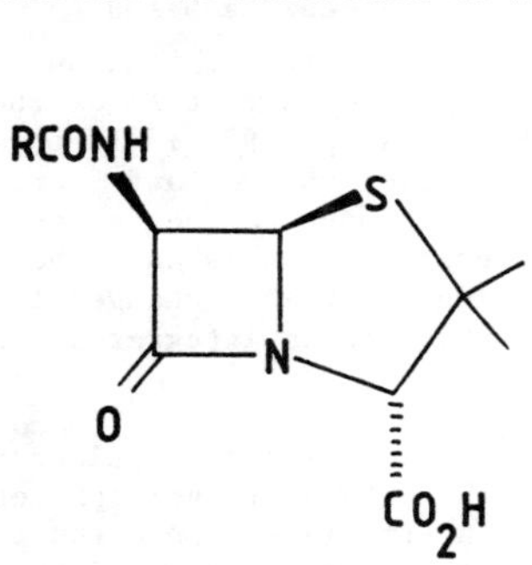

1

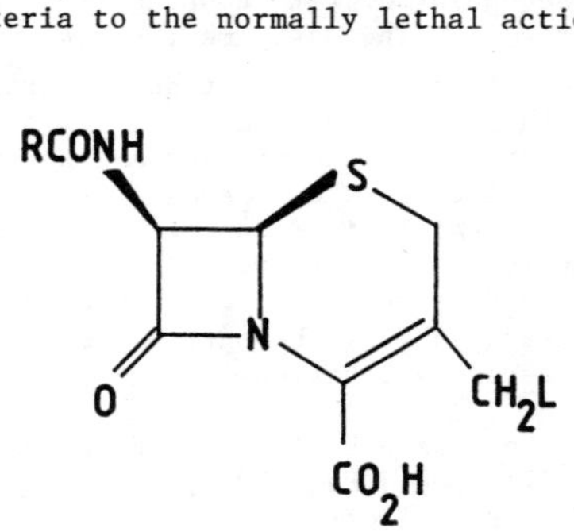

2

© 1987 Pergamon Journals Ltd.
Printed in Great Britain

β-lactam antibiotics(Hamilton-Miller and Smith, 1979). The second class of enzymes that interact with the antibiotic are the transpeptidases which are the inhibitory targets for β-lactams when they kill bacteria. The intermediates or derivatives formed between transpeptidases and the β-lactam antibiotics are relatively stable and are responsible for inhibiting the enzyme and preventing cell wall synthesis(Frère, 1985). Another class of enzymes that indirectly recognise β-lactams are those involved in antibiotic synthesis from acyclic peptide precursors. Some of these enzymes appear quite discrete, for example, gram-positive bacteria excrete a β-lactamase in the culture medium during growth but also have a distinct enzyme bound to the membrane (Johnson and others, 1975; Neilson and Lampen, 1983). In Gram-negative bacteria, β-lactamases are often located in the periplasmic space and appear indistinguishable from the membrane bound enzyme (Mantsala and Suominen, 1981).

β-Lactamases may be regarded as peptidases although at present only β-lactams and not acyclic peptides have been shown to be substrates for these enzymes. Monocyclic β-lactams can however, be substrates for β-lactamases (Pratt, Anderson and Odeh, 1980). Similar to peptidases, there are different classes of enzymes which catalyse the hydrolysis of β-lactams. Mechanistically these are serine and zinc enzymes, similar to, say, α-chymotrypsin and carboxypeptidase respectively. Based on their primary structures there have been three classes recognised, A, B and C. (Ambler, 1980; Jaurin and Grundstrom, 1981). Class B is distinguished by the requirement of zinc ions for activity whereas classes A and C, although structurally distinguishable, act by using an active site serine (Knott-Hunziker and others, 1982; Fisher and others, 1981: Cohen and Pratt, 1980). The serine β-lactamase catalysed hydrolysis of penicillins and cephalosporins is thought to occur by a two step acylation-deacylation mechanism.

In the serine proteases the active site histidine is apparent from the pH-rate profile and shows a pK_a of 7. There is no similar evidence for the β-lactamases. The pH-rate profile shows a very broad bell-shaped pH-dependence which is apparently independent of substrate. The classical kinetic model for describing this is equation (1):

$$\frac{k_{cat}}{K_m} = \frac{k_{max}}{1 + H^+/K_1 + K_2/H^+} \qquad (1)$$

where the enzyme has two ionising groups with ionisation constants K_1 and K_2 for EH_2 and EH respectively. One group is in its unprotonated form and the other is in its protonated form, i.e. EH, for productive binding to occur. The pH-independent second order rate constant for the reaction of EH and the substrate is given by k_{max}. The assignment of pK_a values to the free enzyme based on K_1 and K_2 assumes that (i) proton transfer steps are much faster than formation and breakdown of the enzyme substrate complex and (ii) only the ionised state of the enzyme and substrate can bind (Peller and Alberty, 1959; Knowles, 1976). If this assumption is applied to the *B.cereus* β-lactamase A, pK_a's of about 5 and 9 are obtained. It is not possible to deduce whether these acid/base groups on the enzyme are mechanistically important. Kinetic parameters reported for this and other β-lactamase catalysed reactions were calculated by fitting absorbance changes of the substrate measured over the entire course of the reaction to an integrated form of the Michaelis-Menten equation.

Contrary to common belief, the β-lactam ring of penicillins and cephalosporins is not readily opened(Page 1984; Page, in press). The attack of nucleophiles on the β-lactam carbonyl carbon to form the tetrahedral intermediate 3 is often reversible and expulsion of the attacking group N_u occurs faster than cleavage of the β-lactam C-N bond (Page 1984; Page, in press) (Page and Proctor, 1984; Gensmantel and Page, 1979). If the attacking atom of the nucleophile is bonded to an ionisable proton then general base catalysis is required to facilitate the reaction by removing

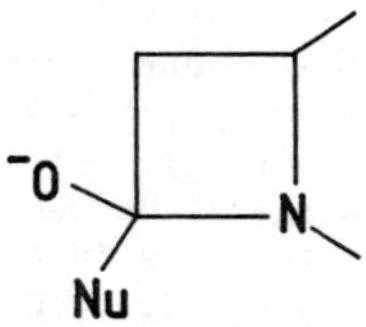

3

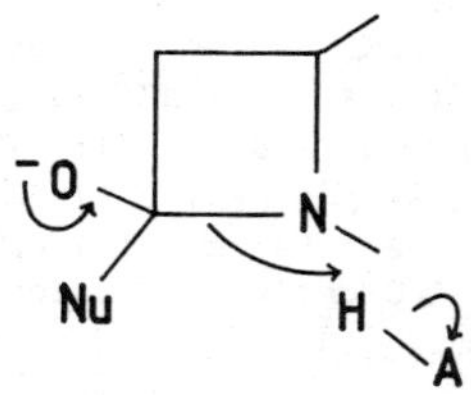

4

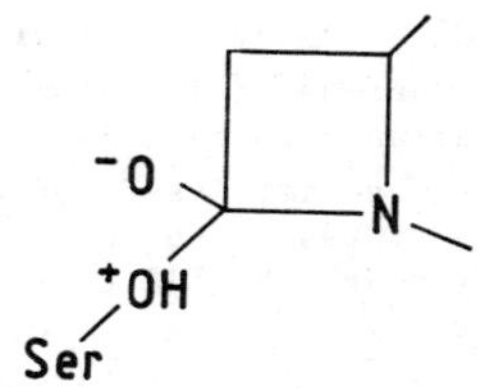

5

this proton. Cleavage of the β-lactam C-N bond is also facilitated by general acid catalysis 4 (Page, 1984; Page, in press; Page and Proctor, 1984; Gensmantel and Page, 1979). So despite the assumed high reactivity of β-lactams and the release of strain upon ring opening it appears that cleavage of the β-lactam requires similar catalytic assistance to that for acyclic peptides. The alcoholysis of penicillin occurs through the alkoxide ion and there is no evidence of the reaction proceeding by general base catalysed addition of the neutral alcohol. The rate limiting step for alcoholysis isbreakdown of the tetrahedral intermediate 3 (Proctor, Davis and Page, unpublished observations). The tetrahedral intermediate 5, formed by attack of the β-lactamase serine on penicillin is therefore likely to require a general base to remove the proton from oxygen and probably a general acid to protonate the β-lactam nitrogen before the acyl enzyme 6 can be formed. In chymotrypsin these two functions are carried out by the same histidine residue but, *a priori* with β-lactamase they could be carried out by two separate residues 7. The logic of serine proteases forming an ester intermediate is usually ascribed to esters being chemically more reactive than amides. In the case of β-lactams this logic no longer applies because penicillin is as reactive as an ester (Page, 1984; Page, in press). It is therefore conceivable that the β-lactamase catalysed

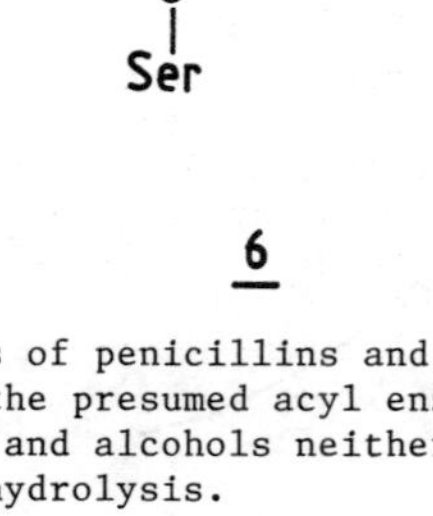

6

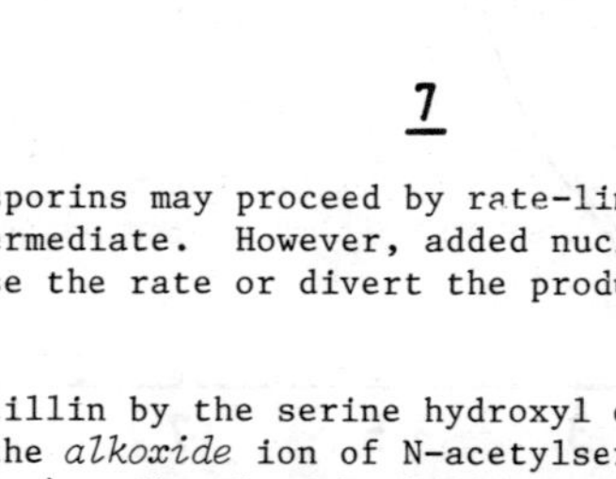

7

hydrolysis of penicillins and cephalosporins may proceed by rate-limiting deacylation of the presumed acyl enzyme intermediate. However, added nucleophiles such as amines and alcohols neither increase the rate or divert the products from being those of hydrolysis.

The rate of alcoholysis of benzylpenicillin by the serine hydroxyl of β-lactamase is *ca.* 10^7 fold faster than that by the *alkoxide* ion of N-acetylserinamide(Proctor, Davis and Page, unpublished observations). Clearly, the binding energy between the enzyme and penicillin is used to affect an alternative mechanism to simple attack by alkoxide ion. The second-order rate constant, k_{cat}/K_m for the β-lactamase catalysed hydrolysis of benzylpenicillin is *ca* 10^8 greater than that for the hydroxide ion catalysed hydrolysis. This rate enhancement is similar to that shown by chymotrypsin for its peptide substrates.

The solvent isotope effect on k_{cat}/K_m for the acylation of α-chymotrypsin is near unity whereas it is between 2 and 3 for the deacylation of acylchymotrypsins (Bender and others 1964; Bender and Kézdy 1975). The effect on k_{cat}/K_m for the β-lactamase catalysed hydrolysis of β-lactam antibiotics is about unity for a variety of substrates. The solvent isotope effects (SIE) on k_{cat} and K_m are between 1 and 2. Although the SIE on k_{cat} is indicative of proton transfer occuring in the rate limiting step it is not of the magnitude to firmly draw this conclusion. Chemical modification studies have so far been unsuccessful in showing unequivocally the presence of an essential group, other than serine, in any β-lactamase. It is still not clear therefore whether the ionisable groups indicated by the pH-rate profile are simply required for binding or are involved mechanistically.

The second order rate constant for the hydrolysis of benzylpenicillin by β-lactamase is $5.0 \times 10^7\ M^{-1}\ s^{-1}$ at 30^0 which is approaching that expected for a diffusion controlled reaction. In an attempt to see if some substrates show higher activity and also to gain an insight to subsidiary binding sites several penicillins and cephalosporins have been synthesised. Various substituents have been introduced at the C6 and C7 acylamido side chains of penicillin and cephalosporin, respectively, and at C3 in cephalosporins.

The effect of increasing the alkyl side chain $CH_3(CH_2)_n$, in penicillins is shown in Figure 1. A maximum in k_{cat}/K_m is shown for octyl penicillin, but this is less reactive than benzylpenicillin.

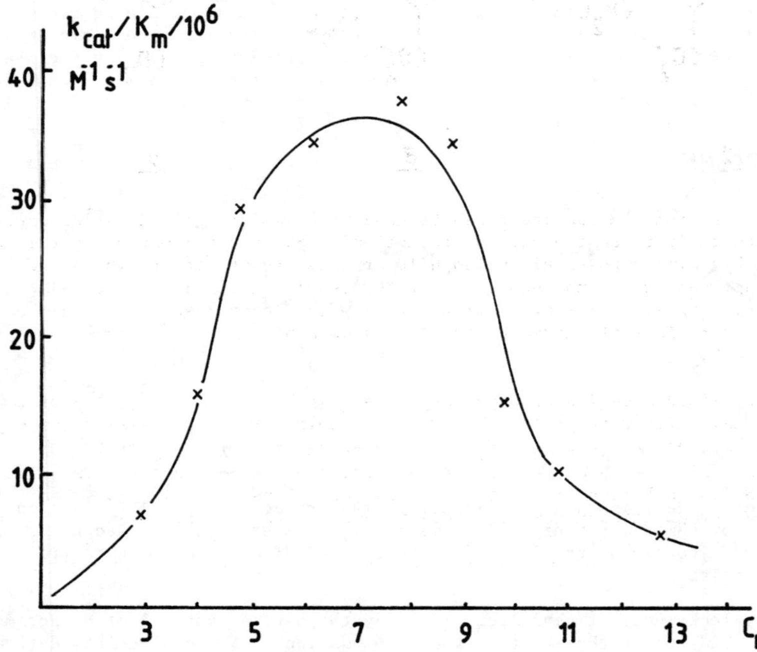

Figure 1. A plot of k_{cat}/K_m for the *B.cereus* β-lactamase catalysed hydrolysis of alkylpenicillins at pH 7.0 and 30^0.

A similar plot is observed for 7- alkyl cephalosporins, except that the maximum occurs with the undecyl derivative. Substitution of a para carboxylate in the phenyl cephalosporin (2, L = OAc) reduces k_{cat}/K_m 150 fold. This could indicate the presence of a negatively charged group on the protein, e.g. a carboxylate, which repels the similarly charged group on the substrate. Such a group on the enzyme could, of course, act as a general base catalyst in the acylation step.

Cephalosporins 2 differ from penicillins by having a leaving group at C3´ which is often expelled during the reactions of these antibiotics with nucleophiles that open the β-lactam ring (Page, 1984; Page, in press). There have been many suggestions that this process is concerted and that this mechanism is important for the lethal action of cephalosporins towards bacteria (Boyd and Lunn, 1979; Bundgaard, 1975; O'Callaghan and others, 1972). However, there have been several reports suggesting that β-lactam ring opening and fission of the C3´-L group are separate processes and an intermediate enamine 8 is formed (Page and Proctor 1984; Gensmantel and Page, 1979) (Faraci and Pratt, 1984; Agathocleous and others 1985; Grabowski, Douglas and Smith, 1985).

Despite this experimental evidence, theoretical arguments are still used to justify a concerted process (Boyd, 1985). There is in fact, an equilibrium between the intermediate enamine 8 and the product α,β-unsaturated imine 9. Expulsion of the

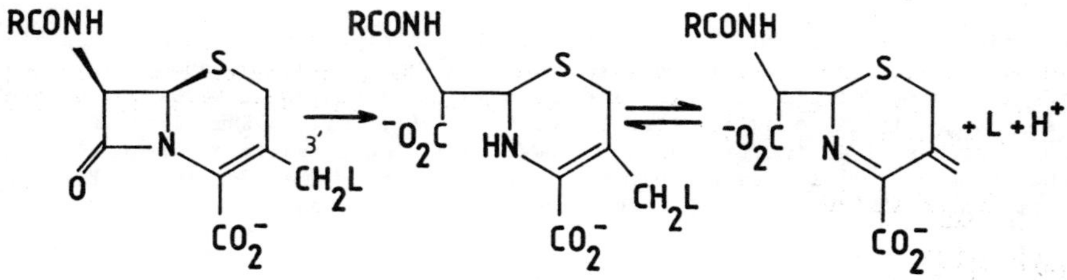

2 8 9

leaving group L from the enamine 8 can be made rate limiting by adjusting experimental conditions to increase the rate of formation of 8. Non-enzymatically this can be achieved by using a powerful nucleophile or substituent effects which facilitate β-lactam ring opening but not expulsion of the leaving group L. Using β-lactamase as a catalyst for β-lactam ring opening, it is possible to detect the enamine 8 at high concentrations of enzyme as the latter does not catalyse expulsion of the leaving group L.

We have synthesised a series of cephalosporins 2 with thiol leaving groups at C3´ (L=SR) that have different pK_a's. The *B.cereus* β-lactamase I catalysed hydrolysis of these cephalosporins show k_{cat}/K_m to be independent of the pK_a of the C3´ substituents, which is inconsistent with expulsion occurring in the rate limiting step. At pH 7.0 cephalosporins with good leaving groups at C3´e.g. pyridine give the α,β-unsaturated imine 9 whereas those with poor leaving groups at C3´ e.g. n-butanethiol give the enamine 8. These two products have different uv spectra, notably between 260 and 270 nm, and nmr spectra. The addition of 10^{-2} M n-butane thiol to an aqueous solution of 9 at 10^{-4} M at pH 7.0 generates 8. This can also be observed if the α,β-unsaturated imine 9 is generated from cephaloridine (L=pyridine). The addition of up to 10^{-2} M pyridine to solutions of 9 does not regenerate 8.

By a step-wise addition of a thiol to aqueous solutions of 9 uv spectra intermediate between those corresponding to 9 and 8 are observed. The equilibrium constant between 7 and 8 is given by equation (2):

$$K = \frac{[9][RS^-]}{[8]} = \frac{[9]}{[8]} \frac{K_a\ [RS]_{tot}}{1 + K_a/H^+} \qquad (2)$$

where K_a is the dissociation constant of the thiol RSH. The equilibrium constant K can be determined by measuring the change in absorbance at 260 nm as a function of thiol concentration.

If the pK_a of the thiol is > pH, the ratio of imine 9 to enamine 8 is pH independent. However, if pH > pK_a of RSH the ratio decreases with decreasing pH. The reversibility of this action is readily demonstrated by increasing the pH of a solution of 8 which regenerates 9.

The equilibrium constants for other thiols have been measured and there is a linear relationship between log. K and the pK_a of the thiol which gives a Brönsted β_{lg} of 0.71. The formation of 8 is favoured with basic thiol anions.

It is not inconceivable that a nucleophilic group on the enzyme, which is the killing target for cephalosporin antibacterial activity, could undergo the Michael type addition to the α,β-unsaturated imine to irreversibly inactivate the enzyme.

The kinetic behaviour of the *B.cereus* β-lactamase catalysed hydrolysis of cephalosporins shows some unusual characteristics. 'Lag' and 'burst' phenomena have been previously reported for penicillin substrates and interpreted in terms of conformational changes in the enzyme (Citri and others, 1984; Citri, Samuni and Zyk, 1976). Our observations of an initial burst followed by a slower reaction were due to the presence of some β-lactamase II, the zinc metalloenzyme. The addition of EDTA or purification of the enzyme completely removed this phenomenon.

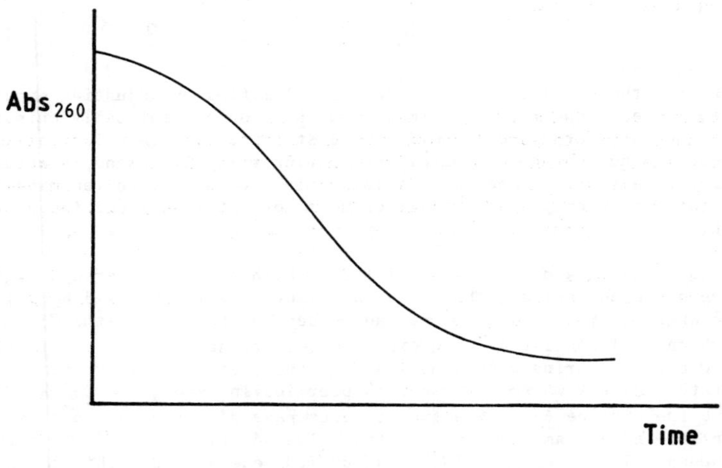

Figure 2. Change in absorbance with time for the *B.cereus* β-lactamase catalysed hydrolysis of some cephalosporins at pH 7.0 and 30°.

More unusual is the initial slower reaction followed by a faster one. Under saturation conditions changes in UV absorbance are as shown in Figure 2. With cephaloridine as substrate this observation can only be seen under flow reaction conditions. However, with thiol groups at C3′ (2 L=SR) the slower reaction is seen under normal kinetic conditions. The addition of thiols to the reaction does not remove this behaviour. The readdition of substrate during or at the end of the reaction regenerates similar behaviour. This is not consistent with a substrate or product induced conformational change in the enzyme. It is also not consistent with one of the products catalysing the breakdown of an intermediate.

β-Lactamase catalyses the hydrolysis of alkyl esters of benzylpenicilloic acid to benzyl penicilloic acid. Some of these esters are better substrates than some of the cephalosporins. If the serine enzyme forms an acyl enzyme intermediate with benzylpenicillin as substrate this same intermediate will be generated from the penicilloyl esters. The effect of substituents on the β-lactamase catalysed hydrolysis of α-alkyl benzylpenicilloates is being investigated to elucidate whether acylation or deacylation is rate-limiting.

References

Agathocleous, D, Buckwell, S, Proctor, P, and Page, M. I. (1985). Recent advances in the chemistry of β-Lactam antibiotics. Ed. Brown, A. G. and Roberts, S. M. Roy. Soc. Chem., London, p.18.

Ambler, R.P. (1980). Philos. Trans. Roy. Soc. London, Ser. B. 289, 321.

Bender, M. L, Clement, C.E, Kézdy, F. J, and Heck, H. D, (1964). J.Am.Chem.Soc. 86 3680.

Bender, M. L, Kézdy, F. J, (1975). Proton transfer reactions. Ed. Caldin, E and Gold, V, Chapman and Hall, London, p.385.

Boyd, D. B, and Lunn, W. H. W, (1979). J.Med.Chem. 22, 778.

Boyd, D. B, (1985). J.Org. Chem, 50 886.

Bundgaard, H, (1975). Arch. Pharm. Chemi. Sci. Ed., 3 94.

Citri, N, Samuni, A and Zyk, N, (1976). Proc. Natl. Acad. Sci., USA, 73, 1048.

Citri, N, Kalkstein, A, Samuni,A, and Zyk, N, (1984). Eur.J.Biochem., 144 333.

Cohen, S. A. and Pratt, R. F, (1980). Biochemistry, 19 3996.

Faraci, W. S. and Pratt, R. F, (1984). J.Am. Chem. Soc., 106, 1489.

Fisher, J, Charnas,, R. L, Bradley, S. M. and Knowles, J.R, (1981). Biochemistry 20, 2726.

Frère, J. M, (1985). CRC Crit. Rev. Microbiol. 11, 299.

Gensmantel, N. P. and Page, M. I, (1979). J.Chem.Soc. Perkin Trans. 2. 137.

Grabowski, E. J. J, Douglas, A. W, and Smith, G. B, (1985). J.Am.Chem.Soc., 107, 267.

Hamilton-Miller, J.M. T. and Smith, J. T, (eds) (1979). β-Lactamases Academic Press New York.

Jaurin, B. and Grundstrom, T, (1981). Proc. Natl. Acad. Sci. USA , 78, 4897.

Johnson, K, Duez, C, Frère, J. M, and Ghuysen, (1975). Methods Enzymol., 43 687.

Knott-Hunziker, V, Petursson, S, Waley, S. G, Jaurin, B, and Grundstrom, T.(1982) Biochem. J., 207, 315.

Knowles, J.R, (1976). CRC Crit. Rev. Biochem. 4, 165.

Mantsala, P. and Suominen, I, (1981). Acta Chem., Scand., Ser. B, 35, 567.

Nielsen, J. B. K. and Lampen, J.O, (1983). Biochemistry, 22, 4652.

O'Callaghan, C. H, Kirby, S. M, Morris, A, Waller, R. E. and Duncombe, R. E. (1972). J.Bacteriol. 110, 988.

Page, M. I, (1984). Acc. Chem. Res., 17, 144.

Page, M. I, (in press). Adv. Phys. Org. Chem.

Page, M. I. and Proctor, P. (1984). J.Am.Chem.Soc. 106, 3820.

Peller, L and Alberty, R. A, (1959). J.Am.Chem. Soc. 81, 5907.

Pratt, R. F, Anderson, E. G, and Odeh, I, (1980). Biochem. Biophys. Res. Comm. 93, 1266.

Proctor, P, Davis, A. and Page, M. I, (unpublished observations) .

Nuclear Magnetic Resonance Studies of Amidase Catalysis

K. H. Röhm
Phillips University, Institute of Physiological Chemistry,
Marburg (Lahn), Federal Republic of Germany

ABSTRACT

The ^{18}O induced isotope shift in ^{13}C nuclear magnetic resonance spectroscopy was utilized to follow oxygen exchange reactions catalyzed by hog kidney aminoacylase I (EC 3.5.1.14) and by *E.coli* L-asparaginase (EC 3.5.1.1). With asparaginase the experiments clearly demonstrated the involvement of a covalent acyl enzyme in catalysis, while with aminoacylase no evidence in favour of such an intermediate was obtained.

KEYWORDS

Hydrolase reactions; intermediates; oxygen exchange; ^{13}C NMR; ^{18}O induced isotope shifts.

INTRODUCTION

Within the last two decades nuclear magnetic resonance spectroscopy (NMR) has gained considerable importance as a tool in biochemical research. Recent developments in instrumentation and methodology have extended the field of application of NMR from simple molecules to very complex systems. Today even intact cells or whole organisms can be studied by magnetic resonance techniques.

NMR spectroscopy is especially attractive for enzymologists as it provides both structural and dynamic information on enzymes and enzyme-catalyzed reactions in solution. NMR methods were utilized

© 1987 Pergamon Journals Ltd.
Printed in Great Britain

to study enzyme structure and conformation, to characterize enzyme-ligand complexes, and to identify intermediates and reaction pathways. Illustrative examples of such uses of NMR may be found elsewhere in this volume.

In the present contribution we will outline a NMR technique that permits one to follow oxygen exchange reactions of any kind, including those catalyzed by hydrolytic enzymes. Results obtained with L-asparaginase from E.coli and with hog kidney aminoacylase I will be discussed in some detail.

THE ^{18}O ISOTOPE SHIFT IN ^{13}C NMR SPECTROSCOPY

In 1979 Risley and Van Etten described an effect that had been predicted for theoretical reasons but apparently not observed before: When normal ^{16}O bound to carbon is replaced by ^{18}O the ^{13}C NMR resonance of the adjacent carbon is shifted to higher field. ^{18}O isotope shifts are small (0.015 - 0.055 ppm, depending on the type of bonding) but easily observable with most NMR spectrometers. When two or more chemically equivalent oxygens are bound to the same carbon their isotope shifts are additive. As an example, Fig. 1 shows the ^{13}C carboxyl signal of partially ^{18}O labeled acetate ion.

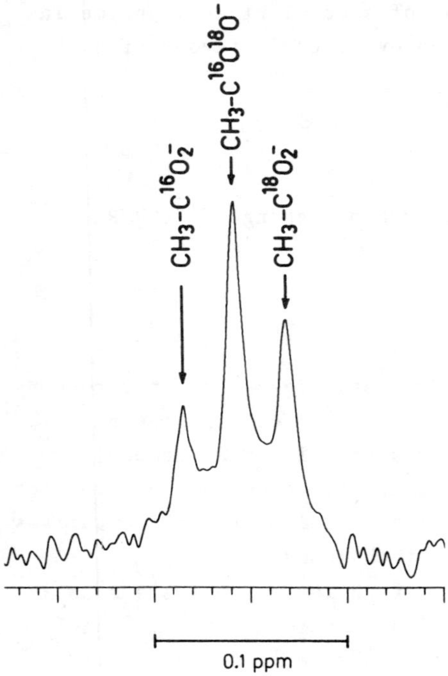

Fig. 1. ^{13}C NMR spectrum of ^{18}O labeled acetate. Only the carboxyl carbon resonance is shown. The spectrum was obtained with the Nicolet NTC-200 spectrometer operating at 50.3 MHz. A 500 Hz sweep width, a 90° pulse angle, 8.2 s acquisition time and a 16 K data block were used. 100 transients were accumulated, and an exponential lide broadening factor was applied before Fourier transformation

Three different signals separated by about 0.027 ppm are visible. Substitution of ^{16}O by ^{18}O does not change the relaxation times of the directly bonded carbon, the relative amounts of the various isotopic forms may, therefore, be calculated from the respective peak areas or peak heights.

Until recently, the usual way to follow oxygen exchange reactions was to allow a reaction to proceed in ^{18}O-enriched water, to isolate reactants or products, to convert them into volatile compounds like CO_2, and to measure the $^{16}O/^{18}O$ isotope ratio of that compound by mass spectrometry. Obviously, such a procedure is cumbersome and time-consuming and does not encourage systematic kinetic studies of oxygen exchange. Moreover, all positional information is lost, e.g. it remains obscure where in the original molecule oxygen had been incorporated. Utilization of the ^{18}O isotope shift on ^{13}C NMR offers a more elegant approach, and in the years following the discovery of the effect its usefulness was borne out by experiment, mainly in Van Etten's laboratory. Among the first systems examined were the acid-catalyzed oxygen exchange between acetic acid and water (Risley and Van Etten, 1981), oxygen exchange at the anomeric carbons of glucose and erythrose (Risley and Van Etten, 1982), the hydration of an epoxide by a microsomal hydratase (Risley, Kuo and Van Etten, 1983), and the hydrolysis of labeled benzyl phosphate by acid phosphatase (Parente, Risley and Van Etten, 1984).

In principle, natural abundance ^{13}C NMR is sufficient to measure ^{18}O isotope shifts. However, positional enrichment of the compounds with ^{13}C is often advantageous. Using highly ^{13}C-enriched reactants oxygen exchange can be monitored almost in real time. As the cost of $H_2{}^{18}O$ is considerable, it is usually more economical to follow the loss of heavy oxygen from prelabeled substrates rather than to measure the incorporation of ^{18}O from bulk water into unlabeled compounds.

OXYGEN EXCHANGE REACTIONS AS PROBES OF HYDROLASE CATALYSIS

The enzymes catalyzing amide bond cleavage may be classified within two main groups according to their mechanism of action: In many cases the existence of covalent acyl-enzyme intermediates was demonstrated or at least suggested; with many other amidases no such intermediate was detected. Among the findings indicative of covalent catalysis are a 'ping-pong' type of steady-state kinetics, 'initial bursts' of product release, or the trapping of intermediates by reaction with strong nucleophiles like hydroxylamine. In certain cases acylation of the enzyme by highly reactive ester substrates yields covalent intermediates stable enough to be isolated. It is still a matter of controversy whether the hydrolysis of amide substrates by these enzymes involves acylated stable intermediates as well.

A rather straightforward approach to distinguish double-displacement mechanisms from linear sequential schemes (i.e. reactions with and without stable intermediates) is based on the analysis of oxygen exchange reactions. The rationale behind this approach is schematically depicted by Fig. 2.

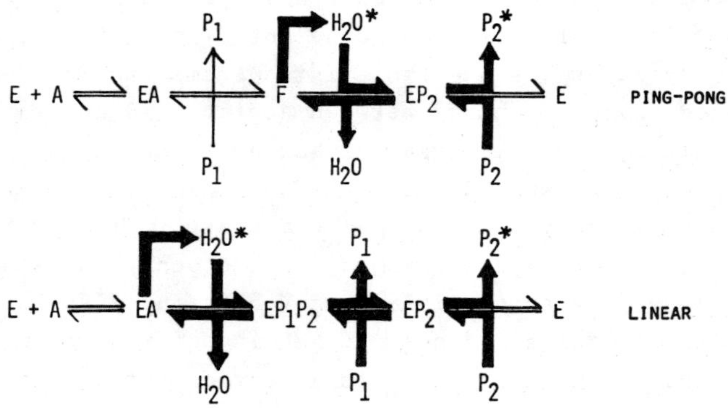

Fig. 2. Pathways of oxygen exchange for 'ping-pong' and sequential mechanisms

Whenever a stable intermediate F exists, the second product P_2 (usually the acyl part of the substrate amide A) should undergo enzyme-catalyzed oxygen exchange with water even in the absence of P_1. The linear mechanisms with a central EP_1P_2 complex requires that both P_1 and P_2 be present for exchange to take place. The reason ist that, in the latter case, the exchange reaction has to proceed all the way to EA and back, while in the double-displacement case exchange is mediated by formation and rehydrolysis of F which do not involve P_1.

AMINOACYLASE I

Aminoacylase I (EC 3.5.1.14) catalyzes the hydrolysis of acyl amino acids. The enzyme from hog kidney is still widely used to resolve amino acid racemates (Greenstein and Winitz, 1961). Early kinetic studies on its mechanism of action yielded contradictory results. Ötvös and his colleagues (1971) proposed a double- displacement scheme, a view supported by results of Chukrai and others (1979) who observed an initial burst of amino acid release during hydrolysis of acetyl methionine. On the other hand, Galaev and Svedas (1982) interpreted the results of a careful study of acetyl methionine synthesis/hydrolysis in terms of a sequential Uni-Bi mechanism. In view of this situation we decided to examine the aminoacylase-catalyzed oxygen exchange between acetate and water utilizing the ^{18}O isotope shift. Commercial sodium [1-^{13}C] acetate was converted to the acid chloride which was then hydolyzed in $H_2{}^{18}O$ to yield $^{13}C,{}^{18}O$-labeled acetate. No oxygen exchange between this compound and water was detected even after prolonged incubation with large amounts of aminoacylase I. However, addition of alanine immediately initiated exchange. The time course of the exchange reaction could be continuously monitored by ^{13}C NMR, and its rate constants were determined under a variety of conditions. The measured velocities and the pH-dependence of exchange were in excellent agreement with the values calculated on the basis of a linear sequential scheme (Röhm and Van Etten, 1986a). In addition, we studied acetyl alanine synthesis from alanine and acetate. Again ^{13}C NMR spectroscopy considerably facilitated the experiments. As shown in Fig. 3 , the carbonyl resonances of acetate, alanine and acetyl alanine are easily separated. Using [1-^{13}C] alanine and unlabeled acetate we were able to measure as a function of pH the rate of acetyl alanine formation and its equilibrium concentrations in the same assay. Unfortunately, the high ionic strengths required for such measurements alter the enzyme's kinetic behaviour in fashion not yet understood.Therefore, our data on acetyl alanine synthesis as well as those of Galaev and Svedas (1982) are difficult to interpret in a consistent way. However, they did not provide any evidence in favour of a covalent intermediate.

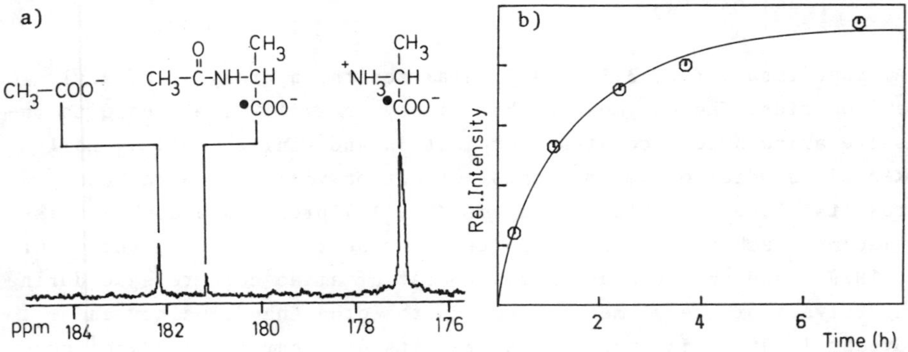

Fig. 3. Detection of aminoacylase-catalyzed acetyl alanine formation by ^{13}C NMR. The assay contained acetate (1 M), [1-^{13}C]alanine (75 mM) and enzyme (10 mg/ml) in phosphate/borate buffer, pH 8.5. ^{13}C NMR signals of reactants and products are shown in a), b) shows the time course of acetyl alanine formation as measured by the intensity of the respective signal.

ASPARAGINASE

L-Asparaginase from E.coli was shown to exhibit 'ping-pong'-kinetics (Ehrman and colleagues, 1972; Röhm and Schneider; 1971) and two alternative mechanisms, both consistent with the kinetic evidence, were put forward. One of these (Röhm and Schneider; 1971) as an uncommon feature, proposed enzyme-bound intramolecular aspartic acid anhydride as the covalent intermediate of asparaginase catalysis. The feasibility of such a scheme was backed by studies of model compounds the nonenzymatic hydrolysis of which nicely followed the suggested sequence of steps. The model provoked a certain amount of debate, however, until recently direct evidence for or against the existence of an anhydride intermediate was unavailable.

Again, studies of oxygen exchange appeared to offer an approach to test the anhydride hypothesis by experiment. Anhydrides may undergo hydrolysis at either carbonyl group which are of comparable reactivity. Thus with aspartic acid anhydride as a stable intermediate of asparaginase catalysis one would expect that during the hydrolysis of substrates in $H_2{}^{18}O$ both carboxyl groups of the resulting aspartic acid should incorporate ^{18}O unless one of them is entirely unaccessible by solvent (an unlikely assumption in view of the known properties of the enzyme). One further anticipates that asparaginase should catalyze oxygen exchange between aspartic acid and water in the absence of ammonia, creating an ^{18}O labeling pattern similar to that obtained upon substrate hydrolysis.

In order to facilitate detection of small amounts of incorporated ^{18}O we synthesized L-aspartic acid $^{13}C,^{18}O$-labled in both carboxyls and in the ß-carboxyl exclusively (Röhm and Van Etten, 1985). Our results on the regiospecificity of ^{18}O incorporation and oxygen exchange (Röhm and Van Etten, 1986b) were quite clear: Asparaginase catalyzed a relatively rapid oxygen exchange between the ß-carboxyl of aspartic acid and water. Exchange at C-1 was not detectable, even after periods of time long enough to permit many cycles of complete exchange at C-4. Similarily, during hydrolysis of aspartic acid-4-methyl ester in $H_2{}^{18}O$ heavy oxygen was incorporated exclusively into the ß-carboxyl group of aspartic acid.

In conclusion, while our results do not lend support to the anhydride hypothesis, they are fully consistent with a ß-aspartyl enzyme intermediate in asparaginase catalysis. Results of Handschumacher and his colleagues (Peterson, Richards and Handschumacher, 1977) suggest that Ser-109 of asparaginase may be the residue involved in covalent catalysis

ACKNOWLEDGEMENTS

My thanks to Robert L. Van Etten for advice, encouragement and warm hospitality. This work was supported by a research fellowship awarded by the Deutsche Forschungsgemeinschaft.

REFERENCES

Chukrai, E.S., D.Lauceniec A. Arens, and O.M. Poltorak (1979) Vestn.Mosk.Univ., Ser. Khim., 20, 118-122.

Ehrman, M., H. Cedar, and J. H. Schwartz (1972). J.Biol.Chem., 246, 88-94.

Galaev, I. Y., and V. K. Svedas (1982). Biochim.Biophys.Acta, 701, 389-394.

Greenstein, J. P., and M. Winitz (1961). Chemistry of the Amino Acids. Wiley, New York

Ötvös, L., E. Moravcsik, and G. Mady (1971). Biochem. Biophys. Res. Commun., 44, 1056-1063.

Parente, J. E., J. M. Risley, and R.L. Van Etten (1984). J.Amer.Chem. Soc., 106, 8156-8160.

Peterson, R. G., F. F. Richards, and R. E. Handschumacher (1977) J.Biol.Chem., 252, 2072-2076.

Risley, J. M., and R. L. Van Etten (1979). J.Amer.Chem.Soc., 101, 252-253.

Risley, J. M., and R. L. Van Etten (1981). J.Amer.Chem.Soc., 103, 4389-4392.

Risley, J. M., and R. L. Van Etten (1982). Biochemistry, 21, 6360-6365.

Risley, J. M., F.Kuo, and R.L. Van Etten (1983). J.Amer.Chem.Soc., 105, 1647-1652.

Röhm, K. H., and F. Schneider (1971). Hoppe-Seyler's Z.Physiol.Chem., 352, 1739-1743.

Röhm, K. H., and R. L. Van Etten (1985).J.Labelled Compds.Radiopharm., 12, 909-915.

Röhm, K. H., and R. L. Van Etten (1986a) Eur.J.Biochem., in press

Röhm, K. H., and R. L. Van Etten (1986b). Arch.Biochem.Biophys., 244, 128-136.

En Epoxide-reactive Residue in *Scytalidium lignicolum* Acid Protease B: Possible Involvement of Glutamic Acid Residue in the Catalytic Activity

D. Tsuru
Faculty of Pharmaceutical Sciences, Nagasaki University,
Bunkyo-machi 1–14, Nagasaki 852, Japan

ABSTRACT

The complete amino acid sequence of Scytalidium lignicolum acid protease B was established and compared with those of other acid proteases. The acid protease B was composed of 204 amino acid residues with threonine and valine as its amino- and carboxyl- termini, respectively. Only a slight homology was found in the sequences of the acid protease B and other acid proteases hitherto reported. The enzyme was modified by 1,2-epoxy-3-(p-nitrophenoxy) propane (EPNP) with the concomitant loss of its enzyme activity, and the EPNP-labeled peptides were isolated from the thermolysin-digest of the modified enzyme by HPLC. The amino acid sequence of the major peptide was determined to be Ile-Leu-Glu-Thr-Gly, which corresponds to the sequence of residue No 51-55 of the acid protease B. The amino acid sequence around Glu-53 of the enzyme showed a high homology with those in the active site aspartic acid residues of calf chymosin and porcine pepsin. These results suggest a possible involvement of Glu-53 of the acid protease B in its catalytic activity.

KEYWORDS

Acid protease; aspartic protease; Scytalidium lignicolum acid protease B; 1,2-epoxy-3-(p-nitrophenoxy) propane; EPNP; pepstatin; diazoacetyl-DL-norleucine methyl ester; DAN; amino acid sequence; active site; epoxide-reactive residue.

INTRODUCTION

Murao and co-workers (1972) reported that a strain of Scytalidium lignicolum produces three types of acid proteases. The enzymes designated acid proteases A-1, A-2 and B were highly purified , and their properties were characterized (Oda and Murao, 1974; Oda et al., 1975). Interestingly, the former two enzymes were insensitive to pepstatin, diazoacetyl-DL-norleucine methyl ester (DAN) and EPNP, all of which were known to be inhibitors specific for acid proteases (aspartic proteases) such as pepsin, chymosin and those of microbial origin. The acid protease B was also insensitive to pepstatin and DAN, while it was inhibited by EP-NP. The present report deals with the amino acid sequence of the acid protease B and that of EPNP-labeled peptides purified from the thermolysin-digest of the modified enzyme, and compares them with those around the DAN- and EPNP- reactive aspartic acid residues of other acid proteases.

© 1987 Pergamon Journals Ltd.
Printed in Great Britain

MATERIALS AND METHODS

Materials: *Scytalidium lignicolum* acid protease B (SLB) was purified by Oda and Murao (1974), and its homogeneity was confirmed by disc gel electrophoresis. Porcine pepsin, bovine trypsin and α-chymotrypsin were purchased from Sigma Chemical Co., U.S.A. Thermolysin was from Nakarai Chemical Co., Kyoto, Japan.

Assay of Proteolytic Activity: One ml of enzyme was added to 3 ml of 1.3% Hammersten casein in 20 mM lactate buffer, pH 2.5, at 37°C and incubated for 10 min. The reaction was stopped by the addition of 4 ml of 0.44M trichloroacetic acid, and after 20 min at 37°C, the reaction mixture was filtered. One ml of the filtrate was mixed with 5 ml of 0.44M Na_2CO_3 and 1 ml of Folin reagent at 37°C. The optical density at 660 nm was measured after 20 min.

Sequence Determination of Acid Protease B: The enzyme (SLB) was reduced and S-carboxymethylated (Cm-) by the method of Crestfield et al. (1963). Digestion of the Cm-SLB with trypsin and of further fragmentation of larger peptides with α-chymotrypsin, thermolysin, subtilisin and dilute acetic acid were carried out by the method of Maita et al. (1979). A denatured and unmodified SLB was also digested with thermolysin and pepsin to locate the disulfide bridges. Each peptide was separated and purified by column chromatography on Chromo Beads P, DE-52 or Biogel p-4 by monitoring through absorbance at 220 nm and 280 nm, reaction with ninhydrin and/or fluorescamine after alkali hydrolysis. Cystine-containing peptides were followed by CN-nitroprussid method or reaction with 5,5'-dithiobis (2-nitrobenzoic acid) after hydrolysis with alkali. Amino acid analyses of protein and peptides were carried out with a Jeol 6AH amino acid analyzer after hydrolysis with 6N HCl in evacuated, sealed tubes at 110°C for 22-72 hr. Tryptophan was detected by Ehrlich color reaction (Smith, 1953) and determined by amino acid analysis according to Matsubara and Sasaki (1969). The sequence of each peptide was determined by Manual Edman method (Blombäck et al., 1966), subtractive Edman method (Koningsberg and Hill, 1962) and dansyl-Edman procedure(Hartley, 1970). Automated Edman degradation was performed in a Jeol Sequencer (Model JAS 47K) and the phenylthiohydantoin (PTH) amino acids were identified by HPLC using a Shimadzu type LC-3A with a column (4.6 x 250 mm) of Zorbax ODS equilibrated with acetonitrile-sodium acetate buffer system. Digestions of Cm-SLB with carboxypeptidases A and B were carried out in 0.1M Tris-HCl buffer, pH 8.0, and those of peptides, in 20 mM borate buffer, pH 8.1, respectively.

Modification of SLB with EPNP and Identification of the Epoxide-reactive Residue. SLB was modified with a 50-fold molar excess of EPNP at pH 4.7 and 25°C for 72 hr under gentle stirring. After removal of undissolved reagent by centrifugation, the modified enzyme was purified by gel filtration on Sephadex G-25 previously equilibrated with 10 mM acetic acid. The number of epoxide molecule incorporated into protein was calculated from the absorbance at 280 nm and 315 nm (Tang, 1971), assuming that the molar extinction coeficient of SLB is 34,000 (Oda et al., 1975) and that of EPNP at 315 nm is 11,000 (Tang, 1971; Chen and Tang, 1972). The EPNP-modified SLB was digested with thermolysin at pH 6.7 and 37°C for 1 hr. The digest was then subjected to HPLC with a Toyo Soda HLC-803D apparatus equipped with a TSK-GEL ODS-120A (4.16 x 250 mm) column. The elution and separation of peptides were performed with an increasing linear gradient of acetonitrile, from 1 to 50%, containing 0.1% trifluoroacetic acid at a flow rate of 0.7 ml per min at 40°C. Elutions of peptides and EPNP-derivatives were monitored by following the absorbance at 220 nm and 315 nm. The fractions showing the absorbance at 315 nm were further purified repeatedly by HPLC with an increasing linear gradient of acetonitrile, from 1 to 50%, containing 10 mM ammonium formate at a flow rate of 0.4 ml per min. The EPNP-labeled peptides were subjected to automated Edman degradation with an Applied Biosystems 470A sequencer and the resulting PTH-amino acids were identified by HPLC. The amino terminal amino acids of the labeled peptides were determined by the dansyl method (Gray, 1967).

RESULTS AND DISCUSSION

Complete Amino Acid Sequence of the Acid Protease B.

Tryptic peptides of Cm-SLB were separated by chromatography on a column of DE-52. Five fragments were isolated and their amino acid sequences were determined. Overlapping peptides were obtained by the enzymatic digestion of Cm-SLB, and their sequences were determined. By combination of these results, the whole amino acid sequence of SLB was established. To locate disulfide bridges, the heat-denatured SLB was digested by thermolysin or pepsin and the cystine-containing peptides were purified by column chromatography and paper electrophoresis. Three to five peptides were obtained and subjected to performic acid oxidation (Gray, 1967). From the comparisons of their amino acid compositions and sequences with the whole sequence of SLB, the location of three disulfide bridges was established. The complete covalent structure of SLB is shown in Fig. 1. The enzyme is a single polypeptide composed of 204 amino acid residues with threonine and valine as its amino- and carboxyl-termini, respectively.

```
                                   10                                      20
Thr-Val-Glu-Ser-Asn-Trp-Gly-Gly-Ala-Ile-Leu-Ile-Gly-Ser-Asp-Phe-Asp-Thr-Val-Ser-
                                   30                                      40
Ala-Thr-Ala-Asn-Val-Pro-Ser-Ala-Thr-Gly-Ala-Ser-Gly-Gly-Ser-Ser-Ala-Ala-Trp-Val-
                                   50                                      60
Gly-Ile-Asp-Gly-Asp-Thr-Cys-Gln-Thr-Ala-Ile-Leu-Glu-Thr-Gly-Phe-Asp-Trp-Tyr-Gly-
                                   70                                      80
Asp-Gly-Thr-Tyr-Asp-Ala-Trp-Tyr-Glu-Trp-Tyr-Pro-Glu-Val-Ser-Asp-Asp-Phe-Ser-Gly-
                                   90                                     100
Ile-Thr-Ile-Ser-Glu-Gly-Asp-Ser-Ile-Gln-Met-Ser-Val-Thr-Ala-Thr-Ser-Asp-Thr-Ser-
                                  110                                     120
Gly-Ser-Ala-Thr-Leu-Glu-Asn-Leu-Thr-Thr-Gly-Gln-Lys-Val-Ser-Lys-Ser-Phe-Ser-Asn-
                                  130
Glu-Ser-Ser-Gly-Leu-Cys-Arg-Thr-Asn-Ala-Glu-Phe-Ile-Ile-Glu-Asp-Phe-Glu-Glu-Cys-
                                  150                                     160
Asn-Ser-Asp-Gly-Ser-Asp-Glu-Phe-Val-Pro-Phe-Ala-Ser-Phe-Ser-Pro-Ala-Val-Glu-Phe-

Thr-Asp-Cys-Ser-Val-Thr-Ser-Asp-Gly-Glu-Ser-Val-Ser-Leu-Asp-Asp-Ala-Gln-Ile-Thr-
                                                                          200
Gln-Val-Ile-Ile-Asn-Asn-Gln-Asp-Val-Thr-Asp-Cys-Ser-Val-Ser-Gly-Thr-Thr-Val-Ser-
        204
Cys-Ser-Tyr-Val
```

Fig. 1 Complete covalent structure of *Scytalidium lignicolum* acid protease B

It has been reported that pepstatin-sensitive acid proteases show high homology in their three dimensional structures as well as in their amino acid sequences, regardless of their origin (Tang, 1977). On the contrary, the amino acid sequence of SLB is markedly different from those of pepstatin-sensitive acid proteases. Only slight homology was, however, observed in the sequences between SLB and other acid proteases. The sequence around Asp-98 in SLB showed some homology to those around EPNP-reactive Asp-32 of pepstatin-sensitive acid proteases. Furthermore, the sequence Nos 47 to 55 in SLB was homologous to those around active site Asp-215 of calf chymosin and porcine pepsin, although Asp-215 in the animal enzymes was replaced by Glu-53 in SLB.

Epoxide-reactive Residues of the Acid Protease B.

Modification of SLB with EPNP resulted in 90% reduction of the enzyme activity. The number of epoxide molecules incorporated into protein was calculated to be 1 from the absorbance at 315 nm and 280 nm. The EPNP-labeled SLB was digested with thermolysin and then subjected to HPLC (Fig. 2), the EPNP-labeled peptides being repeatedly purified on HPLC to homogeneity. The analytical data of these fractions were shown in Table 1. Peak ET1 contained neither amino acid nor peptide and was concluded to be 1,2-dihydroxy-3-(p-nitrophenoxy) propane (EPNP-OH), a hydrolysate of EPNP, from its retention time on HPLC. Peak ET2 was separated to two fractions, a minor peak, ET2a, and major peak, ET2b. The sequence of ET2a

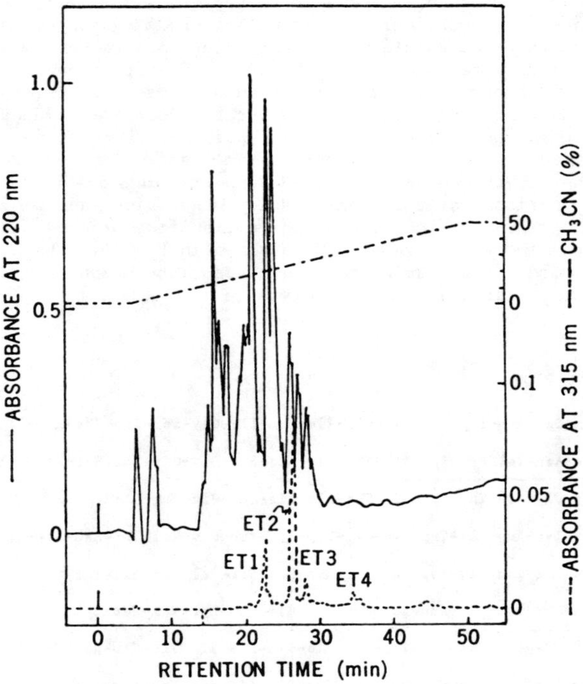

Fig. 2. Elution profile of the thermolysin-digest of EPNP-modified SLB on HPLC

TABLE 1 Chemical Structure of 315 nm-Positive Fractions Purified from Thermolysin-digest of EPNP-labeled SLB

Fractions	Amino acid composition	Chemical structure	Amino-terminal	Yield (%)
ET1	Not detected	EPNP-OH	/	/
ET2a	Asx(1.1),Glx(1.0) Ile(0.9)	Ile-Glu-Asp 136	Ile	4
ET2b	Thr(0.9),Glx(0.9),Gly (1.0),Ile(0.9),Leu(1.0)	Ile-Leu-Glu-Thr-Gly 53	Ile	21
ET3	Not detected	Unknown	/	/
ET4	Not detected	EPNP	/	/

( ): relative molar ratio

corresponded to that of residue Nos 134-136, and the sequence of ET2b coincided to that of residue Nos 51-55 of SLB. The molar ratio of EPNP to ET2b was calculated to be just 1.0 on the basis of absorbance and amino acid analysis.

If EPNP was ester-linked to carboxyl group of acidic residues of SLB, as reported in other acid proteases, the modifier would be released from peptides by alkali treatment, resulting in disappearance of absorbance at 315 nm. To confirm this, ET2b was treated with 0.3N NH_2OH at pH 8.5 and 37°C for 24 hr. Peak ET2b disappeared on this treatment as checked by absorbance at 315 nm, and a new peak corresponding to EPNP-OH appeared. These results indicate that the dissociation of EPNP from the peptide takes place by alkali-treatment and that the binding of the modifier to ET2b is base-labile, presumably the ester-linkage between epoxide and carboxyl group of the peptide. Thus, Glu-53 in SLB was concluded to be the sole amino acid residue reactive to EPNP in the peptide.

No amino acid was detected in peak ET4, and the alkali-treatment of this peak led to the appearance of peak ET1, EPNP-OH. From these results and from its retention time on HPLC, peak ET4 was identified to be EPNP. The amount of peak ET3 was too small for its identification.

To confirm that our experimental procedures are correct, porcine pepsin was also modified by EPNP under the same conditions as above. About 3 moles of EPNP were incorporated into pepsin with a 90% loss of the activity. Analyses of EPNP-labeled sites in pepsin indicated that the modifier was incorporated into Asp-32, Asp-215 and Met-290 of pepsin. These results are in agreement with those reported by Tang and co-workers (Chen and Tang, 1972; Hartsuck and Tang, 1972).

In Fig. 3, the sequence around Glu-53 in SLB and those around active site aspartic acid residues in pepstatin-sensitive acid proteases (aspartic proteases) are compared. There are significant homologies among them; especially high homology is observed in the sequence around DAN-reactive aspartic acid residue of calf chymosin and that around Glu-53 of SLB. These results suggest a high possibility that Glu-53 is one of the acidic amino acid residues involved in the catalytic activity of SLB.

	53
SLB	-Cys-Gln-Thr-Ala-Ile-Leu-Glu-Thr-Gly-Phe-
	215
Calf chymosin	-Cys-Gln-----Ala-Ile-Leu-Asp-Thr-Gly-Thr-
Porcine pepsin A	-Cys-Gln-----Ala-Ile-Val-Asp-Thr-Gly-Thr-
Penicillopepsin	-Ser-Gly-Ile-Ala-Asp-Thr-Gly-Thr-
Rhizopus pepsin	-Asp-Thr-Gly-Thr-
	32
Calf chymosin	-Leu-Phe-Asp-Thr-Gly-Ser-
Porcine pepsin A	-Phe-Thr-Val-Ile-Phe-Asp-Thr-Gly-Ser-
Penicillopepsin	-Asp-Phe-Asp-Thr-Gly-Ser-
Rhizopus pepsin	-Asp-Thr-Gly-Ser-

Fig. 3. Comparison of the amino acid sequence around Glu-53 of SLB with those around the active site aspartic acid residues of some acid proteases.

Chemical modification and X-ray crystallographic analysis of pepstatin-sensitive acid proteases indicated that two acidic amino acid residues participate in the catalytic reaction (Tang, 1977; James et al., 1982; Subramanian and Davis, 1982). It remains ambiguous as to whether this is also true for SLB. To clarify this problem, the nucleotide-directed site specific mutagenesis for SLB has been undertaken in our laboratory.

REFERENCES

Blombäck, B., M. Blombäck, P. Edman, and B. Hassel (1966). Biochim. Biophys. Acta, 115, 371-396.
Chen, K. C. S., and J. Tang (1972). J. Biol. Chem., 245, 2566-2574.
Crestfield, A. M., S. Moore, and W. H. Stein (1963). J. Biol. Chem., 238, 622-627.
Gray, W. R. (1967). Methods in Enzymol., (Hirs, C. H. W., ed). 11, 469-475, Academic Press, New York.
Hartley, B. S. (1970). Biochem. J., 119, 805-822.
Hartsuck, J. A., and J. Tang (1972). J. Biol. Chem., 247, 2575-2580.
James, M. N. G., A. Sielecki, F. Salituro. D. H. Rich, and T. Hofmann (1982). Proc. Natl. Acad. Sci. U.S.A., 79, 6137-6141.
Koningsberg, W., and R. J. Hill (1962). J. Biol. Chem., 237, 2547-2561.
Maita, T., K. Morokuma, and G. Matsuda (1979). Hoppe-Seyler's Z. Physiol. Chem., 360, 1483-1495.
Matsubara, H., and R. M. Sasaki (1969). Biochem. Biophys. Res. Commun., 35, 175-181.
Murao, S., K. Oda, and Y. Matsushita (1972). Agric. Biol. Chem., 36, 1647-1650.
Oda, K., and S. Murao (1974). Agric. Biol. Chem., 38, 2435-2444.
Oda, K., S. Murao, and K. Morihara (1975). Agric. Biol. Chem., 39, 377-485.
Smith, I. (1953). Nature, 171, 43-44.
Subramanian, B. R., and D. R. Davis (1982). Biochemistry, 21, 6956-6962.
Tang, J. (1971). J. Biol. Chem., 246, 4510-4517.
Tang, J. (1977). "Acid Proteases: Structure, Function and Biology", Plenum Press, New York and London.

THE CATALYTIC POWER OF PROTEASES

AB VOL65-R

The Destabilization of Amide Group as the Possible Reason of Proteases Efficiency and Specificity

V. K. Antonov and S. L. Alexandrov
Shemyakin Institute of Bioorganic Chemistry, Academy of Sciences of the USSR, Moscow, USSR

ABSTRACT

The specificity and the efficiency of proteolysis have been analyzed in the framework of concept of the stabilization by enzyme of the most reactive form of the substrate, namely, such form in which the amide group is pyramidalized and resonance destabilized. The dynamic systems, modelling the acylation by serine proteases were calculated and analyzed by the CNDO/2 method.

KEYWORDS

Proteases efficiency and specificity; serine proteases mechanisms; quantum-chemical analysis; amide group pyramidalization and resonance destabilization; proton transfer; nucleophilic attack.

The role of proteolytic enzymes in protein catabolism and regulation of many biochemical reactions is well known. Proteolytic enzymes also serve for the enzymologists as a touchstone in formulation of several concepts of efficiency and specificity of enzymes. These two phenomena discern enzymes from nonbiological catalysts.

It has often been suggested that proteases are enzymes with comparatively low efficiency. However, with the appropriately chosen substrates, the efficiency of these enzymes, as compared with the water catalyzed reaction, comprises a fourteen to fifteen order of magnitude. In comparison with the model reactions, catalyzed by low-molecular weight nucleophiles, the efficiency reaches eight to twelve orders (Antonov and Alexandrov, 1984). Such acceleration has to be explained in terms of enzyme and substrate structure.

Low-molecular weight catalysts are practically deprived of substrate specificity. Among proteases one can find enzymes with very narrow specificity, as for example kidney renin, and enzymes with wide specificity. It is difficult to compare quantitatively the specificity of different enzymes. As the measure of it might be proposed the slope of the graph representing the dependence of the

© 1987 Pergamon Journals Ltd.
Printed in Great Britain

rate constant on some structural parameter for the range of substrates. The structural parameter mentioned may be, for example, the hydrophobicity index (Dorovska and others, 1972) or some empirical specificity index (Zinchenko, Rumsh and Antonov, 1977).

Talking about the quantitative measure of efficiency and specificity it is necessary to define precisely what we are measuring in the enzyme kinetic experiment. Usually the efficiency is determined as the ratio of enzymic catalytic constant to the rate constant of a model, the so called congruent reaction. For comparison of specificity, the second order catalytic constants are used because they are not dependent on the wrong binding (Hein and Niemann, 1962). But experimental k_{cat} and K_m values reflect the rate of substrate chemical transformation and dissociation constant for enzyme-substrate complex, respectively, only in the simplest case of Michaelis-Menten kinetics. In a more complicated cases experimental kinetic parameters become complex values. Particularly, if on the way from substrate to the product two sorption complexes are formed, experimental k_{cat} and K_m values will include effective equilibrium constant between these two complexes and the difference in reactivity in series of substrates may be due to the difference in relative concentrations of the two complexes:

$$E+S \underset{k_{-1}}{\overset{k_1}{\rightleftharpoons}} ES_o \underset{k_{-2}}{\overset{k_2}{\rightleftharpoons}} ES_p \xrightarrow{k_3}, \qquad v=k_3K_p[E]_o[S]_o/\left[K_s+(1+K_p)[S]_o\right],$$

$$K_p=k_2/(k_{-2}+k_3),\ K_s=(k_1+k_3)/k_{-1},\ k_{cat}=k_3K_p/(1+K_p),\ K'_m=K_s/(1+K_p).$$

In case of proteolytic enzymes the two-step formation of enzyme--substrate complex is a very common phenomenon (see Antonov, 1983) and very large variations of the isomerization rate are observed. It is clear that in such cases the chemical transformation leading to the product is realized in the second enzyme-substrate complex which can be called productive one. That is the structure of the productive complex is most important for understanding the efficiency and specificity of catalysis. Among others, two questions are claimed for answer: 1) what is the state of the nucleophile attacking the amide group of substrate? and 2) what is the state of this amide group?

It is well known (Hartley, 1960) that proteolytic enzymes can be classified according to the structure of their catalytic apparatus into four main groups, namely, serine, thiol, carboxylic and metal--dependent proteases. Besides the functional groups the active site of enzymes may include a water molecule which also can act as a nucleophile. So we have first of all differentiated between two possible mechanisms: nucleophilic, when a group of the enzyme itself attacks the carbonyl carbon of the substrate, and the general base mechanism of action of the enzyme group, when the bound water molecule serves as a nucleophile. This problem was solved before only for ester substrates of some proteases (Zerner, Bond and Bender, 1964).

We have developped in our laboratory a few general methods of discrimination between these two mechanisms (Antonov and others, 1981). One of them consists in performing the enzyme catalyzed acyl-transfer or transpeptidation reaction in heavy oxygen water and measuring the ^{18}O content in the product of this reaction:

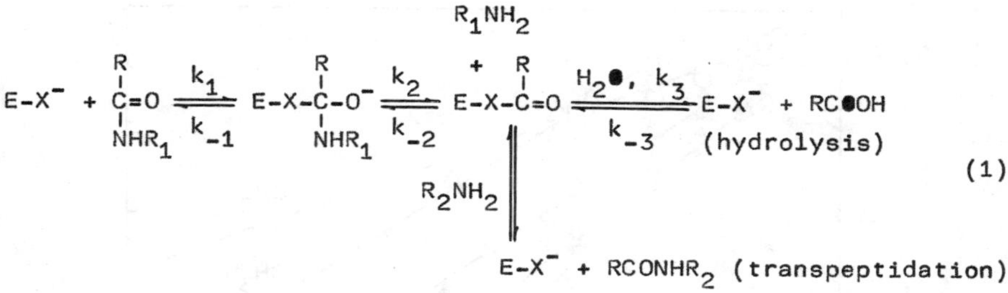

(1)

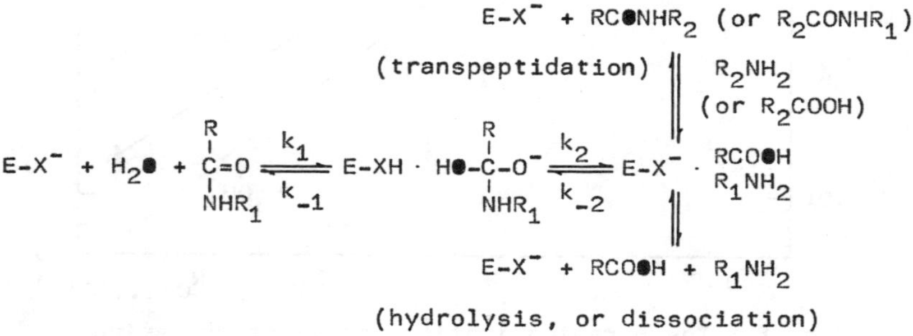

(2)

As follows from our investigations, in the cases of chymotrypsin and papain the reaction proceeds by nucleophilic mechanism (eq.1); the reactions catalyzed by pepsin and some metal-dependent proteases proceed by general base mechanism (eq.2) with a water molecule acting as a nucleophile.

A water molecule as well as serine hydroxyl are very poor nucleophiles (see Antonov, 1983). They are about ten order of magnitude weaker in reaction of amide hydrolysis than hydroxyl ion. This means that catalysis by such weak nucleophiles must be inefficient. On the other hand the well known "charge relay system" (Birktoft, Hartley and Blow, 1969) explaining for a long time the unique reactivity of serine hydroxyl was questioned on the basis of new X-ray (Matthews and others, 1977), neutronographic (Kossiakoff and Spencer, 1981), NMR (Bachovchin and Roberts, 1978) and others data (Umeyama, Nakagawa and Kudo, 1981). What is the principal cause of high efficiency of proteolytic enzymes?

Apparently, the reason of the high efficiency of proteases one must search in the structure peculiarity of amide bond of substrates bound with proteases. It was shown by Fastrez (1977) that the resonance stabilization of amide bond gives the dominant contribution (18-22 kcal/mole) in the total activation energy of the nucleophilic addition reaction of amides. Destabilization of the amide group may proceed by rotation around C'-N bond or by "pyramidalization" of amide group at the carbonyl carbon. The second possibility corresponds to the changes taking place during the nucleophilic addition reaction.

Is the pyramidalization of amide bond a real possibility? First of all, it must be stressed that some degree of pyramidalization exists even in crystals of simple peptides (Kolaskar and others, 1975; Ramachandran and Kolaskar, 1973). Secondly, the X-ray data available (summary see Antonov, 1985) for enzyme-quasisubstrate

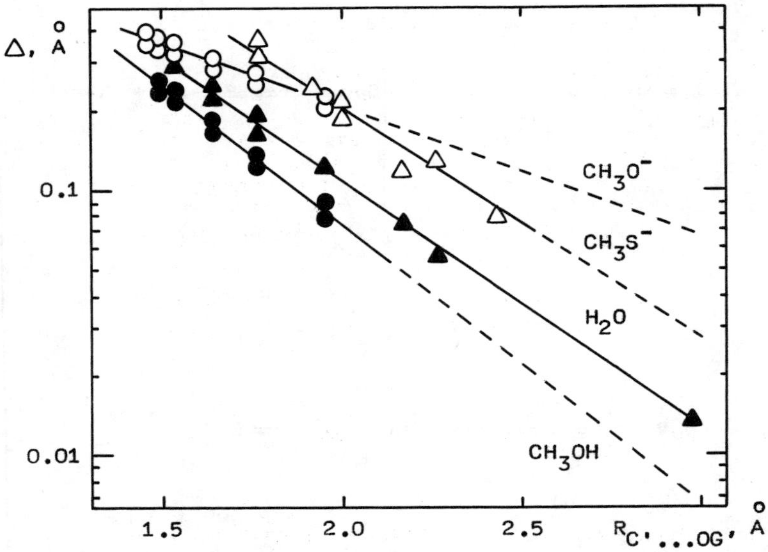

Fig. 1. The pyramidalization degree of the amide group at carbonyl carbon (in N-methylacetamide) by some nucleophiles (Δ is the distance between carbonyl carbon and CA-(=O)-N plane).

complexes show that in many cases pyramidalization of potentially susceptible bond is significant.

Pyramidalization is energetically a very unfavourable process. But, first, it is very possible that secondary enzyme-substrate interactions compensate this unfavourability; second, as has been shown for series of cyclic aminoketones (Bürgi and others, 1974), approach of nucleophile induces pyramidalization. Also, the inductive effects of different nucleophiles, simulated the reactive groups of various types proteases, were calculated quantum-chemically (by CNDO/2 method), which are differed very dramatically (Fig. 1)(Alexandrov and Antonov, 1984).

The quantum-chemical analysis of the acylation stage for serine proteases shown that there is no appreciable activation barrier at the nucleophilic attack of N-methylacetamide by methanol up to the distance between reacting atoms of 1.5 Å. At this point the distance is somewhat higher than that of covalent C'-OG bond and the bond order is about 0.5 at. un. (Alexandrov and Antonov, 1984).

The calculation of proton transfer according to "charge relay system" shown that the proton transfer on the imidazole is to accompany with the dramatical energy decrease due to the acception of the imidazole proton by Asp-102 (Alexandrov and Antonov, 1986). The system falls in some kind of energetic pit and the proton transfer onto the substrate leaving group becomes impossible.

The transfer of the proton onto the leaving group proceeds, apparently, via donor-acceptor water molecule (Fig. 2). This water molecule was found from X-ray data for some trypsin-quasisubstrate complexes (Marquart and others, 1983), which was not displaced by the ligand binding and was localized near the catalytically active groups. Here we did not find any energetically or sterically forbidden situation for the formation of an acyl-enzyme (Alexandrov and Antonov, 1986). This model is attractive because it does not require the addition of new water molecule after the elimination of the leaving group for the deacylation stage. The same water molecule which acts as proton donor-acceptor at the stage of acylation may serve as nucleophile at the deacylation step.

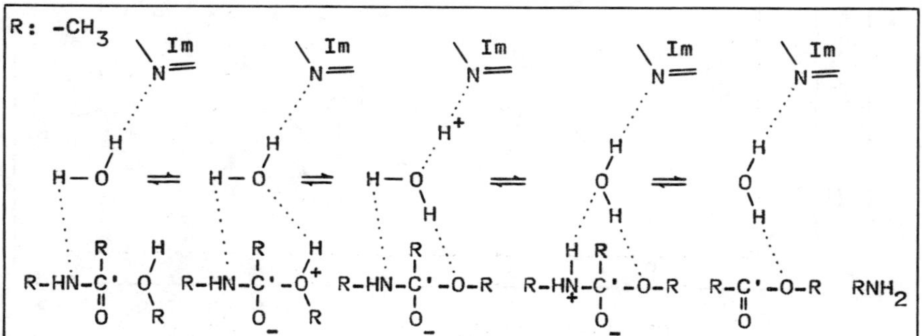

ΔH (kcal/mole)		-90	-70	-77	7
bond orders (at. un.):					
C'-N	1.17	0.94	0.86	0.61	-
C'-OG	-	0.58	0.78	0.94	0.96

Fig. 2. The proton transfer with the assistance of the water molecule as proton donor--acceptor (the energy zero point at the initial state).

The concept of resonance destabilization of amide group to be splitted as the reason of the efficiency of proteases has been discussed in different terms by a number of investigators (Delbaere and Brayer, 1985; Marquart and others, 1983). This concept is similar to the well known strain theory. However, strain, or stress, theory has been criticized (Fersht, 1977) because it joins the efficiency of catalysis with the events which take place in the ground state of a substrate and do not provide the stabilization of the transition state. On the other hand it is difficult to imagine physical and especially chemical mechanisms of the transition state stabilization because this state is not more than the fluctuation of the system.

Analyzing this situation we came to the conclusions (Karpeisky, Yakovlev and Antonov, 1980) which permit to eliminate the paradox mentioned. When we postulate the pyramidalization of the amide group in the productive enzyme-substrate complex, we have to choose the congruent system, for comparison of efficiency, where substrate has the same or very similar structure (Fig. 3). It is obvious that to compare the activation barriers of enzymic and model reactions we must start from the adequates ground states of the reacting molecules. In the solution amide group is nearly planar but always theoretically or experimentally it is possible to share

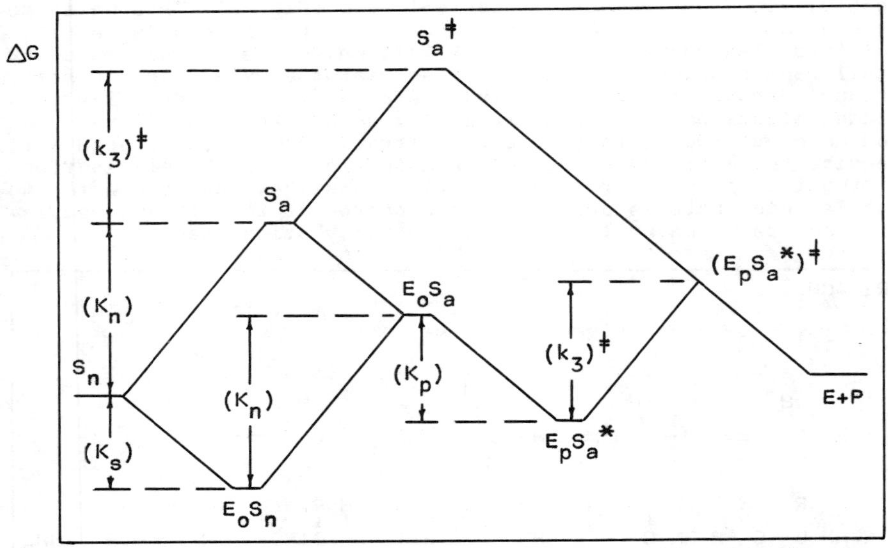

Fig. 3. The diagram of the free energy change along the reaction coordinate.

out part of molecules possesing the same degree of pyramidalization as the molecules in the enzyme active site. If we assume that only these molecules possess full reactivity of the amide in solution and others are not reactive, we can express the reaction rate as the product of the equilibrium constant between these "active" and "inactive" forms and the rate constant of chemical transformation of the "active" form of substrate:

$$S_n \overset{K_n}{\rightleftharpoons} S_a \xrightarrow{k_3} P, \qquad k_{ne}=k_3K_n.$$

The concentration of the "active" form of substrate in solution is very low and enzyme will bind practically exclusively an "inactive" form. But the interconversion between two forms may be the same in complex and in solution. We postulate that once formed in the active site of enzyme the "active" form of substrate is stabilized by the secondary interactions because the minimum energy is reached when the substrate is bound in the distorted "active" form. Thus the first postulate of our hypothesis is: enzyme stabilized the "active" form of the substrate:

$$E+S_n \overset{K_s}{\rightleftharpoons} E_oS_n \overset{K_n}{\rightleftharpoons} E_pS_a \overset{K_p}{\rightleftharpoons} E_pS_a^* \xrightarrow{k_3} E+P.$$

Now comparing the activation barriers of the elementary steps of chemical transformation of a substrate in the enzyme and in the solution we are starting with the same ground state and we can formulate the second postulate: enzyme does not decrease the activation barriers of the elementary step of the substrate chemical transformation. This postulate is admissible because we have already isolated in the solution the form of a substrate with the same reactivity as in the productive enzyme-substrate complex.

Due to the equivalence of the activation energy of chemical transformation of substrate in enzyme and in solution it is not necessary to introduce the concept of the transition state stabilization. Certainly transition state is stabilized relative to the "inactive" form, but not relative to the "active" form. However, "inactive" form does not participate in the chemical transformation of the substrate by enzyme. In the productive complex all substrate molecules are in the "active" form due to its stabilization.

Experimentally determined kinetic constants include in this case the equilibrium constant of the "active" and "inactive" forms interconversion (K_n) and a constant (K_p) which is proportional to the energy of stabilization of the "active" form:

$$k_{cat}=k_3K_nK_p/(1+K_nK_p), \quad K'_m=K_s/(1+K_nK_p).$$

The efficiency of the enzyme is now proportional to the K_p value or to the reciprocal of K_n value:

$$k_{cat}/k_{ne}=K_p/(1+K_nK_p)=K_p(\text{if } K_nK_p<1), \quad =1/K_n(\text{if } K_nK_p>1).$$

If we assume that the enzyme deprives totally the substrate amide group of resonance energy then the efficiency of the enzyme will be 10^{13}-10^{14} which is close to the values of maximal acceleration observed for enzymic hydrolysis of amide substrates.

The proposed scheme permits to describe the specificity of enzyme as its ability to activate the substrate sensitive bond at different degree, depending on the number and strength of enzyme-substrate contacts. It is known for example (Zinchenko, Rumsh and Antonov, 1977) that in case of pepsin the specificity is manifested in k_{cat} whereas K_m remains constant for a wide range of peptide substrates. If we assume that during the two-step productive complex formation only the primary specificity is realized at the first stage we shall have for the series of substrates with the same primary specificity identical K_s values. The K_m values will be constant and k_{cat} will change when the product of K_nK_p is less than unity. For better substrates this product can exceed unity and situation must change: the specificity will be manifested now in the change of K_m. The experimental data correspond to this prediction (Antonov, 1985).

We think that the concept described here is attractive because it tries to explain efficiency and specificity of proteases on the basis of substrate structure in ground state of productive complex and not in terms of experimentally unapproachable transition state. We believe that this is one of the possible ways to the formulation of a general theory of enzymic hydrolysis of the carboxylic acid derivatives.

REFERENCES

Alexandrov, S. L. and V. K. Antonov (1984). Mol. Biol. (Moscow),

18, 1576-1582.
Alexandrov, S. L. and V. K. Antonov (1986). Mol. Biol. (Moscow). Accepted for publication.
Antonov, V. K. (1983). Proteolysis Chemistry. Science, Moscow.
Antonov, V. K. (1985). Proceedings 16th FEBS Meeting, Part A. pp. 31-40.
Antonov, V. K. and S. L. Alexandrov (1984). In Yu. M. Torchinsky (Ed.), Physico-chemical Problems of Enzymatic Catalysis. Science, Moscow. pp. 232-241.
Antonov, V. K., L. M. Ginodman, L. D. Rumsh, Yu. V. Kapitannikov, T. N. Barshevskaya, L. P. Yavashev, A. G. Gurova and L. I. Volkova (1981). Eur. J. Biochem., 117, 195-200.
Bachovchin, W. W. and J. D. Roberts (1978). J. Amer. Chem. Soc., 100, 8041-8047.
Birktoft, J. J., B. S. Hartley and D. M. Blow (1969). Nature, 221, 337-340.
Bürgi, H. B., J. D. Dunitz, J. M. Lehn and G. Wipff (1974). Tetrahedron, 30, 1563-1572.
Delbaere, L. T. G. and G. D. Brayer (1985). J. Mol. Biol., 183, 89-103.
Dorovska, V. N., S. D. Varfolomeyev, N. F. Kazanskaya and A. A. Klyosov (1972). FEBS Lett., 23, 122-124.
Fastrez, J. (1977). J. Amer. Chem. Soc., 99, 7004-7013.
Fersht, A. (1977). Enzyme Structure and Mechanism. W. H. Freeman and Co., San Francisco.
Hartley, B. S. (1960). Ann. Rev. Biochem., 29, 45-72.
Hein, G. E. and C. Niemann (1962). J. Amer. Chem. Soc., 84, 4495-4503.
Karpeisky, M. Ya., G. I. Yakovlev and V. K. Antonov (1980). Bioorg. Chem. (Moscow), 6, 645-654.
Kolaskar, A. S., A. V. Lakshminarayanan, K. P. Sorothy and V. Sasisekharan (1975). Biopolymers, 14, 1081-1084.
Kossiakoff, A. A. and S. A. Spencer (1981). Biochemistry, 20, 6462-6474.
Marquart, M., J. Walter, J. Deisenhofer, W. Bode and R. Huber (1983). Acta Crystallogr., B39, 480-490.
Matthews, D. A., R. A. Alden, J. J. Birktoft, S. T. Freer and J. Kraut (1977). J. Biol.Chem., 252, 8875-8883.
Ramachandran, G. N. and A. S. Kolaskar (1973). Biochim. et biophys. acta, 303, 385-388.
Umeyama, H., S. Nakagawa and T. Kudo (1981). J. Mol. Biol., 150, 409-421.
Zerner, B., R. P. M. Bond and M. L. Bender (1964). J. Amer. Chem. Soc., 86, 3674-3679.
Zinchenko, A. A., L. D. Rumsh and V. K. Antonov (1977). Bioorg. Chem. (Moscow), 3, 1663-1670.

Theoretical Aspects of the Catalytic Power of Serine Proteases: on the Role of Quantum Tunneling

Calvin W. Mordy and Richard L. Schowen
Department of Chemistry, University of Kansas,
Lawrence, Kansas 66045–0046, USA

ABSTRACT

The solvent isotope effect varies with temperature in the hydrolysis of succinyl-L-phenylalanyl p-nitroanilide, catalyzed by α-chymotrypsin. The isotopic difference in enthalpies of activation is consistent with traditonally held views of the origins of isotope effects in protolytic catalysis, and is inconsistent with a resonant tunneling process as the major aspect of catalysis in the action of α-chymotrypsin on this substrate.

KEYWORDS

Enzyme, α-chymotrypsin, protolytic catalysis, solvent isotope effect, temperature dependence, tunneling.

INTRODUCTION

Protolytic Catalysis in the Serine Proteases

The serine proteases (Fersht, 1985) are among the large number of enzymes that employ protolytic catalysis (general acid-base catalysis, i.e., the donation or abstraction of protons at points of high or low electron density in the transition state) in order to assist in the acceleration of their substrate reactions. Among the evidences that this is true is the general observation of solvent isotope effects in the action of serine proteases (Venkatasubban and Schowen, 1984).

Although in principle the donation or abstraction of the proton, not coupled to other events, could determine the rate for these enzymic reactions, this seems not to be the case. The best evidence on this point is the observation of oxygen and nitrogen isotope effects when the substrate leaving group is labeled (reviewed in Hegazi et al., 1978) and the observation of secondary deuterium isotope effects for labels at the α- and β-positions of substrate molecules (Stein et al., 1983). These isotope effects show that bonding changes are

© 1987 Pergamon Journals Ltd.
Printed in Great Britain

occurring at the labeled positions in the catalytic transition state. Thus the formation and fission of heavy-atom bonds at the substrate carbonyl center is almost surely a major component of the reaction coordinate in the rate-determining transition state for both acylation of the enzyme and deacylation of the enzyme, since heavy-atom isotope effects are observed in both cases.

The observation of solvent isotope effects shows that some bonding changes are also occurring at sites which are labeled when the enzyme is placed in deuterium oxide. The most likely candidate site in the acylation of the enzyme is a proton bridge linking the active-site histidine imidazole to either the serine oxygen (formation of the tetrahedral intermediate) or the proximal atom of the leaving group (decomposition of the tetrahedral intermediate). Further sites, perhaps between the histidine imidazole and the nearby carboxylate (the "charge-relay system" or "catalytic triad;" the checkered history of structural work on this entity is recounted by Bachovchin [1986]; for isotope effect studies, see Venkatasubban and Schowen, 1984) or conceivably but less probably in the "oxyanion hole," may become involved when sufficient substrate-enzyme interaction is present (Venkatasubban and Schowen, 1984; Stein et al., 1983) but in the present report we shall be dealing only with cases of "one-proton catalysis." The most likely candidate site for generation of solvent isotope effects in the deacylation step is again the proton involved in general catalysis. This will be the proton of the water molecule which attacks the acyl-enzyme carbonyl group, here also bridging to the active-site histidine imidazole if formation of the tetrahedral adduct is rate-limiting. It will be the protonic bridge from imidazole to the serine oxygen if decomposition of the tetrahedral intermediate is rate-limiting.

The Nature of the Protolytic Bridges

Generally speaking, there have been two schools of thought about the nature of these protonic bridges, both adhering to the more or less traditional viewpoint of transition-state theory. According to one view, the transfer of the proton from donor to acceptor is coupled in the reaction coordinate to the reorganization of the heavy atoms (see calculations and discussion by Engdahl et al., 1983). According to a second view, the proton bridges are more in the nature of specialized hydrogen bonds, with the proton motion itself having a small amplitude in the reaction coordinate (Swain et al., 1965). Both viewpoints attempt to address the fact that the typical magnitude of the isotope effects is intermediate between the expectations for simple hydrogen bonding and for fully coupled reaction-coordinate motion. Ordinary hydrogen bonds commonly give isotope effects of only 10-20 percent or less (Schowen and Schowen, 1982) while fully coupled proton transfer and heavy-atom reorganization, as in E2 elimination reactions not uncommonly generates isotope effects of 5-7 (Melander and Saunders, 1980). The isotope effects for protolytic bridging fall in the middle, being typically in the range of 2-4 for both enzymic and non-enzymic catalysis. Adherents of the view that these reactions are E2-like have generally supposed that greater contributions of heavy-atom motion to the reaction coordinate have led to decreased hydrogen motion and thus to decreased isotope effects compared to E2 reactions. Adherents of the hydrogen-bond viewpoint have imagined that the transition-state hydrogen bonds are of a specialized character, similar to the linkages studied by Kreevoy and his co-workers (1977, 1980). These "AHA^-" hydrogen bonds generate isotope effects in the range of those observed for protolytic catalysis.

The Observation of Temperature-Independent Solvent Isotope Effects for Serine Protease Action

Reports by Khoshtariya and coworkers (1978, 1979) suggest a third and quite novel formulation of protolytic catalysis by the serine proteases. The action of the serine proteases chymotrypsin and trypsin on simple substrates such as generally produce "one-proton catalysis (Venkatasubban and Schowen, 1984) was studied. The dependence of the solvent isotope effects was carefully determined and in all cases, the activation energies for the two isotopic solvents were found to be identical.

This is quite in contrast to what is expected on the basis of either of the customary mechanisms reviewed above. Whether the isotope effect arises from reaction-cooordinate motion of the bridging proton or from the formation of a special type of hydrogen bond, it is the loss of part of the reactant-state zero-point energy difference for the two isotopic sites that is expected to produce the isotope effect (Schowen, 1972). A zero-point energy effect will be essentially wholly enthalpic and will produce a temperature-dependent solvent isotope effect.

The temperature-independent isotope effect can be understood in terms of concepts put forward by Dogonadze and Krishtalik (1975; see also Ulstrup, 1979) and closely related to concepts of Brickmann and Zimmermann (1966) as used by Limbach and his associates (1982). Essentially the same idea has been advocated by Verhoeven et al. (1986). We shall describe here a version of the concept which permits agreement of the model with observations on heavy-atom and secondary-hydrogen isotope effects.

Resonant-Tunneling Model for Temperature-Independent Isotope Effects

The essence of this model, as applied to enzymic catalysis, is that the critical rate-limiting event involves "resonant tunneling," i.e., tunneling from one energy-well to another between which the energy levels are reasonably well matched. An ideal case is a double-minimum potential in which the two wells have similar energies at the minimum and similar force constants. In the case of a symmetrical double-minimum hydrogen-bond potential, tunneling might occur between the ground states of the two wells if the separation of the minima were small (a "thin barrier") and if the energy levels were sufficiently in resonance. We shall describe the model in its application to data on the kinetic parameter k_{cat}, but extension of the argument to k_{cat}/K_m is easy.

In the enzymic reactions of interest, the rate cannot be determined simply by a proton-tunneling event from the ground state of a major form of the enzyme-substrate (ES) complex. Such a process would have no activation energy, while enthalpies of activation around 10 kcal/mol are common for the kinds of cases studied by Khoshtariya et al. (1978, 1979).

Such an enthalpy of activation might be conferred on the reaction if tunneling occurred, not from the ground state of a major form of the ES complex, but instead from the ground state of a low-concentration form of the ES complex. (As we shall see, there are further requirements as well.) The reaction would then exhibit an enthalpy of activation equal to the enthalpy of conversion of the main ES complex to the low-concentration form.

The model in this formulation would require that the isotope effect on the conversion of the main ES complex to the low-concentration form from which tunneling occurs be very small or nil. Then the enthalpies of activation for the two isotopic reactions would be identical.

One possible suggestion, already alluded to, would be that tunneling would now occur directly from the ground state of the minor ES complex. The tunneling probabilities for the two isotopes would be quite different and this difference would in fact generate the entire isotope effect. Since tunneling is postulated to occur from the ground state of the complex, it would be temperature-independent. The enthalpies of activation would be identical for the two isotopic systems and the relative tunneling probabilities would produce an apparent difference in entropies of activation.

However, this model in itself would not account for the heavy-atom and secondary-hydrogen isotope effects for labeled substrates. These indicate that in the critical rate-limiting event, substantial bonding changes about the carbonyl carbon are in progress. If the rate-limiting event were tunneling from the ground state of an ES complex with intact bonds about the carbonyl center, the isotope effects would not have been observed. Thus a further level of complexity must be introduced.

Consider that a minor ES complex - in a sense - might be formed as follows. As conceived in the customary mechanisms, the major ES complex proceeds toward the transition state for, say, formation of the serine-oxygen-to-carbonyl-carbon bond. The proton bound to the serine is not, however, simultaneously transferred to the histidine imidazole. Instead it remains in a double-minimum hydrogen bond between the oxygen of the serine and the nitrogen of the imidazole. Only a small isotope effect is generated as the O-C bond begins to form. At some point, the hydrogen-bond potential becomes symmetrical: this is when the acidity of the O-H bond matches the acidity of the imidazolium N-H bond. Then resonant tunneling can occur. Tunneling can in effect carry the system <u>under</u> a transition state for the proton transfer from one hydrogen-bond minimum to the other. At the same time, the tunneling is occurring at a point where the bonding changes at the heavy-atom centers correspond to a classical transition state.

This model thus incorporates bonding changes at heavy-atom centers with the maintenance of a double-minimum hydrogen-bond potential. Whether resonant tunneling actually comes into play will be a matter of the character of the double mimimum potential. If the "barrier is thin" and the energy levels strongly in resonance, it may play a large role. Then the isotope effect might be wholly or nearly temperature-independent. If these conditions are not both met, then resonant tunneling should be unimportant.

Scope of the Present Work

Because of the great intrinsic interest of the question of whether quantum tunneling is a feature of enzymic protolytic catalysis, we undertook a parallel study to those reported by Khoshtariya et al. (1978, 1979). Demuth et al. (1987) had already shown that the solvent isotope effects are temperature-dependent for inactivation of subtilisin and elastase by diisopropyl fluorophosphate, a reaction in which the catalytic power of the enzymes is substantially "recruited" by the inhibitor. The results were completely consistent with the traditional view of protolytic bridging. We were concerned, however, that resonant tunneling might be an aspect of the enzyme catalytic power not recruited by this inhibitor. In the present work, we have examined the situation with a catalytic substrate. The conditions are different from those employed in the studies of Khoshtariya et al. (1978, 1979).

RESULTS

System Employed

The action of α-chymotrypsin (EC 3.4.21.1; obtained from Sigma Chemical Company, salt-free, three times crystallized, from bovine pancreas) on succinyl-L-phenylalanyl p-nitroanilide (Sigma) was studied spectrophotometrically at 410 nm. Data were collected in digital form from a Cary 118 spectrophotometer on a

Heath H-11A computer. pH was controlled by 0.1 M phosphate buffers and the temperature maintained by a Lauda K4R circulating bath. Rates were measured as a function of substrate concentration and pH (6.4 - 8.2) or pD (6.7 - 8.3) in protium oxide (HOH, distilled and purified by ion-exchange) and deuterium oxide (DOD, from KOR, 99.7% D by nmr). The kinetic constants were calculated by least-squares procedures. Measurements were made at eight temperatures from 6.4° to 40.2°, with linear and reproducible initial-rate measurements being obtained at all temperatures.

Proton Inventory

The rates of reaction at 16 μM substrate and 25°, pH 7.5 in HOH and pL equivalent (Schowen and Schowen, 1982) in mixtures of HOH an DOD, 0.1 M phosphate buffer, were determined in 7 mixtures of HOH and DOD. The dependence of the velocity on the atom fraction n of deuterium in these mixtures was found to be linear, consistent with "one-proton catalysis," as expected (Venkatasubban and Schowen, 1984) for this substrate of simple and restricted structure. This confirms the more detailed studies of Fujihara (unpublished work; Demuth et al., 1987) with this system, in which both kinetic parameters were found to be linearly dependent on n.

Solvent Isotope Effects as a Function of Temperature

The Table presents the main findings of our work. The overall solvent isotope effects (SIE) are shown for each temperature and for each of the two kinetic parameters. Although the errors associated with the isotope effects are larger than would be wished, it is completely apparent that the magnitude of the isotope effect for each of the parameters decreases with increasing temperature. The decrease in both cases very substantially exceeds the errors in the data.

Temperature Dependence of the Rate Constants

Eyring or Arrhenius plots (not shown) for k_{cat} and k_{cat}/K_m are, as previously observed (Khoshtariya et al., 1978, 1979), non-linear, the apparent slope decreasing with increasing temperature. The precision of the data is insufficient to distinguish the several mechanistic models for such non-linearity (e.g., shift in rate-determining step with temperature; shift among conformational isomers of the enzyme, substrate or complexes of the two with temperature; change in heat capacity upon activation). The approximate magnitude of ΔH^* near room temperature is 10 kcal/mol, about the same as previous observations (e.g., Khoshtariya et al., 1978, 1979).

Fits of Solvent Isotope Effect Temperature Dependences

The best-fit Arrhenius representations of the solvent isotope effects are:

$$k_{cat}(HOH)/k_{cat}(DOD) = (0.4 \pm 0.3) \exp\{(1200 \pm 400)/RT\}$$

$$[k^{cat}/K^{m}(HOH)]/[k^{cat}/K^{m}(DOD)] = (0.09 \pm 0.03) \exp\{(1200 \pm 200)/RT\}$$

Table Temperature dependence of k_c and k_c/K_m and their solvent isotope effects for α-chymotrypsin with Suc-Phe-pNA

Solvent	Temp	10^3 k_c	k_c/K_m	k_c SIE	k_c/K_m SIE
H_2O	6.4	1.82±.05	3.04±.2	3.51±.2	3.14±.5
D_2O	6.4	.518±.03	0.966±.2		
H_2O	10.4	3.66±.05	4.99±.2	3.67±.06	2.96±.2
D_2O	10.4	0.589±.021	1.69±.06		
H_2O	15.8	6.18±.09	8.42±.3	2.92±.09	2.88±.2
D_2O	15.8	2.12±.06	2.92±.2		
H_2O	20.2	11.5±.4	10.6±.7	3.41±.1	2.07±.2
D_2O	20.2	3.37±.05	5.13±.2		
H_2O	25.0	19.3±.4	15.7±.7	3.13±.08	2.34±.1
D_2C	25.0	6.17±.08	6.70±.2		
H_2O	30.7	34.1±.8	18.5±.9	3.31±.2	2.41±.3
D_2O	30.7	10.3±.4	7.69±.8		
H_2O	35.7	48.0±.9	21.9±.7	2.78±.1	2.25±.2
D_2O	35.7	17.3±.5	9.75±.6		
H_2O	40.2	60.8±3	27.1±2	2.71±.2	2.16±.2
D_2O	40.2	22.4±.7	12.5±.8		

DISCUSSION

The simplest way to appreciate the significance of the data is to inspect the Table, which exhibits an unmistakable downward trend in SIE as the temperature rises. This is what is expected for a simple zero-point energy origin for an isotope effect arising from bonding changes between reactant state and transition state. It is fully consistent with the traditional viewpoints about protolytic catalysis discussed above and it provides no evidence in favor of a resonant tunneling process in the action of chymotrypsin on this substrate.

It is not impossible that resonant tunneling is at work and is being concealed by other mechanistic complexity. The non-linear Eyring plots could signal a shift among rate-limiting steps, each of which involve resonant tunneling with temperature-independent isotope effects. This would be a strained interpretation, however, and nothing in the data suggests it.

In fact, the apparent Arrhenius factors exhibit a difference in activation energies that is *larger* than that expected from zero-point energy differences, rather than smaller. This is combined with the apparent ratios A^H/A^D which are substantially less than unity (which is their expected magnitude on the zero-point energy model).

This kind of temperature dependence is commonly associated with a different type of tunneling, namely non-resonant tunneling at or near the transition state (Bell, 1980). The conclusion cannot be drawn from the data that this kind of tunneling is involved, because the Arrhenius representation is inappropriate for the isotope effects when the Arrhenius or Eyring plots for the individual rate constants are curved, as is the case here. It is worth noting, however, that when the data are constrained to fits with $A^H/A^D = 1.00$, the resulting values of ΔE^a remain well away from the value of zero expected for resonant tunneling (680 ± 20 cal/mol for k^{cat} and 560 ± 25 cal/mol for k^{cat}/K^m).

CONCLUSION

The present study adds to the evidence from serine-protease inhibition (Demuth et al., 1987) that resonant tunneling is not a universal, major feature of serine protease action. It remains a fascinating question what the differences are that lead to temperature-dependent solvent isotope effects in our work and to temperature-independent solvent isotope effects in the cases studied by Khoshtariya et al. (1978, 1979).

ACKNOWLEDGMENT

This work was supported through research grant number GM-20198 from the National Institute of General Medical Sciences of the National Institutes of Health (USA).

REFERENCES

Bachovchin, W. W. (1986). *Biochemistry*, 25, 7751.

Bell, R.P. (1980). *The Tunnel Effect in Chemistry*. Chapman & Hall, London.

Brickmann and Zimmermann (1966). *Ber. Bunsenges. Phys. Chem.* 70, 157, 521.

Demuth, H.-U., J. Heins, H. Fujihara, C. W. Mordy, G. Fischer, A. Barth and R.L. Schowen (1987). *Proc. 8th Conference Phys. Org. Chem. IUPAC*. Elsevier, Amsterdam.

Dogonadze, R.R., and Krishtalik, L.I. (1975). *Russian Chem. Rev. (translated from Uspekhi Khimii)*, 44, 11.

Fersht, A. (1985) *Enzyme Structure and Mechanism*. Freeman, San Francisco.

Hegazi, M.F., D.M. Quinn and R.L. Schowen (1978). In R.D. Gandour and R.L. Schowen (Eds.), *Transition States of Biochemical Processes*. Plenum, New York.

Khoshtariya, D.E., V.V. Topolev and L.I. Krishtalik (1978). *Bioorg. Khim.*, 4, 1341.

Khoshtariya, D.E. (1978). *Bioorg. Khim.*, 4, 1673.

Khoshtariya, D.E., V.V. Topolev, L.I. Krishtalik, I.L. Reizer and V.P. Torchilin (1979). *Bioorg. Khim.*, 5, 1243.

Kreevoy, M.M., T.-M. Liang and K.-C. Chang (1977). *J. Am. Chem. Soc.*, 99, 5207.

Kreevoy, M.M. and T.-M. Liang (1980). *J. Am. Chem. Soc.*, 102, 3315.

Limbach, H.-H., J. Hennig, D. Gerritzen and H. Rumpel (1982). *Faraday Discuss. Chem. Soc.*, 74, 229.

Melander, L., and W.H. Saunders, Jr. (1985). *Reaction Rates of Isotopic Molecules*. Wiley, New York.

Schowen, K.B., and R.L. Schowen (1982). *Meth. Enzymol.*, 87C, 551.

Schowen, R.L. (1972). *Prog. Phys. Org. Chem.*, 9, 275.

Stein, R.L., J.P. Elrod and R.L. Schowen (1983). *J. Am. Chem. Soc.*, 105, 2446.

Swain, C.G., D.A. Kuhn and R.L. Schowen (1965). *J. Am. Chem. Soc.*, 87, 1553.

Ulstrup, J. (1979). *Charge Transfer Processes in Condensed Media*. Springer, Berlin.

Venkatasubban, K.S., and R.L. Schowen (1984). *CRC Crit. Rev. Biochem.*, 17, 1.

Verhoeven, J.W., G.J. Koomen and S.M. van der Kerk (1986). *Rec. trav. chim. Pays-Bas*, 105, 343.

Thermodynamic and Microscopic Aspects of Stability and Functional Interactions of Biomolecules in Solution

M. B. Palma-Vitorelli and M. U. Palma
Department of Physics, University of Palermo, and Istituto per le Applicazioni Interdisciplonari della Fisica, Consiglio delle Nazionale Ricerche, via Archirafi 36, I–90123 Palermo, Italy

ABSTRACT

This review, based on recent work by the Authors and their colleagues, points out the possibility for stereodynamic (microscopic) and thermodynamic instabilities, and for noisy fluctuations of a relevant parameter, to concur to the stability and orderly function of biomolecules in solution.

KEYWORDS

Biologic order; biomolecular stability in solution; biosystems: instabilities and phase separation; spinodal decomposition; percolation; site-bond correlated-percolation; noise-induced order; water structure; microviscosity.

INTRODUCTION

Two conceptual steps mark the advances made in the last decade or so in the study of biosystems at molecular level, that is:

i) Recognition of the need of extending to the entire phase space the study of the structure-to-function relationship, to encompass dynamic, spatio-temporal structures. A synoptic table of developments related to this step, starting from early suggestions (Linderstrom-Lang and Schellman, 1959; Frohlich, 1973; Aiello and others, 1973; Davidov, 1982) can be found elsewhere (Palma, 1983).
ii) Recognition that biomolecular functionally should not be expected to correspond to one specific microscopic structural/dynamic order or state. (Mc Connell and von Hippel, 1970; Careri, 1974; Austin and others, 1975; Nakanishi and Tsuboi, 1978; Mandal and others, 1979; Englander and others, 1980). A great number of microscopic and equally functional stereodynamic states of both biomolecules and solvent molecules is instead required for the stable and orderly deterministic operation of biomolecular systems -objects lying midway between quantum and classical physics (Fornili and others, 1983; Palma-Vittorelli and Palma, 1985).

All in one, greater efficiency and flexibility, and access to a much ampler volume in the phase space, accounting for a functional stability close to that of macroscopic machines, can now be traced in water-biomolecule systems to a loose

© 1987 Pergamon Journals Ltd.
Printed in Great Britain

stereodynamic order and chaotic flickering from one to another configuration of an ample set, within a strict topological and functional context. The statistically most relevant flickering configurations were shown to have the larger share in solvent-mediated biomolecular interactions (Fornili and others, 1983; Palma-Vittorelli and Palma, 1985). The question poses whether instabilities of thermodynamic nature, or fluctuations in the value of a relevant parameter, may have, too, a functional role. Experiments that we shall here review and comment were designed as a first attempt to answer this question.

THE POSSIBLE ROLE OF A THERMODYNAMIC INSTABILITY

We first summarize the underlying standard thermodynamics and then proceed discussing the results and implications of recent experiments. A phase diagram for a two-component symmetrical system is schematically shown in Fig. 1. Above a critical temperature the system is stable for any proportion of the two components, and thermodynamic stability corresponds to the condition $\delta^2G/(\delta c)^2 > 0$, where c and G are respectively the concentration of one component and the free energy of the system. The so-called spinodal line (S in Fig. 1) encompassing the instability region is defined by the condition $\delta^2G/(\delta c)^2 = 0$. In terms of μ_1 and μ_2 (the chemical potentials of species 1 and 2), the same condition is written as $\delta\mu_1/\delta c_1 = \delta\mu_2/\delta c_2 = 0$.

From anywhere in the instability region, the system spontaneously evolves towards its equilibrium condition by what is called a spinodal decomposition. In the case of Fig. 2, there will be two coexisting compositions, having different concentrations $c^{(\alpha)}$ and $c^{(\beta)}$ such that $\mu_1^\alpha = \mu_1^\beta$ and $\mu_2^\alpha = \mu_2^\beta$. These conditions define another boundary, that of the binodal line (B in Fig. 1). A region of

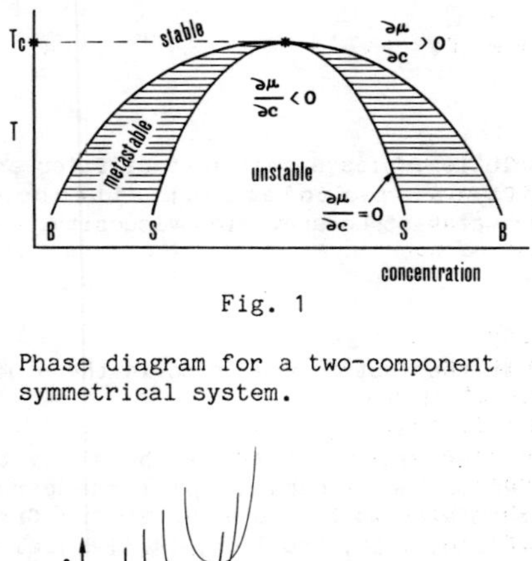

Fig. 1

Phase diagram for a two-component symmetrical system.

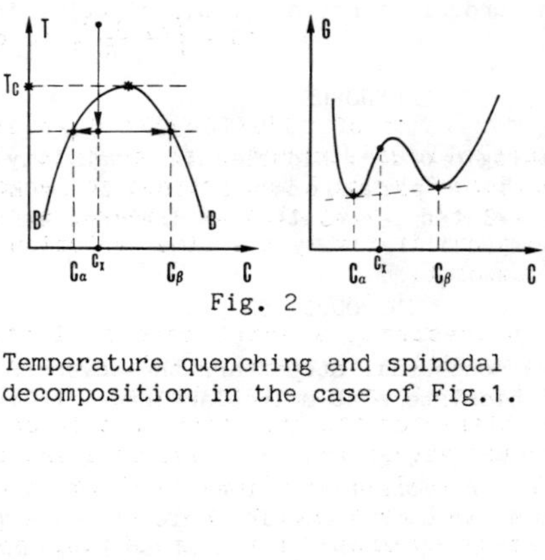

Fig. 2

Temperature quenching and spinodal decomposition in the case of Fig.1.

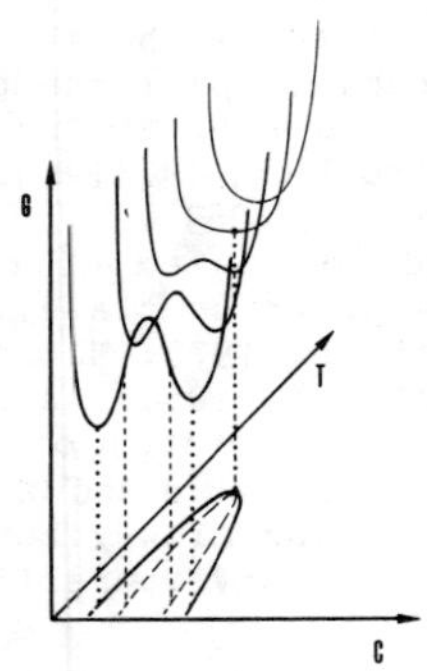

Fig. 3

Dependence of the free energy upon concentration and temperature, in the case of Fig.1.

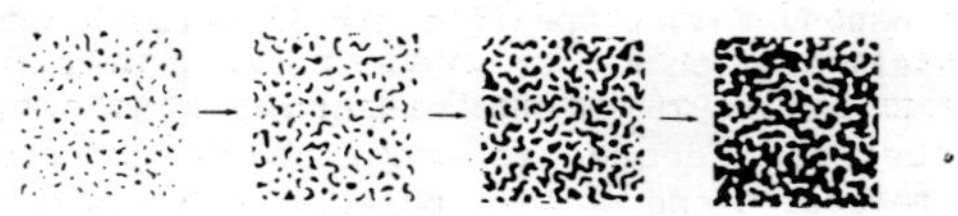

Fig.4

Pictorial illustration of a spinodal decomposition.

metastability exists between the spinodal and the binodal line. The situation is better illustrated in Fig. 3, which shows the behavior of the free energy, as the temperature is varied across the critical value. The same figure also shows how the binodal and spinodal lines are derived.

For a solution, the actual positions and shapes of the spinodal and binodal lines will depend upon interactions among solute(s) and solvent in all possible combinations, including solvent-mediated solute-solute, and solute-perturbed solvent-solvent. Changes in these interactions due to pH/ co-solutes/ co-solvents, will alter positions and shapes of both lines, so that very complex multi-dimensional diagrams will be needed. Symmetrical phase diagrams are expected for symmetrical systems only. For a simple solution, if P is the solute-to-solvent ratio of molar volumes, the critical concentration will progressively shift towards smaller values for increasing P values, according with the expression:

$$\Phi_C = (1+P^{1/2})^{-1} \tag{1}$$

The distribution of phases in space due to spinodal decomposition has long been studied in various fields, such as metals and inorganic glasses, and in more recent years it has stimulated a great deal of theoretical (Cahn, 1965; Van Aartsen, 1970; Kawasaki and Ohta, 1978; de Gennes, 1980) experimental (Tanaka and others, 1979; Tohyama and Miller, 1981;Snyder and others, 1983; Hashimoto and others, 1984; San Biagio and others 1986) and computer experimental (Abraham, 1986 and references therein) work. This distribution is easily understood by considering the free energy contribution due to fluctuations of concentration. For a symmetric system and a positive value of the appropriate transport coefficient κ, and retaining the first non-vanishing term, this becomes:

$$\Delta G = \int [\tfrac{1}{2} \frac{\delta^2 G}{\delta c^2} (c-c_0)^2 + \kappa (\nabla c)^2] \, dc \tag{2}$$

In the region of $\delta^2 G/(\delta c)^2 < 0$ we expect the system to be unstable against fluctuations occurring on a scale large enough to make the gradient term sufficiently small and yet not so large as to meet a diffusion barrier. These will grow, giving rise to a peaked structure function. The cartoon of Fig. 4 shows qualitatively the time evolution of one structure resulting from spinodal decomposition.

In biosystems, a shift into an instability region and subsequent spinodal decomposition may easily be due to synthesis/degradation of some species. The question of the possible biological relevance of spinodal decomposition has not yet attracted the attention that it seems to deserve, although it has already started attracting an active interest in the case of cell membranes (Knoll and others, 1983, and references therein; Wonnenberg, 1986). Here we shall recall the results of some recent experiments at our laboratories, pointing out the role of this mechanism in the self-assembly of the supramolecular order of a biostructural gel (San Biagio and others, 1986 a, and b).

The biopolymer considered in these experiments is Agarose. Aqueous Agarose systems are ideally suited for studying the effects of thermodynamic instabilities on the self-assembly of biostructural order, for the following reasons: i) they are simple two-component systems (apart for some polydispersity), having a notable biological relevance; ii) small increases of concentration, e.g. due to synthesis in vivo, can possibly throw them into the instability region; iii) being gelation of these systems markedly concentration-dependent, the polymer clustering due to spinodal decomposition is expected to promote gelation just where the clustering occurs; iv) at concentrations beyond the threshold for formation of a percolative structure of strings of islets in contact with each-other, spinodal decomposition would not only promote gelation, but also, ipso facto the very occurrence of the percolation along polymer double helices. It is this very connectivity property that provides the self-supporting capacity of the gel as biostructure; v) a shift of these systems to the instability region can alternatively be obtained by a temperature quench (see Fig. 2), which makes feasible a simple study <u>in vitro</u>.

The study has in fact been performed as it follows. Aqueous sols of Agarose (0.5% w/v) were quenched to a temperature (about 48 °C) at which gelation takes weeks to months to occur. After a temperature-dependent waiting time, a structural transition became observable through a low-angle light diffraction pattern (San Biagio and others, 1986, a and b). Concurrent photon correlation spectrometric measurements showed that the transition has no measurable effects on polysaccharide size dispersity or shape. Also, experiments with 0.04 μm radii polystyrene latex spheres performed to monitor zero-shear microviscosity (Madonia and others, 1983) showed no measurable effects accompanying the transition. The latter must thus be due to a spatially modulated polymer clustering, typical of the non-nucleated spinodal decomposition. Further, low-angle turbidometric experiments showed that the growth of the spinodal decomposition could in fact be followed in the course of time, as Fig. 5 shows (Leone and others, 1986). A parallel study of gelation processes and structures showed that at least at those temperatures and concentrations at which a spinodal decomposition is observed to occur well before gelation, the gel structure was observed to reflect the one resulting from the spinodal decomposition of the sol (Leone and others, 1986; San Biagio and others, 1986), see Fig. 6.

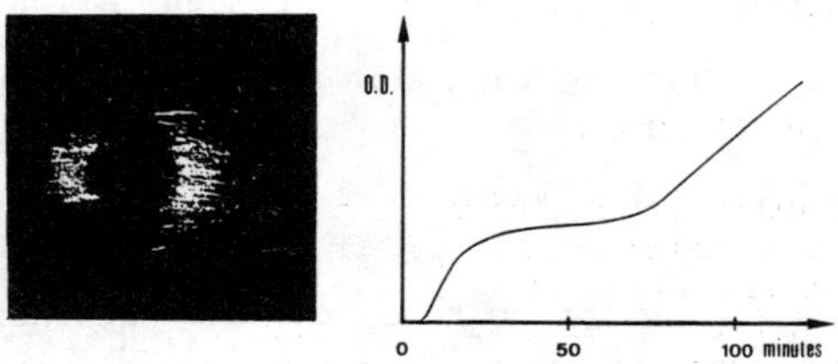

Fig. 5
Low-angle scattered light (left), and low-angle turbidity (right) monitoring the kinetics of spinodal decomposition and of subsequent start of gelation.

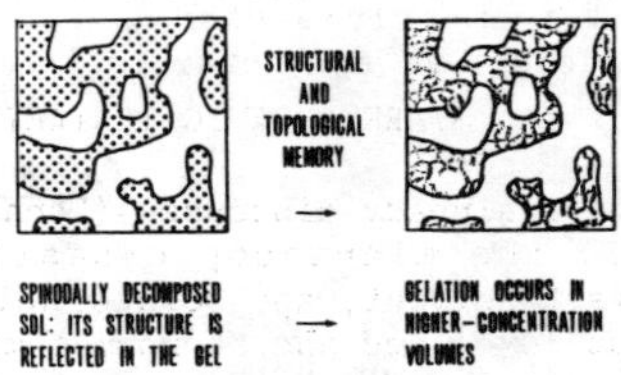

Fig. 6
Illustration of gelation occurring in regions of higher Agarose concentration, resulting from spinodal decomposition

These results are in excellent agreement with what could have been expected in terms of site-bond correlated-percolation theory (Coniglio and others, 1982). In Fig. 7 a hypothetical phase diagram is shown for this type of systems. As a consequence of a temperature quench or of an increased polymer concentration due to biosynthesis, the system can be driven into the region of instability, from where the spinodal decomposition will start. As illustrated by Fig. 7 (see also Fig. 2), this will allow the higher concentration phase to cross the gelation boundary and become a gel (see Fig. 6). The interrelation between the spinodal and gelation curves in Fig. 7 has a notable interest of its own, that may also have biological implications: although the gelation line is primarily determined by the topological properties of the system, the spinodal line is concurrently determined by solute-solute, solvent-solvent, and solute-solvent interactions. Any change in these variables (as due e.g. to co-solutes) is expected to alter the relative positions of the two curves, with consequences on aggregation processes and final structures. More experimental work is in progress in this direction.

It seems reasonable to infer that in the case of solute biomolecules capable of joining to each other in strings or bundles, or fibers, spinodal decomposition might promote the formation of structures of the type of those found e.g. in the cytoskeleton or in sickle cell Hemoglobin gels.Such structure(s) would assemble

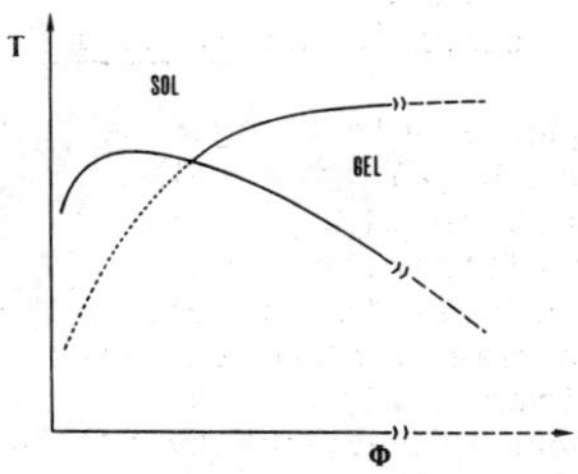

Fig. 7
Hypothetical spinodal and sol-gel phase boundaries.

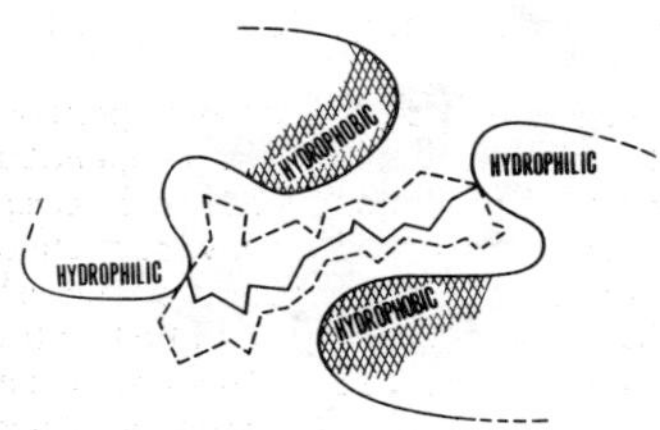

Fig. 8
Amphiphilic reinforcement of flickering H-bond pathways.

and disassemble depending e.g. upon pH or temperature changes, or solute(s) synthesis, as it is in fact observed. We must observe in closing that the potential relevance of spinodal decomposition is not necessarily limited to structure formation, as the very preferential clustering that it causes can be of relevance e.g. to temporary two- or three-body interactions in solution, as it is the case with enzymology or immunology. In this way, the onset of a thermodynamic instability, e.g. in a cell, would cause the onset of microheterogeneities capable of enhancing specific functional interactions.

STEREODYNAMIC INSTABILITIES IN THE SOLVENT

We have already discussed (Fornili and others, 1983; Palma-Vittorelli and Palma, 1985) the microscopic basis for the interpretation of solute-solute, solvent-solvent, and solute-solvent interactions. On a lifetime shorter than about 10^{-11} sec and at room temperature, H-bonds make a percolating network such that, at this temperature and timescale, pure liquid water can be considered as a gel. Solutes and co-solutes, however, are expected to alter both the topology and timescale of this network in a way that is significant to structure, function and interaction of biomolecules. We recall that this alteration is due to proton donor/acceptor sites acting as pinning points for otherwise flickering H-bond pathways, and to the (possibly synergetic) action of hydrophobic areas. The mechanism illustrated in Fig. 8 (amphiphilic synergism) may concern either different parts of one molecule or different molecules, and it is expected to be relevant to biomolecular conformation in the former case, and to biomolecular (functional) interaction in the latter (Palma-Vittorelli and Palma, 1985).

A discussion of the role of H-bond pathways in proton transfer processes through the solvent, and a comparison with existing data had been explicitly omitted in previous discussions (Fornili and others, 1983; Palma-Vittorelli and Palma, 1985). Here we shall briefly consider such processes, limitedly to specific cases where the same H-bond pathways are expected to be relevant to biomolecular structure/interaction. One such case is that of a pathway of H-bonded water molecules originating and terminating at proton acceptor/donor sites, either on the same biomolecule or on two biomolecules at reasonable distance. Large protonic conductivities will be expected along such pathways, because the two conditions of high polarizability and of molecules forming at least two H-bonds (Zundel, 1983 and references therein) are fulfilled.

The close interdependence of conformational changes, proton-transfer processes in easily polarizable proton-transfer hydrogen bonds, and the degree of hydration has been evidenced by I.R. spectroscopy, and it was anticipated that "this should be of great significance for functions of proteins with which proton-transfer processes occur" (Zundel, 1983 and references therein). It should be noted, however, that the stabilizing role of longer-lived (although still flickering) H-bond pathways in the solvent, originating and terminating at solute biomolecule sites (Fornili and others, 1983; Palma-Vittorelli and Palma, 1985), easily allows predicting that the interdependence of biomolecular conformation (or mutual presentation) and proton-transfer should be extended to proton transfer along pathways through the solvent, outside the solute. In other words, H-bond pathways through the solvent, promoted by acceptor/donor sites of solute biomolecules, are expected to be of great potential significance not only for biomolecular conformation (Zundel, 1983) but also for functions that require proton-transfer processes.

To be more specific, we may think of active centers of enzymes, where hydrogen bonds with great proton polarizability are often present. Experimental evidence suggests that an interplay between solvent reaction field, cooperative proton polarizability, and localized charges at the surface of the substrate, results in an increased reactivity of specific groups at the active center (Zundel, 1983). Here again, the microscopic view of the role of solvent should be extended to include H-bond pathways connecting the active center to sites on the enzyme itself or on a nearby substrate molecule. Proton transfer along these pathways is expected to help and extend the same mechanism, thereby increasing or even triggering the reactivity of specific groups at the enzyme active center. A set of recent results on protonic conductivity of lysozyme powders (with and without substrate) at low hydration level (Careri and others, 1985 and 1986) supports this view. In the actual case of a fully hydrated enzyme in solution, the same mechanism will of course be even more efficiently operative, as a consequence of the larger available number of H-bond pathways making the required connection between sites. The volume of phase space corresponding to functionality will be correspondigly increased, so that a better stability will result (Fornili and others, 1983; Palma-Vittorelli and Palma, 1985).

Additional evidence supporting our model comes from ongoing simulation work on the role of H-bond pathways in solvent-mediated solute-solute interaction and on the effect of pinning points for H-bond pathways at the surface of fixed or slowly moving solutes (Noto, 1986; Fornili and others, 1986). More evidence is provided by recent quasi-elastic light scattering experiments on the onset and progress of the spinodal decomposition discussed in the foregoing section. Perturbations of the solvent by substitution of hydrogen with deuterium or by addition of small amphiphilic molecules such as monohydric alcohols, give indeed the results predictable from the model (Fornili and others, 1983; Cordone and others, 1982; Palma-Vittorelli and Palma, 1985 and references therein).

Addition of small quantities of alcohols to water causes effects that are not always univocally interpreted, because stabilizations and distortions of H-bonds by hydrophilic pinning and amphiphilic reinforcement (see Fig. 8) have contributions of opposite signs on many properties. To sort them out, measurements of macro- and microviscosity down to the supercooled state, have been performed in these systems. These quantities are much more sensitive to stabilization than to distortion of hydrogen bonds, so that experiments have permitted a unified interpretation of data concerning a variety of thermodynamic and transport properties of these systems, as well as the alteration of biomolecular structure and function induced by small amounts of alcohols, in terms of the model (Spinnato, 1985; Spinnato and others, 1986).

STOCHASTIC FLUCTUATIONS OF A RELEVANT PARAMETER

Let us consider the time evolution of a systems towards its state of thermodynamic equilibrium. If the evolution depends non-linearly upon some parameters, one of which (at least) fluctuates e.g. between two fixed values, the possibility of observing an ordering effect of these fluctuations, or a "noise-induced order" is predictable (Lefever and others, 1983). The prediction is of particular interest in the case of biosystems, since these operate far from thermodynamic equilibrium and under not strictly fixed, and therefore "noisy" values of relevant parameters such as temperature. It appears particulary promising, in this context, a study of the effects of noise in a biosystem undergoing a micro-decomposition of the spinodal type, starting from a region of thermodynamical instability as described in one of the foregoing sections.

The systems selected for this type of studies was again Agarose in water, and the role of a fluctuating temperature on the final order obtained upon gelation was studied (Sciortino and others, 1986; Carollo, 1986). Reasons for the choice of this systems were in part those outlined in the foregoing sections. In addition, the order in the gel had already been characterized in terms of an order parameter and associated correlation length, and measured in the case of isothermal gelation (Leone and others, 1986), which made quantitative comparisons possible. Also, the role of spinodal decomposition in gelation had been elicited (San Biagio and others, 1986). Finally, the long gelation kinetics helped, since temperature fluctuations could be short compared with kinetics.

In the experiments, the temperature of cells where the self-assembly of order occurred, fluctuated in a dicotomous markovian fashion between two fixed extremes. Computer-controlled Peltier modules were used to this purpose. Both the extremes and the central (average) value, as well as a noise correlation time were selectable. The solid curve in Fig. 9 shows diagrammatically the dependence of the order parameter upon temperature, for the case of isothermal gelation (Leone and others, 1986). In the same figure, a dicotomous fluctuation of temperature is also shown, together with the corresponding interval of the order parameter expected in the case of no major effects of noise-induced order. Experiments have shown that in the presence of noise (temperature fluctuations) the final order parameter is always higher than that corresponding to the average temperature, and in many cases higher than any value in the interval shown in Fig. 9, and revealed the effects of noise amplitude and correlation time (Sciortino and others, 1986). More recent preliminary results (Carollo, 1986) seem to suggest a marked enhancement of this effect when the temperature fluctuates in the region where a spinodal decomposition precedes gelation, and triggers it (region between the "gel" broken line and the spinodal line, in the left part of Fig. 7). If confirmed, this would be an interesting example of an improved order, resulting in a biologically significant structure from a non-linear interaction of a thermodynamic instability causing microheterogeneities, and the noisy character of a relevant parameter.

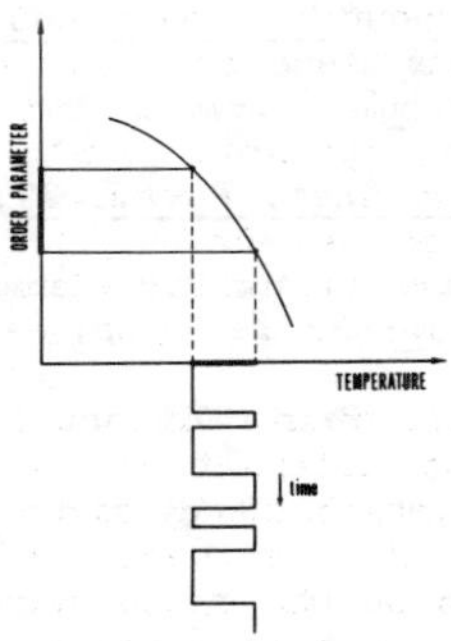

Fig. 9
Illustration of the philosophy of the experiment.

CONCLUSIONS

The experiments here reviewed suggest that for a deeper understanding of the orderly and deterministically, fuctional interactions of biomolecules, the picture should be extended, to include: i) thermodynamic instabilities leading to local redistribution(s) of concentration(s) and triggered by biosynthesis/degradation; ii) solvent stereodynamics, chaotically flickering within a well-defined topology, and iii) the noisy character of environmental parameters. Each of these disturbances can in fact contribute to the extraordinary functional stability and flexibility of biosystems.

References

Abraham, F.F. (1986). Adv. in Physics, 35, 1-111 and references therein.
Aiello, G., M.S. Giammarinaro-Micciancio, M.B. Palma-Vittorelli and M.U. Palma (1973). In H. Haken and M. Wagner (Eds.), Cooperative Phenomena. Springer-Verlag, Berlin. pp. 395-402.
Austin, B.H., K.W. Beeson, L. Eisenstein, H. Frauenfelder and I.C. Gunsalus (1975). Biochemistry, 14, 3555-3559.
Cahn, J.W.(1965). J. Chem. Phys., 42, 93-99 and references therein.
Careri, G. (1974). In B. Kursunoglu, S.L. Mintz and S.M. Widmayer (Eds.), Quantum Statistical Mechanics in the Natural Sciences. Plenum Press, N.Y. p. 54-57.
Careri, G., Geraci, M., Giansanti, A., and Rupley, J.A. (1985). Proc. Nat. Ac. Sci. USA, 82, 5342-5346.
Careri, G. A. Giansanti, and J.A. Rupley (1986). Preprint.
Carollo, C.M. (1986). Thesis.
Coniglio, A., H.E. Stanley and W.Klein (1982). Phys. Rev. B., 25, 6805-6821.
Cordone L.,A.Cupane and G.De Stefano (1982) Farad. Symp. Chem. Soc. 17, 161-169.
Davidov, A.S.(1982). Biology and Quantum Mechanics. Pergamon Press, Oxford.
de Gennes, P.G. (1980). J.Chem. Phys., 72, 4750-4763, and references therein.
Fornili, S.L., M. Leone, F. Madonia, M. Migliore, M.B. Palma-Vittorelli, M.U. Palma and P.L. San Biagio (1983). J. Biomol. Str. Dyn., 1, 473-486.
Fornili, S.L. and Noto, R. (1986). Unpublished.
Frohlich, M.H. (1973). In H. Haken and M. Wagner (Eds.), Cooperative Phenomena. Springer-Verlag, Berlin Hidelberg. p. VII-XII and references therein.
Hashimoto, T., K. Sasaki and H. Kawai (1984). Macromolecules, 16, 757-762.
Kawasaki, K. and T. Ohta (1978). Progr. Theor. Phys., 59, 362-374.
Knoll, W., G. Schmidt, E. Sackmann and K. Ibel (1983). J. Chem. Phys., 79, 3439-3442.
Lefever, R., W. Horsthemke and J.C. Michean (1983). In E. Clementi and R.H. Sarma (Eds.), Structures and Dynamics: Nucleic Acids and Proteins. Adenine Press, New York. p.107-112 and references therein.
Leone, M., F. Sciortino, M. Migliore, S.L. Fornili and M.B. Palma-Vittorelli (1986). Biopolymers. In press.
Linderstrom-Lang, K. U. and J. A. Schellman (1959). In D. Boyer, H. Lardy and K. Myrbach (Eds). The Enzymes, Vol. 1, Academic, Press, N.Y.
Madonia, F., P.L. San Biagio, M.U. Palma, G. Schiliro`, S. Musumeci and G. Russo (1983). Nature, 302, 412-415.
Mandal, H., N.R. Kallenbach and W. Englander (1979). J. Mol. Biol., 135,391-397.
Mc Connell, B. and P.H. Von Hippel (1970). J. Mol. Biol.,50, 297-300.
Nakanishi, M. and M. Tsuboi (1978). J. Mol. Biol., 124, 61-64.
Noto, R. (1986). Thesis.
Palma, M.U. (1983). In E. Clementi and R.H. Sarma (Eds.), Structure and Dynamics: Nucleic Acids and Proteins. Adenine Press, New York. pp. 125-132.
Palma-Vittorelli, M.B. and M.U. Palma (1985). In E. Clementi, G. Corongiu, M.H. Sarma and R.H. Sarma (Eds.), Structure and Motion: Nucleic Acids and Proteins. Adenine Press, Guilderland, N.Y. p. 9-36.
San Biagio, P.L., J. Newman, F. Madonia and M.U. Palma (1986, a). In R. H. Sarma and M.H. Sarma (Eds.), Biomolecular Stereodynamics III, Adenine Press, Guilderland, N.Y. p. 277-285.
San Biagio,P.L., J.Newman, F.Madonia and M.U.Palma (1986, b). Biopolym. In press.
Sciortino, F., M. Lapis, C.M. Carollo, S.L. Fornili and M.B. Palma-Vittorelli (1986). In R.H. Sarma and M.H. Sarma (Eds.), Biomolecular Stereodynamics III. Adenine Press, Guilderland, N.Y. p. 287-297.
Snyder, H.L., P. Meakin and S. Reich (1983). Macromolecules, 16, 757-762.
Spinnato, C. (1985) Thesis.
Spinnato, C. and Bulone, D. (1986). To be published.
Tanaka, T., Y. Swislow and I. Ohmine (1979). Phys. Rev. Lett., 42, 1556-1559.
Tohyama, K. and W.G. Miller (1981). Nature, 283, 813-814.
Van Aartsen, J.J. (1970). Eur. Polym. J., 6, 919-924, and ibidem 1105-1112.
Wonnenberg, W. (1986). Preprint.
Zundel, G. (1983). In W. Hoppe, W. Lohmann, H. Markl, H. Ziegler (Eds.) Biophysics. Springer-Verlag, Berlin. pp.243-254, and references therein.

Author Index

Subject Index